BARRON'S

IB

MATH STUDIES

Allison Paige Bruner
Garner Magnet High School

BARRON'S

All inquiries should be addressed to:
Barron's Educational Series, Inc.
250 Wireless Boulevard
Hauppauge, New York 11788
www.barronseduc.com

Library of Congress Catalog Number: 2014945458

ISBN: 978-1-4380-0335-1

PRINTED IN THE UNITED STATES OF AMERICA

9 8 7 6 5 4 3 2 1

CONTENTS

AUTHOR'S NOTE

The International Baccalaureate was founded in 1968, and the Diploma Programme administered its first examinations in 1970. In December 2013 there were 2457 IB World Schools offering the Diploma Programme.

There are four courses offered through the Diploma Programme: Math Studies Standard Level, Mathematics Standard Level, Mathematics Higher Level, and Further Math Higher Level. Math Studies is geared toward students who have varying backgrounds in mathematics, and the course is based in real-world application. Math Studies is a wonderful course to prepare for introductory college math. It also builds confidence in, and appreciation for, mathematics.

Writing this book has been quite a journey, and I would like to thank my wonderful husband, Ray, and fabulous daughters Mackenzie and Megan for putting up with the long hours working on the computer and for doing more than their fair share around the house. Thank you to my dad and mom, Bob and Vicki, along with my sister-in-law, Kristen, for providing love and support—and of course, my best friend, Shenny, who was always there with a listening ear. Finally, to all my students, past, present, and future, thank you for all you have taught me. The teacher I am today is because of all of you.

Paige Bruner
Garner Magnet High School
2014

INTRODUCTION

This book is designed around the new curriculum that began examinations in 2014. Each chapter gives a brief explanation of the material with supporting examples followed by practice questions in both the "short" and "long" question style. All examples are fully explained and enhanced with calculator directions. After the seven review chapters, there are three full-length practice examinations. Each practice examination contains both a Paper 1 and Paper 2 section followed by explanations and a marking scheme.

When taking the IB Math Studies exam, you will be provided with a clean formula packet. Formulas available in the packet are referenced in this book and utilized when appropriate. All formulas given in the formula packet may be found in the appendix. A calculator is also required for the examination. The IBO allows the use of the TI-83 and TI-84 families, along with the TI-Nspire (non CAS) using the 84 faceplate. The TI-Nspire (non CAS) versions 1.3 or higher must be in PRESS-TO-TEST mode, blocking certain features. There are many Casio calculators also allowed, but the TI-84 is the most commonly used. In this book, all calculator directions and images are from a TI-84 Plus Silver Edition with operating system 2.55. To upgrade the OS on a TI calculator, visit the Texas Instruments website. There is a chapter at the end of this book focusing on the TI-Nspire CX (non CAS) with operation system 3.6.0.546 in PRESS-TO-TEST mode.

THE EXAM

The IB Math Studies exam is held over two days. Paper 1 is given on day one, and Paper 2 is given on day two.

	Paper 1 Examination	Paper 2 Examination
Exam Design	• 15 short questions, limited in scope • Calculator is required • Formula packet provided • Questions often focus on one topic • Questions may contain words, symbols, diagrams/tables, or any combination • Difficulty level will vary for each question • All work, answers, and graphs are written in exam booklet	• 6 long questions, deeper in scope • Calculator is required • Formula packet provided • Questions often incorporate more than one topic • Questions may contain words, symbols, diagrams/tables, or any combination • Difficulty increases within each question with the beginning being easier than the end • All work, answers, and graphs are written on separate paper
Time	• Exam is 95 minutes: 5 minutes to read through exam and then 90 minutes to complete the work • Roughly 6 minutes for each question	• Exam is 95 minutes: 5 minutes to read through exam and then 90 minutes to complete the work • Roughly 15 minutes for each question

Marks	• Exam is worth a total of 90 marks • Each question is worth 6 marks • Correct answers earn full marks • Incorrect answers can earn partial marks for correct method shown	• Exam is worth a total of 90 marks • Each question is worth different marks, often between 12 and 20 marks • Correct answers must be supported with correct method to earn full marks
Overall Exam Grade	• Assigned a level based on total number of marks earned; see boundaries below • Marks earned are 40% of IB grade	• Assigned a level based on total number of marks earned; see boundaries below • Marks earned are 40% of IB grade

Each exam is worth 40% of the overall grade while the project is worth 20%. A perfect score of 90 marks on either paper will contribute 40 marks towards the overall grade.

Therefore, to compute your overall grade for IB Math Studies, use the formula:

$$\text{Overall Grade} = \frac{4}{9}(\text{Paper 1 marks}) + \frac{4}{9}(\text{Paper 2 marks}) + \text{Project marks}$$

For example, a student who earns 75 marks on Paper 1, 60 marks on Paper 2, and 15 marks on the project would have an overall grade of:

$$\text{Overall Grade} = \frac{4}{9}(75) + \frac{4}{9}(60) + 15$$

$$\text{Overall Grade} = 75$$

Each exam along with the project is assigned a grade based on the number of marks earned. The grade boundaries are set by IB and change from year to year. The last three years are shown below for each component and overall grade.

Paper 1 Grade Boundaries (out of 90 marks)

Grade	1	2	3	4	5	6	7
May 2013	0–13	14–26	27–35	36–48	49–60	61–73	74–90
May 2012	0–13	14–27	28–40	41–52	53–65	66–77	78–90
May 2011	0–14	15–29	30–41	42–53	54–65	66–77	78–90

Paper 2 Grade Boundaries (out of 90 marks)

Grade	1	2	3	4	5	6	7
May 2013	0–15	16–30	31–42	43–52	53–62	63–72	73–90
May 2012	0–13	14–27	28–41	42–52	53–63	64–74	75–90
May 2011	0–14	15–28	29–42	43–53	54–63	64–74	75–90

Project Boundaries (out of 20 marks)

Grade	1	2	3	4	5	6	7
May 2013	0–4	5–6	7–8	9–11	12–14	15–16	17–20
May 2012	0–4	5–6	7–8	9–11	12–14	15–16	17–20
May 2011	0–4	5–6	7–8	9–11	12–14	15–16	17–20

Overall Grade Boundaries

Grade	1	2	3	4	5	6	7
May 2013	0–16	17–31	32–42	43–55	56–68	69–80	81–100
May 2012	0–16	17–30	31–44	45–57	58–71	72–83	84–100
May 2011	0–16	17–31	32–45	46–58	59–71	72–83	84–100

Source: International Baccalaureate Organization, Subject Reports May 2013, May 2012, and May 2011.

Using the previous example of the student who earns 75 marks on Paper 1, 60 marks on Paper 2, and 15 marks on the project, the student would have earned either level 6 or 7 on Paper 1, level 5 on Paper 2, level 6 on the project, and level 6 for the overall grade.

STRATEGIES AND STUDY TIPS

Proper review is critical to earning a high score on the Math Studies examination. Do not wait until the night before to look through this book. Begin studying months before and work out as many examples as possible. Be comfortable with the formula packet, calculator, and especially the IB graph paper. With ample preparation, you can do well.

1. **Check the settings of your calculator.**
   Your calculator should always be in degree mode and the diagnostics should be turned on. To quickly check the mode, press [MODE] and make sure "degree" is darkened. To turn the diagnostics on, press 2nd 0 (Catalog) and then press enter twice on [DIAGNOSTIC ON].

2. **Become good friends with the formula packet.**
   If you are familiar with the formula packet and know where to find the formulas you need, you will save time on exam day. Since the exam is only 90 minutes, each second wasted searching through the formula packet is time that could have been spent working. The formula packet is a very helpful and useful tool.

3. **When in doubt, 3 sig fig it out.**
   Remember, you will lose one accuracy point per question if your answer is not written either exactly or as three significant figures, unless the problem states otherwise. If you are unsure if an answer is exact, then round to three significant figures. You could lose an accuracy mark on each question if you round incorrectly, and these points will add up quickly.

4. **Label and scale ALL graphs.**

   Every time you are asked to draw a graph, you must label the horizontal axis and vertical axis, and you must scale appropriately. If a scale is given, you must use that scale. Remember, one block of 5 small tick marks is 1 centimeter. Scale is always given in terms of those centimeter blocks. Graphs will either earn one mark for label and scale OR you could earn one mark for labeling and one mark for the scale. Even if you do not know how to graph the problem, you can at least set up the axes and earn one to two marks.

5. **Stuck? Make up a number.**

   There are times when the answer to one part of a question is to be used in the next part. If you are stuck and do not know how to solve the first part of a question, make up a *reasonable* answer. Then use this answer to continue the problem. You will earn no marks for the wrong answer, but you can earn full marks for the rest of the question if you used that made-up solution correctly. Instead of missing an entire question, you will only lose a few marks!

6. **Become a graphing expert.**

   Graphing equations using the graphics display calculator can be super helpful. You can solve two equations, solve one equation by graphing both sides, find the maximum or minimum function value, determine intervals of increasing and/or decreasing, find the value of the first derivative at a certain point, find function values, see the overall shape, and much more. Be comfortable and confident with the graphing options of your calculator. You should easily be able to change the window to see a graph, calculate key features of graphs, and access the table of values.

COURSE OBJECTIVES

Math Studies is a standard level course and has a focus of application of mathematics. The largest portion of the course covers statistics. Math Studies offers a great sampling of different mathematics from algebra, functions, triangle trigonometry, statistical applications, and introductory calculus. Problem solving is a major component of this course.

Course Topics

Topic	Description
1. Number and Algebra	Number sets, order of operations, scientific notation, rounding, percent error, SI conversions, currency conversions, solving a system of linear equations, solving a quadratic, arithmetic and geometric sequences and series, and compound interest
2. Descriptive Statistics	Discrete and continuous data, frequency tables, histograms, cumulative frequency, box and whisker plots, measures of central tendency, and measures of dispersion
3. Logic, Sets, and Probability	Symbolic logic, compound statements, truth tables, converse/inverse/contrapositive, set theory, Venn diagrams, sample space, probability of an event, expected value, compound events, mutually exclusive events, independent events, tree diagrams, and conditional probability
4. Statistical Applications	Normal distribution, scatter plot, correlation, linear regression, and chi-squared test of independence
5. Geometry and Trigonometry	Equation of a line, parallel and perpendicular lines; sine, cosine, and tangent ratios; angles of elevation and depression, sine rule, cosine rule, geometry of three-dimensional shapes, and surface area and volume of three-dimensional shapes
6. Mathematical Models	Function, domain and range, linear models, quadratic models, exponential models, polynomial models, rational models, drawing accurate graphs, and solving equations
7. Introduction to Differential Calculus	First derivative, gradient of a tangent line, tangent line equations, increasing and decreasing functions, stationary points, and optimization

Numbers and Algebra

- **NUMBER SETS:** Real, Rational, Integer, Natural

- **PERCENT ERROR:** $\varepsilon = \left| \dfrac{v_A - v_E}{v_E} \right| \times 100\%$

- **SCIENTIFIC NOTATION:** $a \times 10^k$, where $1 \le a < 10$, and $k \in \mathbb{Z}$

- **CONVERSIONS:** SI units (meters, kilometers, etc.) and currency (euro to USD, etc.)

- **LINEAR EQUATIONS:** solving a system of 2 linear equations (intersection)

- **QUADRATIC EQUATION:** solving a quadratic (zeros)

- **ARITHMETIC SEQUENCES AND SERIES:** $u_n = u_1 + (n-1)d$ and

  $S_n = \dfrac{n}{2}(2u_1 + (n-1)d) = \dfrac{n}{2}(u_1 + u_n)$

- **GEOMETRIC SEQUENCES AND SERIES:** $u_n = u_1 r^{n-1}$ and

  $S_n = \dfrac{u_1(r^n - 1)}{r - 1} = \dfrac{u_1(1 - r^n)}{1 - r}$

- **COMPOUND INTEREST:** $FV = PV \times \left(1 + \dfrac{r}{100k}\right)^{kn}$

1.1 NUMBER SETS

We classify numbers into the different number sets: Real ($\mathbb{R}$), Rational ($\mathbb{Q}$), Irrational ($\mathbb{Q}'$), Integer ($\mathbb{Z}$), and Natural ($\mathbb{N}$). Numbers can be an element, or member, of more than one set.

In IB Math Studies, all numbers used are elements of the **REAL** number set, which means the number can be graphed on a number line. We will classify all numbers as REAL, since we can plot them on a number line (they **real**ly exist).

All numbers are also elements of either the **RATIONAL** or **IRRATIONAL** number set. Rational numbers are any numbers than can be written as fractions. For example, $\dfrac{7}{5}$, $-\dfrac{2}{3}$, and 5 are all rational numbers but so are 1.2, $0.\overline{6}$, and $3.\overline{546}$. Remember, whole numbers such as 5 can be written as $\dfrac{5}{1}$, and all decimals that end, such as 1.2, or repeat, such as $0.\overline{6}$ and $3.\overline{546}$, can also be written as fractions. Irrational numbers are numbers that cannot be written as fractions. Irrational numbers are decimals that do not end and do not repeat. Examples include: π, e, $\sqrt{15}$, and $-1.3452\ldots$

Rational numbers can also be elements of the **INTEGER** number set, if the number does not contain a fraction or decimal. Integers are the set of the counting numbers, their opposites, and zero. $\mathbb{Z} = \{\ldots, -3, -2, -1, 0, 1, 2, 3\ldots\}$

Finally, integers can be elements of the **NATURAL** number set, if the number is a positive counting number or zero. $\mathbb{N} = \{0, 1, 2, 3,\ldots\}$

In the U.S., there is a slightly different classification system. Whole numbers are the set of zero and positive counting numbers, and the natural numbers are simply the positive counting numbers. For the IB exam, you disregard whole numbers and use the natural number set.

➡ EXAMPLE 1.10

Complete the following table by placing an X in the box if the number is an element of the set.

Number	Number Set			
	$\mathbb{R}$	$\mathbb{Q}$	$\mathbb{Z}$	$\mathbb{N}$
$-\dfrac{6}{7}$				
$\sqrt{9}$				
$\sqrt{47}$				
-2π				
$-\dfrac{40}{10}$				

Answer Explanations

Number	Number Set			
	$\mathbb{R}$	$\mathbb{Q}$	$\mathbb{Z}$	$\mathbb{N}$
$-\dfrac{6}{7}$	X	X		
$\sqrt{9}$	X	X	X	X
$\sqrt{47}$	X			
-2π	X			
$-\dfrac{40}{10}$	X	X	X	

- $-\dfrac{6}{7}$: All numbers are real and the number is a fraction.

- $\sqrt{9}$: This can be simplified to 3, which is real (all numbers are real). 3 is also rational, since it can be written as $\dfrac{3}{1}$. It is also an integer and a natural number, since 3 is a positive counting number.

- $\sqrt{47}$: Since 47 is not a perfect square, the radical cannot be reduced. It is only real. The decimal for $\sqrt{47}$ is approximately $6.8556546\ldots$. Since it does not end or repeat, it is not a rational number.

- -2π: All numbers are real, but π is irrational. Therefore, all multiples of π are also irrational.

- $-\dfrac{40}{10}$: This can be simplified to -4, which is real (all numbers are real), rational (it was already a fraction), and an integer (it is the opposite of the counting number 4). This cannot be an element of the natural set, since it is a negative number.

The number sets are often represented as a Venn diagram because the diagram illustrates the relationship between the sets.

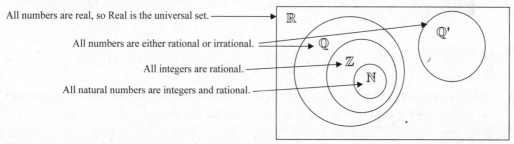

All numbers are real, so Real is the universal set. —— $\mathbb{R}$

All numbers are either rational or irrational. —— $\mathbb{Q}$ $\mathbb{Q}'$

All integers are rational. —— $\mathbb{Z}$

All natural numbers are integers and rational. —— $\mathbb{N}$

1.2 APPROXIMATION, ROUNDING, AND ESTIMATION

Rounding a number is when you approximate the number to a given degree of accuracy. This alters the exact value of the number. You round to the nearest whole number (e.g. ones, tens, hundreds) or to a specified decimal place (e.g. tenths, hundredths, thousandths).

The number of significant figures in a solution gives a degree of accuracy in the answer. Being able to count the number of significant figures (sig figs) involves mastering a few rules.

1. All non-zero digits are counted as significant.
2. For any number **smaller** than 1, the leading zeros are **not** significant.
3. For integers **larger** than 1, zeros at the end are **not** significant.
4. Once you begin counting sig figs, all digits are significant.

An easy way to know where to begin counting digits is to follow the Pacific/Atlantic decimal analogy.

Pacific Ocean

P = Present

If your decimal is **P**resent, begin counting on the LEFT with the first non-zero number

Your Number Here
(Once you begin counting, you cannot stop)

Atlantic Ocean

A = Absent

If your decimal is **A**bsent, begin counting on the RIGHT with the first non-zero number

➡ **EXAMPLE 1.20**

Identify the number of significant figures for each of the following.

1. 4 035 km

2. 56 000 kg

3. 56 040 mm

4. 0.0060700 g

5. 2.0016 cm

6. 5 000 001 m

7. 234.056 kg

Answer Explanations

1. **4 sig figs**
   The decimal is **A**bsent, so begin on the right. There are no zeros at the end of the number, so start counting right away. There are 4 significant figures.

2. **2 sig figs**
   The decimal is **A**bsent, so begin on the right. There are three zeros at the end of the number, so start counting at the "6". There are 2 significant figures (5 and 6).

3. **4 sig figs**
   The decimal is **A**bsent, so begin on the right. There is one zero at the end of the number, so start counting at the "4" and do not stop until you reach the end. There are 4 significant figures (5, 6, 0, and 4).

4. **5 sig figs**
   The decimal is **P**resent, so begin on the left. There are three leading zeros, so begin counting at the "6" and do not stop until you reach the end. There are 5 significant figures (6, 9, 7, 0, and 0).

5. **5 sig figs**
   The decimal is **P**resent, so begin on the left. There are no leading zeros, so begin counting right away. All digits are significant.

6. **7 sig figs**
   The decimal is **A**bsent, so begin on the right. There are no zeros at the end of the number, so start counting right away and do not stop until you reach the end. All digits are significant.

7. **6 sig figs**
   The decimal is **P**resent, so begin on the left. There are no leading zeros, so begin counting right away. All digits are significant.

QUICK TIP

Absent Decimal—start counting on the RIGHT (Atlantic Ocean)

Present Decimal—start counting on the LEFT (Pacific Ocean)

Once you start counting, you cannot stop until you reach the end!

> Remember, once you start counting significant figures, you CANNOT stop until you have counted to the end. This means zeros in the middle of a number are significant.

Why are significant figures so important in Math Studies? When you write your answers you have two choices, unless the problem states otherwise. You can write your answers as **exact** or **rounded to 3 significant figures**. You would choose to write an answer as exact if you do not have to round the number at all to write it on your paper. If there is any doubt that what you are writing is the exact answer, then round the answer to 3 sig figs.

➡ EXAMPLE 1.21

Write each number correct to the number of significant figures given.

1. 2.546 (1 s.f.)
2. 24 650 (3 s.f.)
3. 0.0429 (1 s.f.)
4. 3.624 (3 s.f.)
5. 45 627 (2 s.f.)
6. 23.456732 (1 s.f.)

Answer Explanations

1. **3**

   Since the decimal is **P**resent, the first significant figure is on the left. You only want one sig fig, so you must keep the first number. Following the rules of rounding, you round the 2 up to a 3. Now double check—Does 3 make a good estimator of 2.546? Yes.

2. **24 700**

   Since the decimal is **A**bsent, the first significant figure is on the right. Remember that zeros at the end do not count when the decimal is absent. You need three sig figs, so you must keep the last three digits. Many students will round the number to 247, but does 247 make a good estimator of 24 650? No! You must fill in the zeros at the end to keep the value of the number. Does 24 700 make a good estimator? Yes.

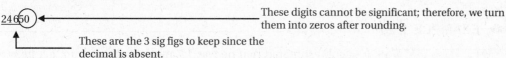

These digits cannot be significant; therefore, we turn them into zeros after rounding.

These are the 3 sig figs to keep since the decimal is absent.

3. **0.04**

   Since the decimal is **P**resent, the first significant figure is on the left. You only want one sig fig, so you must keep the first non-zero number. Following the rules of rounding, you do not round the 4 up to a 5. You must keep the leading zero to maintain the value of the number. Now double check—Does 0.04 make a good estimator of 0.0429? Yes.

4. **3.62**

   Since the decimal is **P**resent, the first significant figure is on the left. You want three sig figs, so you must keep the first three digits. Following the rules of rounding, you do not round the 2. Now double check—Does 3.62 make a good estimator of 3.624? Yes.

> **QUICK TIP**
>
> To round to a specific sig fig:
> 1. Identify the last desired sig fig
> 2. Look at the digit directly behind last sig fig
> 3. Round appropriately
> 4. Remember to fill in zeros if the rounded number is greater than 1 000

5. **46 000**

Since the decimal is **A**bsent, the first significant figure is on the right. Remember that zeros at the end do not count when the decimal is absent. You need two sig figs, so you must keep the last two digits and the 5 must round up to a 6. Many students will round the number to 46, but 46 does not make a good estimator of 45 627. You must fill in the zeros at the end to keep the value of the number. Does 46 000 make a good estimator? Yes.

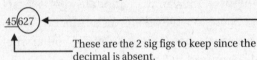

These digits cannot be significant; therefore, we turn them into zeros after rounding.

These are the 2 sig figs to keep since the decimal is absent.

6. **20**

Since the decimal is **P**resent, the first significant figure is on the left. You only want one sig fig, so you must keep the first non-zero number. Following the rules of rounding, we do not round the 2. You cannot keep the decimal because, once you begin counting significant figures you cannot stop until the end. Zeros would become significant with a decimal. Without the decimal, you must put a zero to keep the value of the number without adding a sig fig. Does 20 make a good estimator of 23.456732? Yes.

Estimation is a form of rounding, since you are using approximations of the numbers given in a problem to make them easier to use. Estimation makes the math easier, but you know the answer is still reasonable.

➡ EXAMPLE 1.22

Estimate the cost of 1.8 meters of fabric when the fabric costs $5.13 per meter.

Answer Explanation

The fabric would cost approximately $10.

The exact calculation would be $1.8 \times \$5.13$, but you use the estimates $2 \times \$5$ to make the math more straightforward but the answer still reasonable.

Due to estimation, rounding, and significant figures, your answers will be a little off from the exact answer. This difference creates an error, and you can calculate the percentage error.

The formula for percentage error is $\varepsilon = \left| \dfrac{v_A - v_E}{v_E} \right| \times 100\%$, where v_E is the exact value and v_A is the approximate value of v.

➡ EXAMPLE 1.23

QUICK TIP

When finding percent error:

The order in which the approximate and exact are subtracted does not matter. What does matter is that you **always** divide by the **exact** value. Remember, the answer is always **positive**.

Jacob estimated that he was 1.7 meters tall when in fact he was 1.85 meters tall. Calculate the percentage error Jacob made in his estimate.

Answer Explanation

The percentage error is 8.11%.

Substituting the values into the formula gives $\left| \dfrac{1.7 - 1.85}{1.85} \right| \times 100\%$.

This simplifies to 8.108108108…. Using the IB rule of 3 sig figs, we round our answer to the second decimal place, which gives 8.11%.

1.3 STANDARD FORM

Standard form is the same as scientific notation, which is a way scientists write very large or very small numbers in a more concise form. Standard form is based upon powers of ten. For example 100 is 10^2 and 1 000 is 10^3. A positive exponent moves the decimal to the right, which in turn makes the number larger. Conversely, 0.10 is 10^{-1} and 0.01 is 10^{-2}. The negative exponent moves the decimal to the left, which in turn makes the number smaller.

When you are asked to write a number in the form $a \times 10^k$ where $1 \leq a < 10$ and k is an integer, you are being asked to write a number in standard form. You might also be asked to write a number in the form $a \times 10^k$, where $1 \leq a < 10$ and $k \in \mathbb{Z}$. Both commands are exactly the same, but instead of "k is an integer," you are given set notation.

To write a number in standard form (for example, 0.0034587):

1. Determine the value of a. Remember, $1 \leq a < 10$, so we will place the decimal behind <u>one</u> digit only. In the example of 0.0034587, you will place the decimal in between the 3 and 4, so that $a = 3.4587$. You can only have <u>one</u> number in front of the decimal.

2. Find k by counting the number of places you had to move the decimal from the original position to where it is in a. If your original number was a small decimal, k will be negative. If your original number was a large number, k will be positive. In the example, you moved the decimal 3 times and the original number was a small decimal, so $k = -3$.

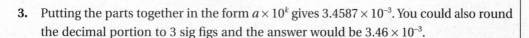

$$0.0\,0\,3\,4\,5\,8\,7 \ (k = -3)$$

3. Putting the parts together in the form $a \times 10^k$ gives 3.4587×10^{-3}. You could also round the decimal portion to 3 sig figs and the answer would be 3.46×10^{-3}.

➥ EXAMPLE 1.30

Write the following numbers in standard form.

1. 35.12
2. 0.041
3. 62 420 000
4. 0.000006132

Answer Explanations

1. **3.512×10 (exact) or 3.51×10 (3 sig figs)**
    The value of a is 3.512, since there can only be one number in front of the decimal. The number is considered large, and you had to move the decimal once so $k = +1$. Remember, you do not need to write an exponent of one since it is understood; however, you can write the answer as 3.512×10^1.

2. **4.1×10^{-2}**
    a is 4.1. The number is small and you moved the decimal two times, so $k = -2$.

3. **6.242×10^7 (exact) or 6.24×10^7 (3 sig figs)**

   a is 6.242, since there can only be one number in front of the decimal. The number is definitely large, and you moved the decimal seven times, so $k = +7$.

4. **6.132×10^{-6} (exact) or 6.13×10^{-6} (3 sig figs)**

   a is 6.132, since there can only be one number in front of the decimal. The number is definitely a small decimal, and you moved the decimal six times, so $k = -6$.

Standard form is also useful when performing calculations with very large or small numbers. The calculator will do the work for you, but you must know how to properly enter values and write the answer.

When entering a standard form number into your calculator, use "E" to represent the $\times 10^k$ portion of standard form. For example, the number 2.34×10^8 would be entered as 2.34E8. In order to enter this number, press 2.34 followed by the 2nd button and then the "," button (note that above the "," button is the symbol "EE". This is the scientific notation "E"). The calculator also displays answers in the same format, but you must use correct standard form notation when writing your answers.

➡ **EXAMPLE 1.31** ―――――――――――――――――――――――――――

Let $x = 7.65 \times 10^{-4}$ and $y = 4.97 \times 10^5$. Calculate the following values, writing the answer in the form $a \times 10^k$, where $1 \le a < 10$ and $k \in \mathbb{Z}$.

QUICK TIP
• Never write solutions using calculator notation
• Always use parentheses when the numerator or denominator has more than 2 numbers

1. $\dfrac{y}{x}$

2. $\dfrac{x}{2y}$

3. $\dfrac{10x - y}{x}$

Answer Explanations

1. **6.50×10^8**

   Entering the problem into the calculator, the answer is given in expanded form. The directions clearly state to write the answer in standard form, so you must follow the previous rules. Round your answer to 3 sig figs, since you cannot write the exact answer.

2. **7.70×10^{-10}**

   Be careful when entering this problem into the calculator as you must put parentheses around the denominator. Round your answer to 3 sig figs, since you cannot write the exact answer.

   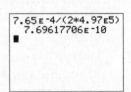

3. **-6.50×10^{-8}**

   Entering the problem into the calculator, you must use parentheses around the numerator. The answer is given in expanded form. The directions clearly state to write the answer in standard form, so you

   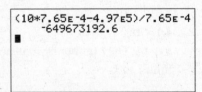

must follow the previous rules. Round your answer to 3 sig figs, since you cannot write the exact answer.

1.4 SI UNITS AND CONVERSIONS

Converting between the *International System of Units*, or SI Units, is based on powers of ten. The most common units of measure are meter, second, liter, meter per second, and gram. These are the base units, and can convert to measurements such as kilogram, centimeter, and milliliter.

When converting between units, you need to know what power of ten corresponds to the prefix. The chart below is helpful in determining what power of ten is needed when performing conversions.

10^3	10^2	10^1	10^0	10^{-1}	10^{-2}	10^{-3}
kilo-	**hecto-**	**deca-**	**Base Unit**	**deci-**	**centi-**	**milli-**
k	**h**	**da**	**(Meter, gram, second, etc.)**	**d**	**c**	**m**

For example, to convert 650 milliseconds to seconds, begin at milli- and move to the base. Thus, you would multiply 6.56 by 10^{-3}. Therefore 650 msec = 0.650 sec.

Conversely, to convert 12.5 meters to kilometers, start at the base and move to the kilo-. This moves against the chart, since you are traveling away from the base. This means you must multiply 12.5 by 10^{-3}. Therefore 12.5 m = 0.0125 km.

Another way to use the chart is to count how many spaces you must move from where you begin to where you end, and move the decimal in the problem in the exact same direction. For the 12.5 meter to kilometer example, begin at the base and move 3 spaces to the left to reach kilo-. This means you would move the decimal three places to the left and end with 0.0125 km.

QUICK TIP

Use the following acronym to remember the order of the SI units:

kilo, hecto, deca, (base), deci, centi, milli

King **H**enry **D**ied **B**y **D**rinking **C**hocolate **M**ilk

➡ EXAMPLE 1.4

Convert the following to the given measurement.

1. 783 cm to hm.
2. 1.09 g to mg
3. 4.52×10^2 km to cm
4. 6540 dl to kl

Answer Explanations

1. **0.0783 hm**

   Starting at centi- and moving towards hecto- means you move four spaces to the left. Move the decimal in the same direction.

kilo-	hecto-	deca-	Base Unit	deci-	centi-	milli-
k	h	da		d	c	m

   ←

   move 4 spaces to left

2. **1 090 mg**

Starting at the base and moving towards milli- means you move three spaces to the right. Move the decimal in the same direction.

kilo- k	hecto- h	deca- da	**Base Unit**	deci- d	centi- c	milli- m

move 3 spaces to right

3. 4.52×10^7 **cm**

Starting at kilo- and moving to centi- means you move five spaces to the right. Move the decimal in the same direction. It is helpful to write the original number as 452, move the decimal five spaces to get 452 000, and then put the number back in standard form.

kilo- **k**	**hecto-** **h**	**deca-** **da**	**Base Unit**	**deci-** **d**	**centi-** **c**	milli- m

move 5 spaces to right

4. **0.654 kl**

Starting at deci- and moving to kilo- means, move four spaces to the left. Move the decimal in the same direction.

kilo- **k**	**hecto-** **h**	**deca-** **da**	**Base Unit**	**deci-** **d**	centi- c	milli- m

move 4 spaces to left

1.5 CURRENCY CONVERSIONS

Countries often have different currencies and these currencies have different values. When converting between currencies, you must use exchange rates. These rates vary from day to day, but they allow us to change between currencies. There are two different methods to convert currencies.

For example, convert 350 US dollars (USD) to euro (EUR) at a rate of 1 EUR = 1.238 USD.

Method 1: Proportions

STEP 1 Set up a proportion with the exchange rate on one side and the problem on the other.

$$\frac{1 \, \text{euro}}{1.238 \, \text{USD}} = \frac{x \, \text{euro}}{350 \, \text{USD}}$$

STEP 2 Cross-multiply and solve.

$$350 = 1.238x$$
$$282.71 = x$$
$$282.71 \, \text{euro}$$

You can write your answer to two decimal places, unless otherwise stated, since that is the standard for money.

Method 2: Conversion Factors

STEP 1 Set up a ratio with the given money amount.

$$\frac{350 \, \text{USD}}{1}$$

STEP 2 Multiply by the exchange rate so that the given currency divides out and the desired currency remains.

$$\frac{350 \, \cancel{\text{USD}}}{1} \times \frac{1 \, \text{euro}}{1.238 \, \cancel{\text{USD}}}$$
$$= 282.71 \, \text{euro}$$

Financial institutions do not exchange currencies at no cost to you. They make money by either different buy/sell rates or by charging commission. When a bank charges a commission, this is deducted before converting the currency because it is a fee for using their services.

➡ EXAMPLE 1.5

Gim is traveling from South Korea to Chile. The bank sells Chilean peso (CLP) at a rate of 1 Korean won (KRW) = 0.46 CLP. A commission is charged at a rate of 1.5% in KRW. If Gim pays 1 million KRW, how many CLP does he receive?

Answer Explanation

Gim receives 472 800 CLP

First, you must subtract the commission from the total KRW Gim will convert since this is a fee. The commission will be $1\,000\,000(0.015) = 15\,000$ KRW.

Gim will convert $1\,000\,000 - 150\,000 = 985\,000$ KRW.

Now, to convert his remaining won.

Method 1

$$\frac{1 \, \text{KRW}}{0.46 \, \text{CLP}} = \frac{985\,000 \, \text{KRW}}{x \, \text{CLP}} \quad \text{so } x = 453\,100 \, \text{CLP}$$

Method 2

$$\frac{985\,000 \, \cancel{\text{KRW}}}{1} \times \frac{0.46 \, \text{CLP}}{1 \, \cancel{\text{KRW}}} = 453\,100 \, \text{CLP}$$

1.6 SOLVING LINEAR AND QUADRATIC EQUATIONS

In IB Math Studies, it is expected that you are able to solve a system of linear equations and quadratic equations by using your graphics display calculator, or GDC.

When solving a pair of linear equations in two variables, you can graph both lines to find the point of intersection. If the two lines are parallel and never cross, the answer is **no solution**. If only one line appears on the graph, then the two lines are the same. The answer is the **equation of the line**.

➥ **EXAMPLE 1.60**

$$\text{Solve } \begin{array}{l} 2x - 3y = 10.5 \\ 6y - x = -16.5 \end{array}.$$

Answer Explanation

(1.5, −2.5)

STEP 1 Rearrange both lines into slope-intercept form.

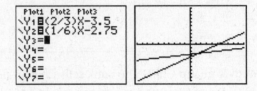

$$2x - 3y = 10.5 \qquad\qquad 6y - x = -16.5$$
$$-3y = -2x + 10.5 \qquad\qquad 6y = x - 16.5$$
$$y = \frac{2}{3}x - 3.5 \qquad\qquad y = \frac{1}{6}x - 2.75$$

STEP 2 Type both equations into your GDC and graph.

STEP 3 Use your GDC to find the intersection.

Press 2ND TRACE. In the CALCULATE menu, choose option 5: INTERSECT. Move the cursor using the left or right arrows to the point of intersection and hit ENTER. Repeat two more times. The solution will be displayed at the bottom of the screen.

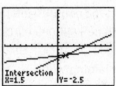

STEP 4 Write down the point of intersection.

(1.5, −2.5)

Solving quadratic equations is also done by graphing, but since there is only one equation, there will be no point of intersection. Instead, you are finding the roots or zeros of the quadratic. This is simply the point or points where the quadratic crosses the x-axis. Remember, at the x-intercept, the value of y is always zero, hence the name "zero".

➥ **EXAMPLE 1.61**

Solve $2x^2 - 5 = 3x$.

Answer Explanation

$x = -1, 2.5$

STEP 1 Set the quadratic equal to zero. This is very important since you need the value of the equation to be zero so you can find the zeros.

$$2x^2 - 3x - 5 = 0$$

STEP 2 Graph the quadratic.

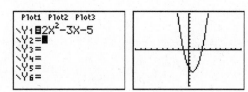

STEP 3 Find the roots or zeros of the graph.

Press (2ND TRACE). In the (CALCULATE) menu, choose option 2: (ZERO). You can only find one zero at a time. Choose an *x*-intercept to find the first and move the cursor using the left or right arrows to the LEFT of the zero and hit (ENTER). Then, move the cursor to the RIGHT of the zero and hit (ENTER). The final step is to just hit (ENTER). The zero will be displayed at the bottom of the screen.

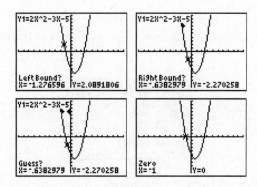

Repeat the process again to find the second root.

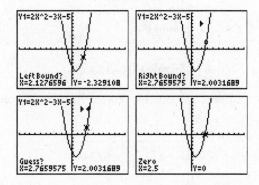

Another way to solve a quadratic equation is through the use of the polynomial root application on the GDC. Depending on your calculator, you may either have (PLYSMLT) or (PLYSMLT2) under the (APPS) menu. If your calculator does not have this app, you should ask your teacher, as calculators can share applications.

In order to utilize the (PLYSMLT) app, the quadratic must equal zero and the terms should be in descending order.

➥ **EXAMPLE 1.62**

Solve $10 - 6x = x^2$.

Answer Explanation

$x = -7.36, 1.36$

STEP 1   Set the quadratic equal to zero and write the terms in descending order.

$$-x^2 - 6x + 10 = 0$$

STEP 2   On your GDC, go to the APPS menu and select PLYSMLT or PLYSMLT2. Choose 1: POLY ROOT FINDER.

```
        MAIN MENU
 1:POLY ROOT FINDER
 2: SIMULT EQN SOLVER
 3: ABOUT
 4: POLY HELP
 5: SIMULT HELP
 6: QUIT POLYSMLT
```

QUICK TIP

To solve any quadratic equation, it MUST equal zero.

Always write a quadratic in the form $ax^2 + bx + c = 0$.

The PLYSMLT app is the quickest way to solve!

STEP 3   Since the highest exponent is "2", the order is 2. The rest of the options can remain in the default settings. If you prefer answers to be expressed as fractions, highlight the FRAC option.

```
 POLY ROOT FINDER MODE
ORDER  1 2 3 4 5 6 7 8 9 10
REAL   a+bi  re^(θi)
DEC    FRAC
NORMAL SCI ENG
FLOAT  0 1 2 3 4 5 6 7 8 9
RADIAN DEGREE
MAIN          HELP NEXT
```

STEP 4   Enter the coefficients from the quadratics.

```
a2x²+a1x+a0=0
 a2 = -1
 a1 = -6
 a0 =10


MAIN MODE CLR LOAD SOLVE
```

STEP 5   Solve and the solutions are on the screen. Be sure to round to 3 sig figs when appropriate.

```
a2x²+a1x+a0=0
 x1 = -7.358898944
 x2 =1.358898944



MAIN MODE COEF STO F◄►D
```

When solving a quadratic equation, there will be two solutions, one solution or no solution. You can determine how many solutions there are by how many times the graph crosses the x-axis.

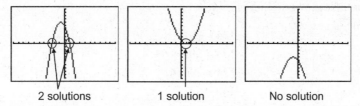

2 solutions   1 solution   No solution

1.7 ARITHMETIC SEQUENCES AND SERIES

An arithmetic sequence is a list of numbers that either increase or decrease by a common difference. For example, 5, 10, 15, 20,... is an arithmetic sequence since each number is 5 more than the previous.

The formula for the nth term of an arithmetic sequence is

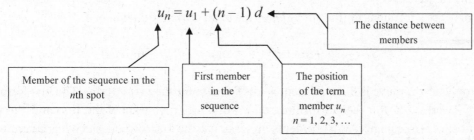

$$u_n = u_1 + (n - 1)\,d$$

The distance between members

Member of the sequence in the nth spot

First member in the sequence

The position of the term member u_n
$n = 1, 2, 3, \ldots$

You can use the arithmetic sequence formula to solve for any of the variables. Just simply substitute in what you are given and solve for the remaining variable.

➡ EXAMPLE 1.70

Given the sequence 20, 16, 12, 8, ... find:

1. the common difference, d;
2. the 81st term;
3. if −124 is a member of the sequence.

Answer Explanations

1. **$d = -4$**

   To find the common difference, subtract <u>two</u> pairs of consecutive members.

   $$16 - 20 = -4;\ 12 - 16 = -4$$

2. **$u_{81} = -300$**

   The information you know from the problem is $n = 81$, $u_1 = 20$, and $d = -4$. Substituting this into the formula gives $u_{81} = 20 + (81 - 1)(-4)$, which equals −300.

> **QUICK TIP**
>
> **For arithmetic sequences:**
>
> - u is **who**—u_n is the value...
>
> - n is **position**—... in the nth seat
>
> - Generated by adding or subtracting the same value (**d**, the common difference)
>
> - If the values are increasing, $d > 0$
>
> - If the values are decreasing, $d < 0$

3. **Yes**

If -124 is a member of the sequence, then it means
$u_n = -124$, $u_1 = 20$, and $d = -4$. You need to find n to know
if the value falls in the sequence.

$$-124 = 20 + (n - 1)(-4)$$
$$-124 = 20 - 4n + 4$$
$$-124 = 24 - 4n$$
$$-148 = -4n$$
$$37 = n$$

Since n is a counting number, this means -124 is the 37th term in the sequence, so it is a member. If you had solved the equation and n was a decimal, then the answer would have been "no," since n must be a counting number.

An arithmetic series is the sum of an arithmetic sequence.

For example, if 4, 7, 10, 13, 16 is an arithmetic sequence, then $4 + 7 + 10 + 13 + 16$ is an arithmetic series.

There are 2 formulas for the sum of n terms of an arithmetic sequence:

$$S_n = \frac{n}{2}[2u_1 + (n-1)d] \qquad \textbf{or} \qquad S_n = \frac{n}{2}(u_1 + u_n)$$

The only new variable is S_n, which represents the sum for the first n terms. The first formula <u>always</u> works, but the second formula only works when you know the first and the last member of the sequence. Most students prefer to use the first formula since it is guaranteed to work.

➡ **EXAMPLE 1.71** ───────────────────────────

A theatre has 21 seats in the first row, 24 seats in the second row, 27 in the third, and so on.

1. What is the **total** number of seats, if the theatre has 20 rows of seats?
2. How many rows of seats would the theatre have, if there was a total of 1935 seats?

Answer Explanations

1. **990 seats**

You should use the first sum formula since you do not know how many seats are in the last row. Substituting into the formula gives $S_{20} = \frac{20}{2}[2(21) + (20 - 1)(3)]$, which simplifies to a solution of 990.

2. **30 rows**

This time you know the actual sum, $S_n = 1935$, while $u_1 = 21$ and $d = 3$. Substituting into the formula gives $1935 = \frac{n}{2}[2(21) + (n-1)(3)]$. To simplify the equation, begin with parentheses.

$$1935 = \frac{n}{2}[42 + 3n - 3]$$

$$1935 = \frac{n}{2}(39 + 3n)$$

Oftentimes, multiplying both sides by 2 makes the algebra easier since the fraction of $\frac{n}{2}$ will become just n.

$$\left[1935 = \frac{n}{2}(39 + 3n)\right] \times 2$$

$$3870 = n(39 + 3n)$$

Now distribute n.

$$3870 = 39n + 3n^2$$

To solve the resulting quadratic, use PLYSMLT. The equation must equal zero and the terms should be listed in descending order.

$$0 = 3n^2 + 39n - 3870$$

Entering the coefficients into the solver application and solving, you get two solutions.

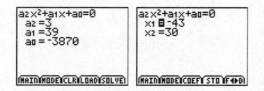

A negative solution is inappropriate since the answers represent the number of rows. Thus, there are 30 rows in the theatre.

1.8 GEOMETRIC SEQUENCES AND SERIES

Geometric sequences are a pattern of numbers like arithmetic, except consecutive numbers are multiplied by the same rate. The list 2, 4, 8, 16, 32, … is a geometric sequence since each number is twice the one before it.

The formula for the nth term of a geometric sequence is

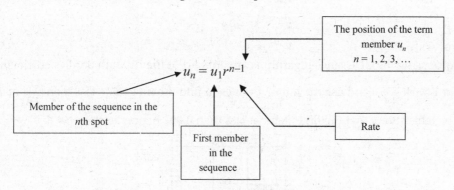

You can use the geometric sequence formula to solve for any of the variables. Just substitute in what you are given, and then solve for the remaining variable.

➡ EXAMPLE 1.80

Given the sequence 64, 16, 4, 1, … find:

1. The rate, r.
2. The 21st term. Write the answer in the form $a \times 10^k$, where $1 \leq a < 10$ and $k \in \mathbb{Z}$.
3. When the sequence will be smaller than $\dfrac{1}{250}$.

Answer Explanations

QUICK TIP

For geometric sequences:

- u is **who**—u_n is the value…
- n is **position**—… in the nth seat
- Generated by multiplying the same value (r, the common ratio)
- If the values are increasing, $r > 1$
- If the values are decreasing, $0 < r < 1$

1. $r = \dfrac{1}{4}$

   To find the rate, simply divide two pairs of consecutive members.

   $$\frac{16}{64} = \frac{1}{4}$$

   $$\frac{4}{16} = \frac{1}{4}$$

   Always check two pairs to ensure it works for both.

2. $u_{21} = 5.82 \times 10^{-11}$

   You know $n = 21$, $u_1 = 64$, and $r = \dfrac{1}{4}$. Substitute into the formula:

   $u_{21} = 64\left(\dfrac{1}{4}\right)^{21-1}$, which equals 5.82×10^{-11}.

3. $n = 8$

   Since you must find which member is smaller than $\dfrac{1}{250}$, this means $u_n = \dfrac{1}{250}$, $u_1 = 64$, and $r = \dfrac{1}{4}$. Substitute into the formula:

   $$\frac{1}{250} > 64\left(\frac{1}{4}\right)^{n-1}$$

   Solve using the GDC since logarithms are not part of the IB Math Studies curriculum. Let $Y_1 = 64\left(\dfrac{1}{4}\right)^{x-1}$ and use the table of values to find the first value less than $\dfrac{1}{250}$ or 0.004.

   The table shows that the first member less than 0.004 happens when $n = 8$.

X	Y1
3	4
4	1
5	.25
6	.0625
7	.01563
8	.00391
9	9.8E-4
X=8	

A geometric series is the sum of a geometric sequence.

For example, if 1, 3, 9, 27, 81 is a geometric sequence, then $1 + 3 + 9 + 27 + 81$ is a geometric series.

There are 2 formulas for the sum of n terms of a geometric sequence:

$$S_n = \frac{u_1(r^n - 1)}{r - 1} \qquad \text{or} \qquad S_n = \frac{u_1(1 - r^n)}{1 - r} \qquad \text{but } r \neq 1$$

The only new variable is S_n, which represents the sum for the first n terms. The two formulas are virtually the same, however the first is for cases when $r > 1$ and the second for cases when $0 < r < 1$.

➡ EXAMPLE 1.81

Consider the sequence 1, 5, 25, 125, ...

1. Find the rate, r.
2. Calculate the sum of the first 7 members.
3. Find the number of terms in the sequence when the sum is 2 441 406.

Answer Explanations

1. **$r = 5$**

   To find the rate, divide two pairs of consecutive members.

   $$\frac{5}{1} = 5 \text{ and } \frac{25}{5} = 5$$

> **QUICK TIP**
>
> **When solving a geometric equation:**
> - Enter the exponential equation into Y_1 in the GDC
> - Use the table to find the desired value
> - The solution is always the value of x

2. **$S_7 = 19\,531$**

   Since the rate is larger than 1, use the first sum formula.

   Substituting the given information gives $S_7 = \dfrac{1(5^7 - 1)}{5 - 1}$. When entering this into your GDC, you must put parentheses around the denominator.

3. **$n = 10$**

   $S_n = 2\,441\,406$, $u_1 = 1$, and $r = 5$. Substitute into the sum formula: $2\,441\,406 = \dfrac{1(5^n - 1)}{5 - 1}$.

   Let $Y_1 = \dfrac{1(5^x - 1)}{4}$, and then use the table of values to find the first value equal to 2.44×10^6.

   Look for 2.44×10^6 rather than $2\,441\,406$ since the calculator has limited space to display digits. For larger numbers, the table converts to standard form. The table shows that when $n = 10$, the sum is 2.44×10^6.

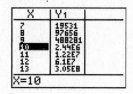

   Since the value of 2.44×10^6 is not exactly $2\,441\,406$, you should check our answer.

   $$\frac{1(5^{10} - 1)}{5 - 1} = 2\,441\,406$$

FEATURED QUESTION

May 2010, Paper 2

Part A

Daniel wants to invest $25000 for a total of three years. There are three investment options.

Option One   pays simple interest at an annual rate of interest of 6%
Option Two   pays compound interest at a nominal annual rate of interest of 5%, compounded **annually**
Option Three   pays compound interest at a nominal annual rate of interest of 4.8%, compounded **monthly**

(a) Calculate the value of his investment at the end of the third year for each investment option, **correct to two decimal places**.
(b) Determine Daniel's best investment option.

Part B

An arithmetic sequence is defined as

$$u_n = 135 + 7n \qquad n = 1, 2, 3, \ldots$$

(a) Calculate u_1, the first term in the sequence.
(b) Show that the common difference is 7.

S_n is the sum of the first n terms of the sequence.

(c) Find an expression for S_n. Give your answer in the form $S_n = An^2 + Bn$, where A and B are constants.

The first term, v_1, of a geometric sequence is 20, and its fourth term, v_4, is 67.5.

(d) Show that the common ratio, r, of the geometric sequence is 1.5.

T_n is the sum of the first n terms of the geometric sequence.

(e) Calculate T_7, the sum of the first seven terms of the geometric sequence.
(f) Use your graphic display calculator to find the smallest value of n, for which $T_n > S_n$.

(See page 435 for solutions)

1.9 FINANCIAL APPLICATIONS

Geometric sequences can be applied to financial problems since percentage rates multiply. The thing to watch out for is the percentage rate given in the problem is oftentimes not the one you will use when calculating the answer. For interest rates that appreciate, meaning grow or increase, add the given rate to 100%. For interest rates that depreciate, meaning decay or decrease, subtract the given rate from 100%.

➥ EXAMPLE 1.90

Lainie began a new job with a starting salary of $23 500 and a 5% increase each year after.

1. Find Lainie's salary after working for 6 years.
2. Calculate the total amount of money Lainie will earn if she works for 20 years at this job.

Answer Explanations

1. $u_6 = \$29\,992.62$

    Since the initial salary is $23 500, $u_1 = 23\,500$, but the rate is **not** 0.05. Lainie keeps 100% of her current salary but adds an additional 5% each year, so the rate is 105% (100% + 5%).

    Thus, $r = 1.05$.

    $$u_8 = 23\,500(1.05)^{6-1}$$

2. $S_{20} = \$777\,049.92$

    Since the rate is larger than 1, use the first sum formula. Substitute the given information: $S_{20} = \dfrac{23\,500\left(1.05^{20}-1\right)}{1.05-1}$. When entering this into the GDC, put parentheses around the denominator.

➥ EXAMPLE 1.91

Miller bought a car worth 26 000 euro. The car's value depreciates at a rate of 14.1% per year. Determine the value of the car after 5 years.

Answer Explanation

12 160.16 euro

The car is depreciating in value, which means it loses 14.1% of its value each year.

The rate is 100% – 14.1%, thus $r = 0.859$. Also, the car will depreciate for 5 full years, including the first year, so do not subtract 1 from n when using the formula.

Substitute the values into the sequence formula:
$u_5 = 26\,000(0.859)^5$

The car is worth 12 160.16 euro after 5 years.

> **QUICK TIP**
>
> When the value of an object **appreciates** (or increases), the common ratio is 100% plus the given percent
>
> When the value of an object **depreciates** (or decreases), the common ratio is 100% minus the given percent

You can also solve problems involving compound interest, meaning the interest is added a set number of times per year.

This involves the use of the formula $FV = PV \times \left(1 + \dfrac{r}{100k}\right)^{kn}$.

The Math Studies formula packet explains each variable:

FV = future value, PV = present value, n = number of years, k = number of compounding periods per year, $r\%$ = nominal annual rate of interest.

A savings account with a bank earns interest. If the bank compounds the interest quarterly, interest is added to the account 4 times a year. Monthly means interest is added 12 times a year, and half-yearly would be twice a year. These values would all represent k. The nominal annual rate of interest merely means the interest earned per year, but this is broken into equal amounts by k, the compounding periods.

➡ EXAMPLE 1.92

Bahir invested 5 500 Turkish lira (TL) in a savings account, with a nominal rate of 2.45% p.a. compounded quarterly.

1. Calculate the amount Bahir will have in his account after 6 years.
2. To retire, Bahir needs to double his investment. Determine the number of years before Bahir can retire.

Answer Explanations

> **QUICK TIP**
>
> **For Compounded Interest:**
>
> Annual: $k = 1$
>
> Semi-annual: $k = 2$
>
> Trimesters: $k = 3$
>
> Quarterly: $k = 4$
>
> Monthly: $k = 12$
>
> Daily: $k = 365$

1. **6 368.09 TL**

   Substitute the given information into the compound interest formula:

   $$FV = 5\,500\left(1 + \frac{2.45}{100 \times 4}\right)^{4 \times 6}.$$

   Before using the GDC, simplify some of the simple math.

   $$FV = 5\,500\left(1 + \frac{2.45}{400}\right)^{24}$$

   Thus, Bahir will have 6 368.09 TL.

2. **After 29 years**

   In order for Bahir to retire, he must double his investment, which is 2(5 500) or 11 000. This is his "future value". Substitute into the formula:

   $$11\,000 = 5\,500\left(1 + \frac{2.45}{100 \times 4}\right)^{4n}$$

   Solve using the GDC.

   Let $Y_1 = 5\,500\left(1 + \dfrac{2.45}{400}\right)^{4x}$

Now, use the table to find when the value of Y_1 is greater than 11 000.

X	Y1
27	10636
28	10899
29	11168
30	11444
31	11727
32	12017
33	12315

X=29

The table shows his investment will double after 29 years.

TOPIC 1 PRACTICE

Paper 1 Questions

1. Place the following numbers in the appropriate circle on the Venn diagram.

$$-\pi, \ -\frac{3}{4}, \ 0, \ \sqrt{36}, \ 1.75, \ -1$$

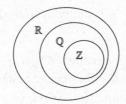

2. Let $A = 6.5 \times 10^{-4}$ and $B = 2.7 \times 10^{-3}$. Find

   (a) $\dfrac{A}{B}$

   (b) $4(A + B)$

   (c) $\dfrac{B}{(A+B)^2}$

   Give your answers in the form $a \times 10^k$, where $1 \le a < 10$ and $k \in \mathbb{Z}$.

3. (a) A rectangular piece of paper has a dimension of 8 cm by 5 cm. Find the area of the paper in **mm**.
   (b) Marco drank 3 liters of water. Write this in **mL**.
   (c) Baskana played soccer for 1 hour 12 minutes. Write this number to the nearest **100 seconds**.
   (d) A rectangular building has dimensions of 76 m × 100 m. Find the perimeter of the building in **km**.

4. Given $p = \sqrt{\dfrac{x^2 - 10}{2x}}$, let $x = 9$:

   (a) Calculate p. Write the answer to the nearest thousandth.
   (b) Write the answer to 2 significant figures.
   (c) Calculate the percentage error between the answer from a and b.

5. The sum of the first 10 terms of an arithmetic series is 165. The common difference is 3.

   (a) Find the value of the first term.
   (b) What is the 75th term of the sequence?
   (c) The sum of the first n terms is 513. Find n.

6. An author begins writing a book. She writes 200 words the first day, and then increases the amount of words she writes by 8% each day.

   (a) On which day does she write 500 words?
   (b) What is the total number of words the author has written after 2 weeks of work?

7. At the local pizza restaurant, Sammy bought 2 slices of pizza and one soda for $9.25, while Robbie bought 5 slices of pizza and two sodas for $22.25.

   (a) Express this information using two equations relating the cost of pizza slices and the cost of sodas.
   (b) Find the cost for one slice of pizza.

8. Camille invests 5 000 euro in an account with a nominal interest rate of 4.3% p.a. compounded half-yearly.

   (a) Find the value of her investment after 15 years.
   (b) Determine the amount Camille would need to invest at the same nominal rate if she wants to have the same amount in 10 years.

9. Eleni is planning a trip from South Africa to England. She converts 45 000 South African Rand (ZAR) at her bank, which sells British pounds (GPB) at a rate of 1 GBP = 15.7 ZAR.

   (a) Calculate how many British pounds Eleni will receive.

   Eleni becomes ill before her trip and she must cancel. She returns to the bank to exchange her GBP back into rand. The bank buys British pounds at a rate of 1 GBP = 15.3 ZAR.

   (b) Determine the amount in ZAR she receives.
   (c) Calculate Eleni's percentage loss from her original 45 000 ZAR.

SOLUTIONS

1. $-\pi, -\dfrac{3}{4}, 0, \sqrt{36}, 1.75, -1$

   - $-\pi$ is an irrational number, so it only belongs in the real number circle.
   - $-\dfrac{3}{4}$ and 1.75 are both rational numbers, but they are not integers since they are fractions/decimals.
   - $0, \sqrt{36}$, and -1 are integers, since they are positive or negative counting numbers plus zero ($\sqrt{36}$ reduces to 6).

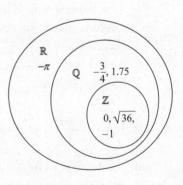

2.   (a) **2.41 × 10⁻¹**

     (b) **1.34 × 10⁻²**

     (c) **2.41 × 10²**

> Each of these was computed using the GDC. Be sure to appropriately use parentheses.

3.   (a) **4 000 mm²**

     8 cm = 80 mm and 5 cm = 50 mm

     Area is 80 × 50 = 4 000 mm²

     (b) **3 000 ml**

     To move from the liters to milli-, you moved 3 places to the right on the chart. Do the same for the decimal.

     (c) **4 300 seconds**

     First: 1 hour = 60 minutes

     1 × 60 = 60 minutes. Each minute has 60 seconds.

     60 × 60 = 3 600 seconds

     Second: 12 minutes × 60 seconds = 720 seconds.

     Add these together to get 4 320. You must round to the nearest hundred.

     (d) **0.352 km**

     76 m = 0.076 km and 100 m = 0.1 km

     Perimeter is the sum of all the sides, so $P = 2(0.076) + 2(0.1)$

4.   (a) $p = 1.986$

     Entering the expression into the calculator involves parentheses around the numerator and denominator.

     ```
     √(9²-10)/(2*9)
              1.986062548
     ```

     The answer must also be rounded to the nearest thousandth, which is the third decimal place.

     (b) **2.0**

     To round to 2 sig figs, you can only have 2 digits, thus you must round off at the tenths place. You must keep the zero in the tenths place so that you have 2 sig figs.

     (c) **0.705%**

     $$\left| \frac{2.0 - 1.986}{1.986} \right| \times 100\% = 0.705\%$$

     Let $v_A = 2.0$, since it is the most rounded, and $v_E = 1.986$, since it is more exact.

5. (a) $u_1 = 3$

Substitute the given information into the arithmetic sum formula and then solve.

$$165 = \frac{10}{2}[2u_1 + (10-1)(3)]$$

$$165 = 5[2u_1 + 27]$$

$$165 = 10u_1 + 135$$

$$30 = 10u_1$$

$$3 = u_1$$

(b) $u_{75} = 225$

Substitute the given information into the arithmetic sequence formula:

$u_{75} = 3 + (75 - 1)(3)$. This simplifies to 225.

(c) $n = 18$

Substitute the given information into the arithmetic sum formula and simplify.

$$513 = \frac{n}{2}[2(3) + (n-1)(3)]$$

$$513 = \frac{n}{2}[6 + 3n - 3]$$

$$1\,026 = n(3n + 3)$$

$$0 = 3n^2 + 3n - 1\,026$$

Now solve by PLYSMLT.

The only solution is $n = 18$, since $n = -19$ does not make contextual sense.

6. (a) $n = 13$

Substitute the given information into the geometric sequence formula.

$$500 = 200(1.08)^{n-1}$$

Let $Y_1 = 200(1.08)^{x-1}$ and use the table of values to find the first value of 500 or greater.

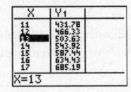

The table shows the first day she writes 500 words will be day 13.

(b) 4 840 words

This is a geometric sum, since you are asked to find a total.

$S_{14} = \dfrac{200(1.08^{14} - 1)}{1.08 - 1}$. You know $n = 14$, since there are 7 days in one week.

The answer must be rounded to 3 significant figures.

7. (a) $\begin{array}{l}2x+y=9.25\\5x+2y=22.25\end{array}$, where $x=$ cost of one slice of pizza and $y=$ cost for one soda.

Sammy bought 2 slices of pizza and one soda for \$9.25. Thus, the equation representing Sammy's cost is $2x+y=9.25$.

Robbie bought 5 slices of pizza and two sodas for \$22.25. Thus, the equation representing Robbie's cost is $5x+2y=22.25$.

(b) $x=3.75$

Solve the system of linear equations by graphing.

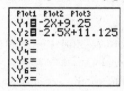 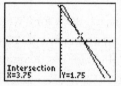

8. (a) **9 465.01 euro**

$$FV=5\,000\left(1+\frac{4.3}{100\times2}\right)^{2\times15}$$

Since Camille invests 5 000 euro, $PV=5\,000$, $r=4.3\%$ p.a., and since the interest is compounded half-yearly, $k=2$.

(b) **6 185.20 euro**

Camille wants the same amount, which means $FV=9\,465.01$ and the interest rate stays the same.

$$9\,465.01=PV\left(1+\frac{4.3}{100\times2}\right)^{2\times10}$$
$$9\,465.01=1.5303PV$$
$$6\,185.20=PV$$

9. (a) **2 866.24 GBP**

There are two methods to convert the 45 000 ZAR to GPB:

Method 1: Multiply Conversion Factor	Method 2: Currency Proportion
$45\,000\,\text{ZAR}\times\dfrac{1\,\text{GBP}}{15.7\,\text{ZAR}}=2\,866.24$	$\dfrac{1\,\text{GBP}}{15.7\,\text{ZAR}}=\dfrac{x\,\text{GBP}}{45\,000\,\text{ZAR}}$ $45\,000=15.7x$ $2\,866.24=x$

(b) **43 853.50 ZAR**

There are two methods to convert the 2 866.24 GPB back to ZAR:

Method 1: Multiply Conversion Factor	Method 2: Currency Proportion
$2\,866.24\,\text{GBP}\times\dfrac{15.3\,\text{ZAR}}{1\,\text{GBP}}=43\,853.50$	$\dfrac{1\,\text{GBP}}{15.3\,\text{ZAR}}=\dfrac{2\,866.24\,\text{GBP}}{x\,\text{ZAR}}$ $x=43\,853.50$

(c) **2.55%**

The percentage error formula can be used to calculate her percentage loss.

v_A = post conversion value, while v_E = original amount

$$\left|\frac{43\,853.50-45\,000}{45\,000}\right|\times100\% = 2.55\%$$

Paper 2 Questions

1. A local town is selling raffle tickets for a large monetary prize. The winner of the raffle may choose one of the following options.

   Option A: $300 each week for 8 weeks.

   Option B: $9 in the first week, $18 in the second week, $36 in the third week, continuing to double for a total of 8 weeks.

   Option C: $75 in the first week, $140 in the second week, $205 in the third week, increasing by $65 each week for a total of 8 weeks.

   (a) Calculate the amount received in the eighth week, if the winner selects:
   - (i) **Option B**
   - (ii) **Option C**

   (b) What is the total payout for **option B**?

   (c) Which option has the greatest total payout? Support your answer with appropriate calculations.

2. (a) Evaluate $34.01\times\dfrac{\sqrt{3.75}}{1.25\times(2.34)^2}$, giving your answer in:

   - (i) 4 significant figures
   - (ii) the nearest hundredth
   - (iii) 1 significant figure

   (b) A large flag is being sewn together from four pieces of fabric. When rounding to the nearest cm, the length of each piece is 1.50 m, but the actual length is 1.48 m.

   - (i) Find the actual length of the flag after the pieces are sewn together.
   - (ii) Calculate the percentage error between the actual length and the rounded length of the fabric.
   - (iii) If the fabric costs $0.05 per centimeter, how much will the flag cost? Use the rounded length.

3. Mirah deposits $3\,000 into an investment account that pays 3.7% nominal interest per annum, compounded monthly.

   (a) Find the value of the account after 7 years.

   (b) Determine the number of years it would take Mirah's investment of $3\,000 to double in value.

   Patrice deposits $300 into an investment account each year on her birthday. The account pays 3.2% nominal interest per annum, compounded yearly.

(c) Find the value of Patrice's account after 5 years.

(d) Determine the number of years it would take for Patrice to have \$2 000 in her account.

SOLUTIONS

1. (a) (i) $u_8 = 9(2)^{8-1} = 1\,152$

This is a geometric sequence, since each member is twice the one before.

(ii) $u_8 = 75 + (8-1)(65) = 530$

This is an arithmetic sequence, since each member is 65 more than the previous.

(b) $S_8 = \dfrac{9(2^8 - 1)}{2 - 1} = 2\,295$

This is a geometric sum, since you are asked to find the **total** payout.

(c) **Option C has the greatest total payout**

Option A total payout: $300 \times 8 = 2\,400$

Option B total payout: $2\,295$

Option C total payout: $S_8 = \dfrac{8}{2}[2(75) + (8-1)(65)] = 2\,420$

2. (a) (i) **9.622**

Entering this into your GDC requires parentheses around the denominator.

```
34.01*√3.75/(1.25*2.34²)
        9.622336446
■
```

(ii) **9.62**

(iii) **10**

Rounding to one sig fig means you can only have one digit. The 6 rounds the 9 up to 10. Remember, when the decimal is absent, zeros at the end do not count.

(b) (i) **4(1.48) = 5.92**

The actual length is 1.48, and there are 4 panels being sewn together.

(ii) $\left|\dfrac{6 - 5.92}{5.92}\right| \times 100\% = 1.35\%$

The answer is the same, if using the given lengths: $\left|\dfrac{1.50 - 1.48}{1.48}\right| \times 100\% = 1.35\%$

(iii) **4(1.50) = 6**

Using the rounded length, you will need 6 meters of fabric.

The price is per centimeter, so 6 m = 600 cm.

$$600(0.05) = \$30$$

3. (a) $FV = 3\,000\left(1 + \dfrac{3.7}{100 \times 12}\right)^{12 \times 7} = 3\,885.35$

Since Mirah deposited \$3 000, $PV = 3\,000$, $r = 3.7\%$, $k = 12$ (compounded monthly), $n = 7$.

(b) $6\,000 = 3\,000\left(1 + \dfrac{3.7}{100 \times 12}\right)^{12n}$

Solve using the GDC.

Let $Y_1 = 3\,000\left(1 + \dfrac{3.7}{1\,200}\right)^{12x}$ and use the table to find when the value is greater than $6\,000$.

X	Y₁
15	5221.4
16	5417.9
17	5621.8
18	5833.3
19	6052.9
20	6280.7
21	6517

X=19

$n = 19$

(c) $S_5 = \dfrac{300(1.032^5 - 1)}{1.032 - 1} = 1\,599.12$

Patrice is **adding** $300 each year on top of earning interest, so the problem is actually a geometric sum. The rate will be $100\% + 3.2\%$.

(d) $n = 7$

$$2\,000 = \dfrac{300(1.032^n - 1)}{1.032 - 1}$$

Solve using the GDC.

Let $Y_1 = \dfrac{300(1.032^x - 1)}{0.032}$ and use the table to find when the value is greater than $2\,000$.

X	Y₁
5	1599.1
6	1950.3
7	2312.7
8	2686.7
9	3072.7
10	3471
11	3882.1

X=7

$n = 7$

CHAPTER OBJECTIVES

Before you move on to the next chapter, you should be able to:

- ☐ Identify to which sets a number belongs (Real, Rational, Integer, Natural)
- ☐ Round a number to a given significant figure
- ☐ Calculate the percentage error between an exact value and an approximate value
- ☐ Write a number in standard form
- ☐ Perform operations (+, −, ×, ÷) involving numbers in standard form
- ☐ Convert between SI measurements
- ☐ Convert between different currencies
- ☐ Determine the commission charged for a currency conversion
- ☐ Solve a system of linear equations by graphing
- ☐ Solve a quadratic equation by graphing or using the PLYMSLT app
- ☐ Identify the common difference or common ratio for an arithmetic or geometric sequence
- ☐ Find a certain term in an arithmetic or geometric sequence
- ☐ Determine if a given value is a member of an arithmetic or geometric sequence
- ☐ Find the first term greater or smaller than a given value for an arithmetic or geometric sequence
- ☐ Determine the number of terms in an arithmetic or geometric sequence
- ☐ Calculate the sum for a set number of terms in an arithmetic or geometric sequence
- ☐ Determine the number of terms in an arithmetic or geometric sum
- ☐ Calculate the final value of an object if the value is appreciating or depreciating
- ☐ Determine the final value of an investment that earns compounded interest
- ☐ Determine the amount of time needed for an investment to reach a certain value if interest is compounded
- ☐ Calculate the rate of an investment whose final value is reached after a certain number of years

Descriptive Statistics

- **DISCRETE AND CONTINUOUS DATA:** counting versus measurement

- **FREQUENCY TABLE:** displays occurrence of data

- **HISTOGRAM:** graph displaying frequency of data intervals

- **CUMULATIVE FREQUENCY:** running total of frequencies

- **BOX AND WHISKER PLOT:** graph displaying minimum, first quartile, median, third quartile, and maximum of a data set

- **MEASURE OF CENTRAL TENDENCY:** mean, median, mode

- **MEASURES OF DISPERSION:** range, interquartile range, standard deviation

2.1 DATA CLASSIFICATION

Collected data can be classified as either *discrete* or *continuous*.

Discrete data are counted, such as the number of red candies in a bag, the number of elementary schools in a certain district, or the number of purchasing customers in a store during one week. Discrete data are exact values, since the data set is gathered by counting.

Continuous data are measured, such as the weights of second graders in a local school, the heights of giraffes at your zoo, or the length of fish caught in a pond. Continuous means the value is never ending, and measurements can always be measured using smaller, more accurate units (for example, kilometer – meter – centimeter – millimeter). Measurements technically never end, hence the term continuous. They are rounded approximations.

➡ **EXAMPLE 2.10**

Determine if the following data examples are discrete or continuous.

1. The time it takes for fifth grade girls to run a mile
2. The marks earned by a class of IB Math Studies students on an exam
3. The number of drivers talking on a cellular phone at a particular stoplight
4. The weights of newborn panda bears
5. The wingspans of birds at a pet store
6. The shoe sizes for a troupe of dancers

> **QUICK TIP**
>
> **Discrete Data:** COUNTED (number of…)
>
> **Continuous Data:** MEASURED (weight, height, etc.)

Answer Explanations

1. **Continuous**
   Time is a measurement and an approximation.
2. **Discrete**
   Marks are counted and are exact.
3. **Discrete**
   The number of drivers would be counted and exact.
4. **Continuous**
   Weight is a measurement and an approximation.
5. **Continuous**
   Wingspan is a measurement from the tip of one wing to the other and is an approximation.
6. **Discrete**
   In this example, the dancers' feet are not being measured, the shoe sizes are being counted.

2.2 FREQUENCY TABLES FOR DISCRETE DATA

When dealing with a large data set, it is often easiest to display the data in a frequency table. *Frequency* is how many times a certain value occurs.

Say you have a bag of red, blue, and green candies. You randomly select a candy, note the color, and then replace it. You do this 12 times. The results could be as follows:

red	red	blue	blue	blue	green
red	green	blue	blue	green	red

To display the data in a frequency table, count the occurrence of each candy.

Color	Frequency
Red	4
Blue	5
Green	3
Total	**12**

Notice the frequencies of each color add up to the total number of data values in the set.

➡ **EXAMPLE 2.20** ———————————————————————

Suppose a fair six-sided die is rolled 20 times and the number that shows face-up is recorded. Construct a frequency table to display the data.

1	5	5	4	6
3	2	3	6	1
2	1	4	4	5
1	6	6	3	3

Answer Explanation

Number	Frequency
1	4
2	2
3	4
4	3
5	3
6	4
Total	**20**

To count the frequency of the data values, list the numbers in ascending order.

1	1	1	1	2	2	3	3	3	3	4
4	4	5	5	5	6	6	6	6	6	6

Now, to construct the chart, write each data member with its occurrence.

2.3 GROUPED DATA

If the data set is spread out over a large range of values, it is best to group the data. For example, if you recorded the ages of people at the movie theater on any given day, the ages might range from 5 to 85 years. The most efficient way to display the data is to group values into **equal** intervals. Continuous data are always grouped, since measurements never end and intervals capture the true, non-rounded values. When creating a frequency table for grouped data, the intervals always include the beginning value but exclude the end.

➡ EXAMPLE 2.30

The following data set represents the time (t), in minutes, 20 high school football players took to run 1.6 kilometers.

10.2	9.5	8.3	8.75	7.2
6.7	11	9.4	9.3	9.1
8.2	7.5	7.1	6	9.3
7	10	12.1	7.3	7.9

Construct a frequency table using the intervals $6 \le t < 7$, $7 \le t < 8$, etc. to display the times.

Answer Explanation

Time	Frequency
$6 \le t < 7$	2
$7 \le t < 8$	6
$8 \le t < 9$	3
$9 \le t < 10$	5
$10 \le t < 11$	2
$11 \le t < 12$	1
$12 \le t < 13$	1
Total	**20**

First, write the data in ascending order.

6	6.7	7	7.1	7.2
7.3	7.5	7.9	8.2	8.3
8.75	9.1	9.3	9.3	9.4
9.5	10	10.2	11	12.1

Then, count and record the values in the provided intervals. Note, 10 was included in the $10 \leq t < 11$ interval, and the value of 11 was included in the $11 \leq t < 12$ interval due to the inequality symbols.

An important part of grouped data frequency tables is the *mid-interval value*, which is the middle of the interval. In example 2.30, the mid-interval values would be 6.5, 7.5, 8.5, 9.5, 10.5, 11.5, and 12.5. To calculate the mid-interval value, average the lower and upper boundaries for each interval.

For continuous data, the ending of one interval is always the beginning of the next. Example 2.30 models this principle. However, if the end of one interval does not begin the next, average the upper boundary of the first interval with the lower boundary of the next interval and continue for all intervals. The beginning value of the frequency chart must decrease and the ending value must increase the same amount as all the other values. This ensures all intervals stay the same width.

➡ **EXAMPLE 2.31** ——————————————————————————

The following frequency chart represents the heights of 100 sunflower seeds 5 weeks after planting.

Height (çm)	Frequency
30–34	9
35–39	22
40–44	56
45–49	10
50–54	3
Total	100

(a) Determine the mid-interval values.

(b) The upper boundaries do not match the lower boundaries of the next interval. Copy and complete the frequency chart so the boundaries are correct.

Answer Explanations

(a) **The mid-interval values would be 32, 37, 42, 47, and 52.**

$(30 + 34) / 2 = 32$

$(35 + 39) / 2 = 37$

$(40 + 44) / 2 = 42$, etc.

(b) **The correct frequency chart would be**

Height (cm)	Frequency
29.5–34.5	9
34.5–39.5	22
39.5–44.5	56
44.5–49.5	10
49.5–54.5	3
Total	100

(34 + 35) / 2 = 34.5

(39 + 40) / 2 = 39.5

etc.

Note, the first value of 30 decreased by 0.5, and the final value of 54 increased by 0.5. This ensures all the intervals are equal width.

Frequency histograms are a graphical way to represent data. Data should be first displayed in a frequency chart in order to draw a histogram. Just like with frequency charts, histograms have equal class intervals. Histograms are constructed by drawing rectangles with a width of the interval and a height of the frequency. The rectangles, or bars, must touch.

In example 2.31, the frequency chart for 100 five-week-old sunflower plants was given. Since the bars of a histogram must touch, the graph must use the corrected frequency chart, where the upper boundaries equal the lower boundaries of the next interval.

Height (cm)	Frequency
29.5–34.5	9
34.5–39.5	22
39.5–44.5	56
44.5–49.5	10
49.5–54.5	3
Total	100

The histogram for the data is shown below. Since the intervals are graphed on the half mark, all bars touch and are equal width.

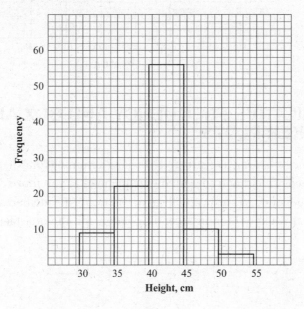

➡ **EXAMPLE 2.32**

Abebe measured the weights of all the four-week-old chicks in his village. The frequency table below displays the weights, in grams, of the 20 chicks.

QUICK TIP

The intervals create the scale for the horizontal axis.

The frequency creates the scale for the vertical axis.

ALWAYS label each axis with its context.

Weights (g)	Frequency
115–120	1
120–125	2
125–130	4
130–135	7
135–140	3
140–145	3
Total	20

Represent the weights of the four-week-old chicks in a histogram. Use a scale of 1 cm to represent 5 grams on the horizontal axis and 2 cm to represent 5 chicks on the vertical axis.

Answer Explanations

Each 1 cm block on the x-axis is scaled by 5, starting at the lowest value of 115 and ending at the highest value of 145. On the y-axis, **two** 1 cm blocks represent 5 chicks, and since the largest frequency is 7, the vertical axis must be labeled up to 10. Each bar of the histogram has a width of 5 and height of the frequency for the interval.

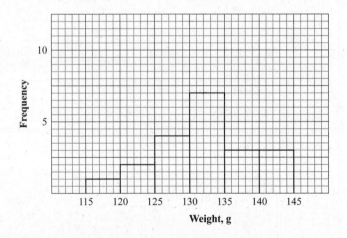

2.4 CUMULATIVE FREQUENCY AND BOX AND WHISKER PLOTS

Cumulative frequency charts are constructed by taking a running total of the frequencies. You start with the frequency chart and sum up the frequencies as you move down the chart.

Cumulative frequency curves are a graphical way to locate the median, which is the middle value of the data set when written in ascending order, and the quartiles. Quartiles are the

Descriptive Statistics

lower 25% of the data and the upper 25%. The lower 25% is called the first quartile, or Q_1, and the upper 25% is called the third quartile, or Q_3. Note, the third quartile represents the upper 25% of the data that is equivalent to the lower 75%.

Refer back to the frequency chart that represented the time in minutes that 20 high school football players took to run 1.6 kilometers (Example 2.30). The cumulative frequency chart is constructed below.

Time	Frequency	Cumulative Frequency	
$6 \le t < 7$	2	2	2
$7 \le t < 8$	6	8	$2 + 6$
$8 \le t < 9$	3	11	$2 + 6 + 3$
$9 \le t < 10$	5	16	$2 + 6 + 3 + 5$
$10 \le t < 11$	2	18	$2 + 6 + 3 + 5 + 2$
$11 \le t < 12$	1	19	$2 + 6 + 3 + 5 + 2 + 1$
$12 \le t < 13$	1	20	$2 + 6 + 3 + 5 + 2 + 1 + 1$
Total	**20**		

Imagine lining up all the data values in ascending order. Your goal is to count each data value. When you start, you have not counted a single value. Imagine walking past the data and counting. By the time you get to 7, you have passed 2 values. Now, continue walking. When you reach 8, you have passed an additional 6 values, which makes 8 in total. Once you walk to 9, you have passed 11 total values, and so on. The chart below illustrates this principal.

	$6 \le t < 7$					$7 \le t < 8$		
Data Value	6	6.7	7	7.1	7.2	7.3	7.5	7.9
Number of Data "seen"	*1*	*2*	*3*	*4*	*5*	*6*	*7*	*8*

	$8 \le t < 9$					$9 \le t < 10$		
Data Value	8.2	8.3	8.75	9.1	9.3	9.3	9.4	9.5
Number of Data "seen"	*9*	*10*	*11*	*12*	*13*	*14*	*15*	*16*

	$10 \le t < 11$		$11 \le t < 12$	$12 \le t < 13$
Data Value	10	10.2	11	12.1
Number of Data "seen"	*17*	*18*	*19*	*20*

When graphing a cumulative frequency curve, the very first point must be on the x-axis. The y-value is zero because when you start the running total, no values have been counted. The cumulative frequency graph for the football player times is shown on page 40.

The graph begins at (6, 0) and each upper boundary is graphed with the cumulative frequency. The end of the graph is (13, 20) because by the time you have counted every data set member, you have reached the last value. Cumulative frequency curves are always drawn as smooth curves.

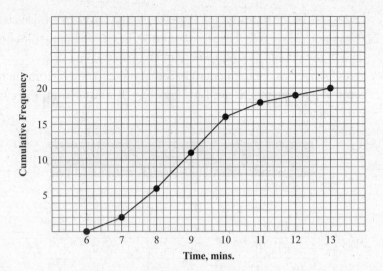

To locate the median, you must find the middle data value. There are 20 data values, thus the median would be located at the 10th value. Using the graph, draw a horizontal line from 10 on the y-axis to the graph, and then down to the x-axis to identify the corresponding data value. The median would be 8.8 minutes.

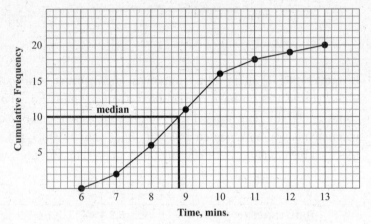

To locate the quartiles, you must find where the lower 25% end and where the upper 25% begin.

There are 20 data values, so the lower 25% would end at $20(0.25) = 5$. The third quartile would begin at 15, since $20 - 5 = 15$.

The first quartile, Q_1, is 7.8 minutes. The third quartile, Q_3, is 9.8 minutes.

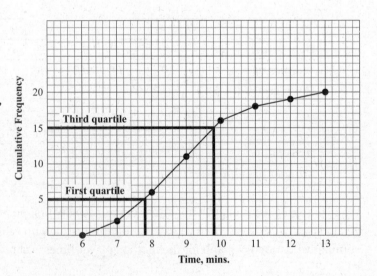

➡ EXAMPLE 2.40

A fish market records the weights (w), in kilograms, of the fish purchased from local fishermen on Monday. The results are displayed in the frequency table below.

Weight w (kg)	Frequency
$0.4 \leq w < 0.6$	4
$0.6 \leq w < 0.8$	8
$0.8 \leq w < 1.0$	12
$1.0 \leq w < 1.2$	9
$1.2 \leq w < 1.4$	5
$1.4 \leq w < 1.6$	2
Total	40

(a) Construct a cumulative frequency chart for the given data.

(b) Draw a cumulative frequency curve. Use a scale of 1 cm to represent 0.2 kg on the horizontal axis, and 1 cm to represent 5 fish on the vertical axis.

(c) Using the graph, find:
  (i)   the median;
  (ii) lower quartile, Q_1;
  (iii) upper quartile, Q_3.

QUICK TIP

To graph a cumulative frequency curve:

1. Make sure the intervals are continuous.

2. The first point on the curve is **ALWAYS** (the very first interval value, 0).

3. The remaining points are: (Each interval's **largest** value, cumulative frequency).

4. Connect the points with a smooth curve. It should look like a stretched out S.

5. **ALWAYS** label each axis with its context.

Answer Explanations

(a)

Weight w (kg)	Frequency	Cumulative Frequency	
$0.4 \leq w < 0.6$	4	4	4
$0.6 \leq w < 0.8$	8	12	4 + 8
$0.8 \leq w < 1.0$	12	24	12 + 12
$1.0 \leq w < 1.2$	9	33	24 + 9
$1.2 \leq w < 1.4$	5	38	33 + 5
$1.4 \leq w < 1.6$	2	40	38 + 2
Total	40		

(b) The graph begins at (0.4, 0). Each upper boundary is paired with the cumulative frequency. The graph ends at (1.6, 40). The cumulative frequency graph is a smooth curve.

QUICK TIP

The cumulative frequency curve is used to find the following descriptors of a data set.

Median: 50%

First Quartile: 25%

Third Quartile: 75%

Percentile: p%

Top Percent: 100% − p%

Multiply the percent and the total number of data. Find this value on the vertical axis, and then locate the corresponding value on the horizontal axis.

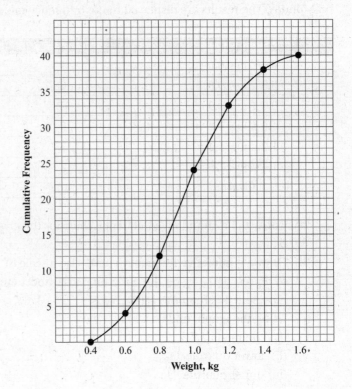

(c) (i) **median: 0.94 kg**

There are 40 fish in the sample, so the median is at the 20th value. Drawing a horizontal line from 20 on the y-axis to the graph, and then down to the x-axis, the weight is 0.94 kg. Each individual tick mark is 0.04 kg. Due to variations in student graphs, answers between 0.92 kg and 0.96 kg would be accepted.

(ii) **$Q_1 = 0.76$ kg**

Since there are 40 fish, the lower 25% would end at the 10th value. Drawing a horizontal line from 10 on the y-axis to the graph, and then down to the x-axis, the weight is 0.76 kg. Answers between 0.74 and 0.78 would be accepted.

(iii) **$Q_3 = 1.14$ kg**

You know 25% of 40 is 10, but you need the value that begins the **top** 10 fish, which is located at the 30th fish (30 = 40 − 10). Drawing a horizontal line from 30 on the y-axis to the graph, and then down to the x-axis, the weight is 1.14 kg. Answers between 1.12 and 1.16 would be accepted.

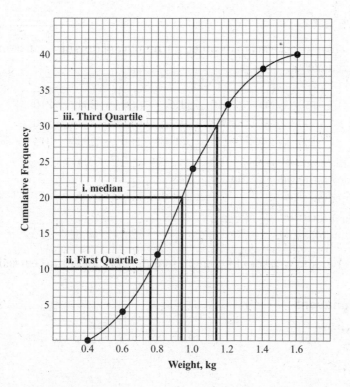

Another graph that displays the median and the quartiles is a box and whisker plot. This graph uses the minimum value, first quartile, median, third quartile, and maximum value to construct a "box" with "whiskers". Each section of the graph represents 25% of the data.

The graph below shows a general box and whisker plot. The "box" is graphed using Q_1, the median, and Q_3 with "whiskers" extending from the middle of the box out to the minimum and the maximum. The percent of data in each section is always 25%. The width of the box is arbitrary.

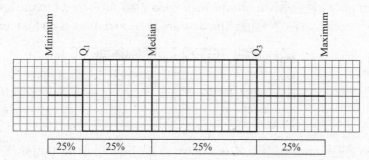

➡ **EXAMPLE 2.41** ────────────────────

The coach of a school's cricket team recorded the number of runs his team earned last season. The data are given below.

150	200	250	195	199	204	210
300	265	288	185	210	232	190
185	240	245	260	288	265	213

(a) Write down the value of the
   (i)   median,
   (ii)   first quartile,
   (iii) third quartile.

(b) Draw a box and whisker plot to represent the number of runs earned.

Answer Explanations

(a) (i) **Median: 213**

    (ii) **Q_1: 197**

    (iii) **Q_3: 262.5**

To quickly identify the median and first and third quartile, enter the data into L_1 in the (STAT) menu.

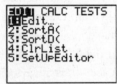 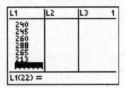

Then press (2ND MODE) (to exit the list). Press (STAT), arrow right to the (CALC) menu, and select option 1: 1–(VAR STATS). The data are in L_1, and there is no frequency list.

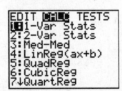 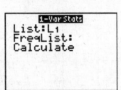

Press (ENTER) on (CALCULATE). Arrow down to the bottom of the output, and the median along with the lower and upper quartile will be found.

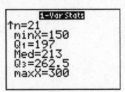

(b) Since the data range from 150 to 300, a larger scale is necessary. Each individual tick mark represents 4 runs earned. The box is constructed by drawing a vertical line at Q_1, the median, and Q_3. The lower whisker extends to the minimum and the upper whisker extends to the maximum, which are also displayed in the calculator output.

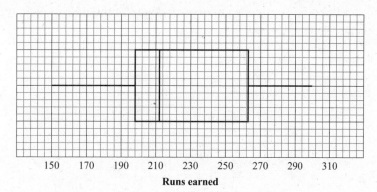

Runs earned

The GDC can also graph box and whisker plots, which is an alternate way to find the minimum, Q_1, median, Q_3, and maximum.

Using Example 2.41, the data are already in L_1.

> **NOTE: To clear any values that may be in L_1, move up so L_1 is highlighted, press CLEAR, and then ENTER.**

To display any type of statistical graph, use the STAT PLOT menu by pressing 2ND GRAPH.

Select PLOT 1, press ENTER on ON and arrow down to the TYPE option. Arrow right until the box and whisker plot is selected, then hit ENTER. Our data are in L_1, so the XLIST remains the same.

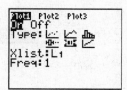

The quickest and most convenient way to generate a statistical graph is to press ZOOM 9: STAT. The graph can identify the 5-number summary. Press TRACE and arrow right and left to see all 5 values.

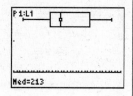

FEATURED QUESTION

May 2009, Paper 2

The diagram shows the cumulative frequency graph for the time t taken to perform a certain task by 2000 men.

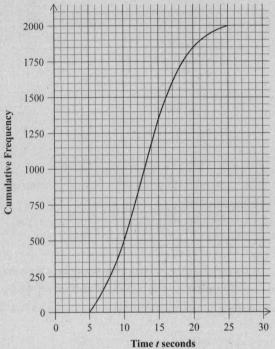

(a) Use the diagram to estimate
  (i) the median time,
  (ii) the upper quartile and the lower quartile,
  (iii) the interquartile range.

(b) Find the number of men who take **more than** 11 seconds to perform the task.

(c) 55% of the men took less than p seconds to perform the task. Find p.

The times taken for the 2000 men were grouped as shown in the table below.

Time	Frequency
$5 \leq t < 10$	500
$10 \leq t < 15$	850
$15 \leq t < 20$	a
$20 \leq t < 25$	b

(d) Write down the value of
  (i) a,
  (ii) b.

(e) Use your graphic display calculator to find an estimate of
  (i) the mean time,
  (ii) the standard deviation of the time.

Everyone who performs the task in **less than** one standard deviation **below** the mean will receive a bonus. Pedro takes 9.5 seconds to perform the task.

(f) Does Pedro receive the bonus? Justify your answer.

(See page 437 for solutions)

2.5 MEASURES OF CENTRAL TENDENCY

With any given data set, the measures of central tendency are a calculated way to describe the "middle" of the data. These values are the mean, median, and mode.

- Mean is the sum of the data divided by total number of data values.
- Median is the middle of the data when in ascending order.
- Mode is the most frequently occurring data value. There can be more than one mode or no mode at all.

As seen previously, the GDC can locate the median. The [1-VAR STATS] output also identifies the mean.

The mode can be found by identifying the data value that occurs the most.

➡ EXAMPLE 2.50

The marks earned on the Paper 1 of the IB Math Studies exam for 24 randomly selected students are recorded below.

85	75	54	77	32	70
44	80	68	53	59	72
81	30	39	47	54	60
77	72	68	71	76	77

Write down the value of the

(a) Mean

(b) Median

(c) Mode

Answer Explanations

(a) **Mean: 63.375 (exact) or 63.4 (3 significant figures)**

Either sum up the values and divide by 24 or use the output of the [1-VAR STATS] in the [STATS] menu. The mean is represented by $\bar{x}$.

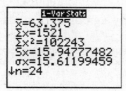

(b) **Median: 69**

Using the [1-VAR STATS] output, scroll down to find the median.

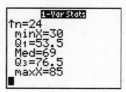

(c) **Mode: 77**

The most frequently occurring data value is 77.

When the data are presented in a frequency table, the frequencies must be taken into consideration.

For example, consider the frequency table below that represents the shoe sizes for a traveling European dance troupe.

Shoe Size	Frequency
38	6
39	5
39.5	4
40	3
40.5	2
Total	**20**

Frequency of 6 means: 38, 38, 38, 38, 38, 38

Frequency of 5 means: 39, 39, 39, 39, 39

Frequency of 4 means: 39.5, 39.5, 39.5, 39.5

Frequency of 3 means: 40, 40, 40

Frequency of 2 means: 40.5, 40.5

There are not 5 data values; there are 20 due to the frequencies. When calculating the central tendencies, the frequencies are critical and must be entered into L_2.

The mode is still the most frequent, which would be a shoe size of 38.

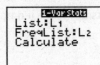

To find the mean and median for the table, enter the shoe sizes into L_1 and frequencies into L_2. You will still select [1-VAR STATS], but the calculator needs to know that the data (L_1) occur with frequency (L_2). Thus, select [1-VAR STATS] and enter L_1 as the [LIST] and L_2 as the [FREQLIST]. This will ensure the calculator takes into account the six values of 38, five values of 39, and so on.

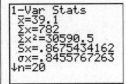

The mean shoe size would be 39.1, and further down the output, the median is shown to be 39.

Grouped discrete or continuous data are also presented in a table format except they are in intervals. The exact mean, median, or mode is unknown, but you can use the data to find an approximation or estimate.

Consider the frequency table below, which represents the length of time spent on the IB Math Studies Internal Assessment by a group of 25 students.

Time, t (in hours)	Frequency
$10 \le t < 15$	3
$15 \le t < 20$	7
$20 \le t < 25$	6
$25 \le t < 30$	5
$30 \le t < 35$	4
Total	**25**

In order to get an approximate mean or median, you need to determine the best representative for each interval. The most logical choice would be the middle. Thus, enter the mid-interval value into L_1, the frequencies into L_2, and proceed as usual.

The approximate mean and median would both be 22.5.

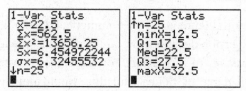

The modal class is $15 \le t < 20$, since it has the highest frequency.

➡ EXAMPLE 2.51

The weights of 80 patients at a pediatric doctor's office were recorded and are displayed in the table below.

Weight (w)	$10 \le w < 20$	$20 \le w < 30$	$30 \le w < 40$	$40 \le w < 50$	$50 \le w < 60$
Frequency	10	18	12	22	18

   (a) Write down the modal class.

   (b) Write down the mid-interval value for the interval $10 \le w < 20$.

   (c) Using a GDC, find an estimate weight for the:

      (i) mean

      (ii) median

Answer Explanations

(a) **$40 \le w < 50$**

    The interval with the highest frequency is the modal class.

(b) **15**

    To calculate the mid-interval value, average the lower boundary with the upper boundary. $(10 + 20) / 2 = 15$

(c) (i) **37.5**

     (ii) **40**

To use the GDC, enter the mid-interval values into L_1 and the frequencies into L_2. Then, calculate (1-VAR STATS) with the (LIST) as L_1 and the (FREQLIST) as L_2.

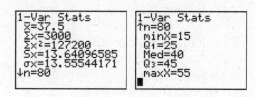

2.6 MEASURES OF DISPERSION

While the measures of central tendencies are a numerical way to describe the center of the data, the measures of dispersion describe how far the data are spread out. Data that are close in values have small measures of dispersion. For example, the weights of newborn babies would be within a small range of values, and thus would not be very spread out. Data that are stretched over a wide range of values will have larger measures of dispersion. The weights of adult men, for example, would be much more spread out than those of newborn babies.

The three measures of dispersion are:

- Range (the maximum – minimum)
- Interquartile range ($Q_3 - Q_1$)

- Standard deviation $\left(\sqrt{\dfrac{\sum (x - \bar{x})^2}{n}} \right)$

When calculating the one-variable statistics, all parts needed to compute the above measures of dispersion are displayed. The standard deviation is represented by σx.

➡ **EXAMPLE 2.60**

The marks earned on the Paper 1 IB Math Studies exam for 24 randomly selected students are recorded below.

85	75	54	77	32	70
44	80	68	53	59	72
81	30	39	47	54	60
77	72	68	71	76	77

Write down the value of the
(a) range,
(b) interquartile range,
(c) standard deviation.

Answer Explanations

(a) **Range: 55**

The range is maximum – minimum: $85 - 30 = 55$

(b) **IQR: 23**

The interquartile range is $\boldsymbol{Q_3} - \boldsymbol{Q_1}$. $76.5 - 53.5 = 23$

(c) **Standard deviation: 15.6**

On the GDC output, standard deviation is $\sigma x = 15.61199$ and this rounds to 15.6.

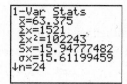

➡ **EXAMPLE 2.61** ─────────────────────────────

The weights of 80 patients at a pediatric doctor's office were recorded and are displayed in the table below.

Weight (w)	$10 \leq w < 20$	$20 \leq w < 30$	$30 \leq w < 40$	$40 \leq w < 50$	$50 \leq w < 60$
Frequency	10	18	12	22	18

(a) Calculate the range.

(b) Write down the approximate value of the
    (i)   first quartile,
    (ii)   third quartile,
    (iii) standard deviation.

(c) Estimate the interquartile range.

Answer Explanations

(a) **Range: 50**

The range is maximum – minimum: $60 - 10 = 50$

(b) (i)   **Q_1: 25**

   (ii)   **Q_3: 45**

   (iii) **standard deviation: 13.6**

    Enter the mid-interval values into L_1 and the frequencies into L_2. Then, calculate 1-VAR STATS L_1, L_2.

(c) **IQR: 20**

The interquartile range is $Q_3 - Q_1$; $45 - 25 = 20$

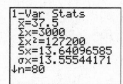 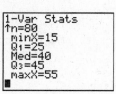

Descriptive Statistics

TOPIC 2 PRACTICE

Paper 1 Questions

1. Marshal kept track of the total hours he worked each week at his part time job. The histogram below shows the hours, h, worked each week over the last 20 weeks.

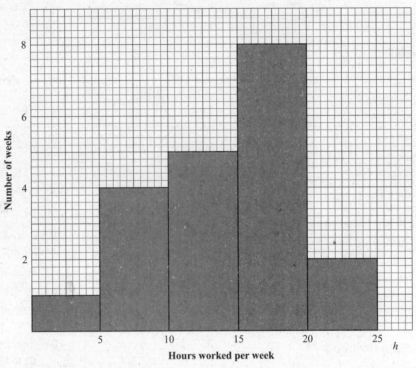

Hours worked per week

    (a) Write down the modal class.

    (b) Write down the mid-interval value for the $20 < h \leq 25$ class.

    (c) Calculate an estimate for the:
        (i)   mean hours worked per week,
        (ii) standard deviation of the hours worked per week.

2. Pamela was interested in the number of minutes her students spent doing math homework during the school week. She asked her 18 students to keep a record of the total minutes they worked on math homework during a certain week. The results are listed below.

50	32	25	60	45	49
44	52	61	65	32	40
46	51	32	44	70	59

    (a) State the value of the
        (i)   median,
        (ii)   lower quartile,
        (iii) upper quartile.

    (b) Calculate the range.

    (c) Draw a box and whisker plot of the minutes spent on math homework.

3. Local school board members wanted to determine how far most students drove to the town's high school. They surveyed 30 seniors and recorded the distance in kilometers driven.

8.1	7.5	9.2	1.2	3.6	3.0	5.6	5.2	6.4	8.0
4.1	7.3	6.2	2.3	3.8	4.1	5.3	6.0	6.7	7.2
4.2	3.3	4.6	6.2	6.6	7.0	2.3	8.0	1.9	7.4

(a) Determine if the data are discrete or continuous.

(b) Complete the frequency table for the distance in kilometers driven.

Distance, km	Frequency
$0 \le d < 2$	
$2 \le d < 4$	
$4 \le d < 6$	
$6 \le d < 8$	
$8 \le d < 10$	

(c) Display the distribution in a histogram using the intervals provided.

4. A local community center held a weight loss program for members. Over 10 weeks, participants made healthy lifestyle choices and increased their activity levels. The total weight loss results are displayed in the graph below.

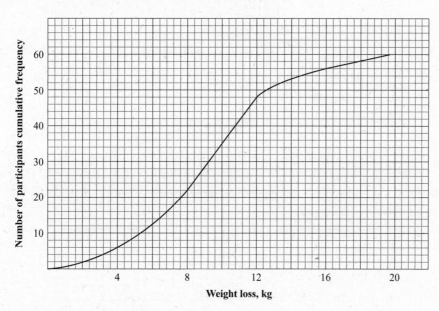

(a) Write down the median weight loss.

(b) Calculate the interquartile range.

(c) Stuartt lost a total of 10 kilograms. How many people lost more weight than Stuartt?

SOLUTIONS

1. (a) $15 < h \le 20$

   This interval has the highest frequency

   (b) **22.5**

   $(20 + 25) / 2 = 22.5$

   (c) (i) **14**

   (ii) **5.27**

   Enter the mid-interval values into L_1 with the frequencies in L_2.

   Then, calculate (1-VAR STATS) L_1, L_2.

   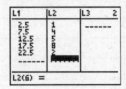 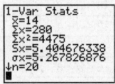

2. (a) (i) **47.5**

   (ii) $Q_1 = 40$

   (iii) $Q_3 = 59$

   Entering the given data into L_1, calculate (1-VAR STATS).

   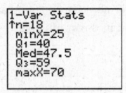

   (b) **45**

   The range is max – min.

   $$70 - 25 = 45$$

   (c)

   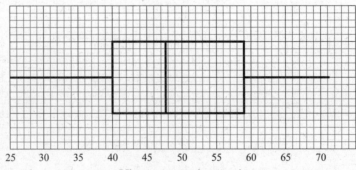

   Minutes spent on homework

The GDC provides a quick look at the box and whisker plot and gives an easy way to find the minimum, first quartile, median, third quartile, and maximum. However, the graph must still be drawn.

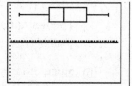

The box and whisker plot displays the minimum (25), the first quartile (40), median (47.5), third quartile (59), and maximum (70). Notice the scale and label on the x-axis.

3. (a) **continuous**

Distance is a continuous variable.

(b)

Distance, km	Frequency
$0 \leq d < 2$	2
$2 \leq d < 4$	6
$4 \leq d < 6$	7
$6 \leq d < 8$	11
$8 \leq d < 10$	4

The frequency table was completed by listing the data in ascending order and counting how many data belong in each interval.

(c)

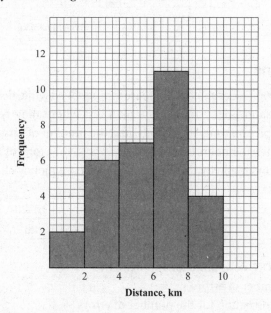

The histogram is graphed using the provided intervals, and the rectangles have heights of the calculated frequencies. The scale is uniform on each axis and labels provided.

4. (a) **9.2 kilograms**

The median occurs at 30 (half of the 60 participants). Draw a line from the y-axis to meet the graph, and then down to the x-axis. This corresponds to three tick marks after eight. Each tick mark is worth 0.4 kg, thus the median is at $8 + (0.4)(3) = 9.2$ kg.

(b) **IQR = 11.4 − 6.6 = 4.8 kg**

The first quartile is found at 15 participants (half of 30) and the third quartile at 45 (half between 30 and 60). Answers within the range of 4.6 to 5.0 kg would be accepted.

(c) Since Stuartt lost 10 kg, she is participant number 36. Thus, 24 people lost more weight than Stuartt (60–36).

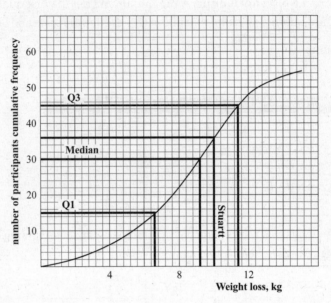

Paper 2 Questions

1. Asif was taking a course designed to increase typing speed while decreasing typing errors. Every Wednesday and Friday the class took a test where they typed a paragraph and recorded the completion time, in seconds, and the number of errors present in the paragraph. There were ten members in the eight-week class for a total of 160 tests. Asif recorded the number of errors made by the entire class in the frequency chart below.

Number of errors, n	1–5	6–10	11–15	16–20	21–25	26–30
Frequency	24	42	38	26	20	10

(a) Using your GDC, determine an approximate:
  (i) mean number of errors,
  (ii) standard deviation for the number of errors.

(b) Construct a cumulative frequency chart using the intervals $0.5 \leq n < 5.5$, $5.5 \leq n < 10.5$, etc.

(c) Graph the cumulative frequency curve. Let 2 cm represent 5 errors on the horizontal axis and 1 cm represent 20 people on the vertical axis.

(d) Hence, find the
  (i) median number of errors,
  (ii) interquartile range.

SOLUTIONS

1. (a) (i) **mean: 13.2**

   (ii) **standard deviation: 7.22**

   To find the approximate mean and standard deviation, use the mid-interval values, which are 3, 8, 13, 18, 23, and 28. These are found finding the average of the interval. For example, $(1 + 5) / 2 = 3$.

   Enter the mid-interval values into L_1 with the frequencies in L_2, and then calculate [1-VAR STATS] L_1, L_2.

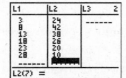 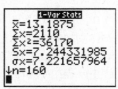

(b)

Number of errors, n	Cumulative Frequency	
$0.5 \leq n < 5.5$	24	
$5.5 \leq n < 10.5$	66	$(24 + 42)$
$10.5 \leq n < 15.5$	104	$(66 + 38)$
$15.5 \leq n < 20.5$	130	$(104 + 26)$
$20.5 \leq n < 25.5$	150	$(130 + 20)$
$25.5 \leq n < 30.5$	160	$(150 + 10)$

The boundaries must be $0.5 \leq n < 5.5$, since the original intervals do not match. Remember, the ending of one interval must be the beginning of the next.

(c)

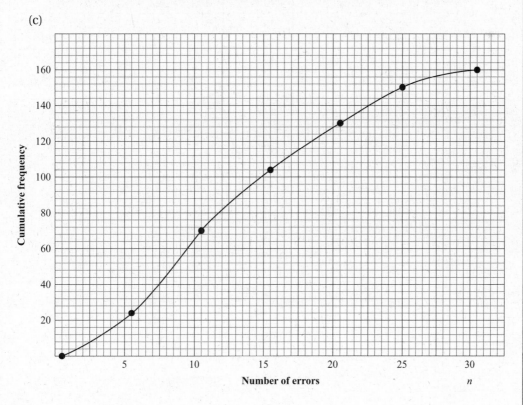

Make sure the scales are correct and the axes are labeled. The points are graphed on the intervals of the cumulative frequency chart. The curve is smooth.

(d) (i) **median: 12 errors**

Using the cumulative frequency graph, the median will be at 40 people, so draw a line from 40 on the *y*-axis over to the graph, and then down to the number of errors. This shows about 12 errors. Due to graphing variations, answers between 11.5 and 12.5 would be accepted.

(ii) **IQR: 13**

Using the graph, the first quartile is 25% of 160, which is 40 people. The third quartile is 75% of 160, which is 120 people.

Finding their corresponding number of errors gives $Q_1 = 7.5$ and $Q_3 = 18.5$. Thus the interquartile range is $18.5 - 7.5 = 11$.

Due to graphing variations, answers between 10 and 12 would be accepted.

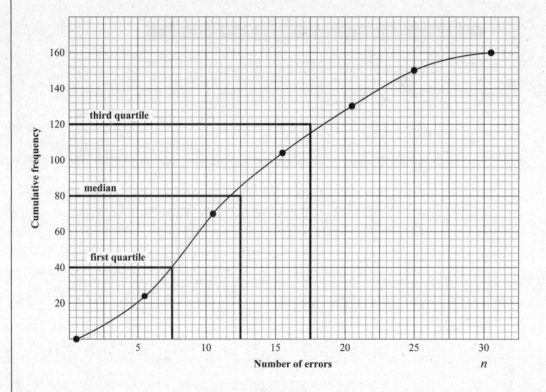

CHAPTER OBJECTIVES

Before you move on to the next chapter, you should be able to:

☐ Determine if data are discrete or continuous

☐ Construct a frequency chart for discrete or continuous data

☐ Determine mid-interval values for continuous data

☐ Rewrite a frequency chart so the intervals are continuous (i.e. the ending of one interval is the beginning of the next)

☐ Draw a histogram

☐ Construct a cumulative frequency chart

☐ Draw a cumulative frequency graph

☐ Using a cumulative frequency graph, identify the median, first quartile, third quartile, number of data values above or below a certain value, or the value corresponding to a certain percent

☐ Identify the five number summary

☐ Draw a box and whisker plot

☐ Given a data set, calculate the mean, median, mode, and standard deviation using the GDC

☐ Given a frequency chart or histogram, calculate an approximate mean, median, mode, and standard deviation using the GDC

☐ Calculate the interquartile range and the range

Logic, Sets, and Probability

- **PROPOSITION:** statement that is either true or false: p, q, or r

- **COMPOUND STATEMENT:** two or more propositions joined with logic connectives: $\wedge$, $\vee$, $\underline{\vee}$, $\neg$, $\Rightarrow$, $\Leftrightarrow$

- **TRUTH TABLES:** displays all possible truth outcomes

- **CONVERSE, INVERSE, CONTRAPOSITIVE:** given $p \Rightarrow q$, the converse is $q \Rightarrow p$, the inverse is $\neg p \Rightarrow \neg q$, and the contrapositive is $\neg q \Rightarrow \neg p$

- **SET THEORY:** sets, subsets, union ($\cup$), intersection ($\cap$), universal set, complement

- **VENN DIAGRAM:**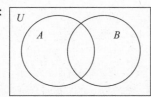

- **PROBABILITY:** $P(A) = \dfrac{\text{number of outcomes in } A}{\text{total number of outcomes}}$

- **EXPECTED VALUE:** how many times you would expect a certain event to occur; probability of event times the number of trials

- **MUTUALLY EXCLUSIVE:** events with no common outcome: $P(A \cap B) = 0$

- **COMBINED EVENTS:** events with a common outcome: $P(A \cup B) = P(A) + P(B) - P(A \cap B)$

- **INDEPENDENT EVENTS:** events that happen together but one does not affect the outcome of the other: $P(A \cap B) = P(A)P(B)$

- **TREE DIAGRAM:** displays all possible outcomes for independent events

- **WITHOUT REPLACEMENT:** probability that involves selecting an item, keeping it, and then selecting another

- **CONDITIONAL PROBABILITY:** a condition is stated in the problem, which affects the sample space: $P(A|B) = \dfrac{P(A \cap B)}{P(B)}$

3.1 SYMBOLIC LOGIC AND PROPOSITIONS

A **proposition** is a statement that is either true or false. Questions, commands, or exclamations would not be propositions, as they are neither true nor false.

Examples of propositions would include:

- It is raining.
- The car is out of gas.
- I wore shorts.

The following are not propositions:

- Clean your room. (*command*)
- Are you hot? (*question*)
- Ouch, that hurt! (*exclamation*)

➡ EXAMPLE 3.10

Determine if each of the following is a proposition.

1. The door is open.
2. Do your homework.
3. The contract is not signed.
4. Are you hungry?
5. Yes, I passed my test!
6. I bought a drink.

Answer Explanations

1. **Proposition.** This is a statement that is either true or false.
2. **Not a proposition.** This is a command and has no truth.
3. **Proposition.** This is a statement that is either true or false.
4. **Not a proposition.** This is a question and has no truth.
5. **Not a proposition.** This is an exclamation and has no truth.
6. **Proposition.** This is a statement that is either true or false.

A **compound statement** is two or more propositions joined with logic connectives.

The table below lists the logic connectives along with their meaning and an example.

Logic Connective	Name	Meaning	Example
$\wedge$	Conjunction	and	$p \wedge q$ "p and q"
$\vee$	Disjunction	one or the other OR both	$p \vee q$ "p or q"
$\underline{\vee}$	Exclusive disjunction	one or the other but not both	$p \underline{\vee} q$ "p or q, but not both"
$\neg$	Negation	not	$\neg p$ "not p"
$\Rightarrow$	Implication	If, then	$p \Rightarrow q$ "If p, then q"
$\Leftrightarrow$	Equivalence	If and only if	$p \Leftrightarrow q$ "p if and only if q"

➡ **EXAMPLE 3.20**

Given the following propositions:

   p: The sun is shining.

   q: I will walk to class.

1. Write the following in words:
   (a) $p \wedge q$
   (b) $p \vee \neg q$
   (c) $\neg p \Rightarrow q$

2. Write the following in symbols:
   (a) I will walk to class, if and only if, the sun is shining.
   (b) The sun is not shining, or I will walk to class, but not both.

Answer Explanations

1. (a) **The sun is shining, and I will walk to class.**
      The symbol $\wedge$ means "and".
   (b) **The sun is shining, or I will not walk to class.**
      The symbol $\vee$ means "or", but notice the proposition q is negated by the symbol $\neg$. The word "not" must be used with the proposition q.
   (c) **If the sun is not shining, then I will walk to class.**
      The symbol $\Rightarrow$ means "If, then" but p is negated.

2. (a) $q \Leftrightarrow p$
      "If and only if" means equivalence, thus the symbol $\Leftrightarrow$ joins q and p.
   (b) $\neg p \underline{\vee} q$
      Since the sun is **not** shining, p must be negated. The phrase "…or…but not both" means exclusive disjunction, $\underline{\vee}$.

3.3 TRUTH TABLES

To determine if an argument is logically valid, use a truth table. A truth table shows all possible truth outcomes of a proposition or compound statement.

In IB Math Studies, three is the maximum number of propositions used for truth tables.

In order to correctly complete truth tables, an understanding of the logic connectives is required. True is denoted by "T" and false by "F". To begin a truth table, the first columns are simply the truth combinations of the given propositions.

p	q
T	T
T	F
F	T
F	F

Those truth values are used to determine the truth values of the compound statements.

p	q	$\neg p$	$p \wedge q$	$p \vee q$	$p \underline{\vee} q$
T	T	F	T	T	F
T	F	F	F	T	T
F	T	T	F	T	T
F	F	T	F	F	F

- $\neg p$: "not"—this switches the truth values of the given proposition.
- $p \wedge q$: "and"—in order for this to be true, both truth values must be true.
- $p \vee q$: "or"—in order for this to be true, one OR the other OR both of the truth values can be true.
- $p \underline{\vee} q$: "one or the other but not both"—in order for this to be true, ONLY one of the truth values can be true.

Let p: I clean my room.
   q: I get a reward.

Using the true/false truth values for p and q, you can determine when "If I clean my room, then I get a reward" will be true.

p	q	$p \Rightarrow q$	Explanation
T	T	T	I DO clean my room, and I DO get a reward, so the implication is true.
T	F	F	I DO clean my room, but I DO NOT get a reward. Thus, "If I clean my room, then I get a reward" cannot be true, since I didn't get a reward.
F	T	T	I DO NOT clean my room, and I DO get a reward, so the implication is true. I might have gotten a reward for a different reason, but the reward still happened, and thus is true.
F	F	T	I DO NOT clean my room, and I DO NOT get a reward is a true implication, since a clean room leads to a reward. A room that is not clean will lead to no reward.

Notice the only way for an implication to be **false** is for "If TRUE, then FALSE".

Logic/Sets/Probability

Equivalence works in a similar way, except one truth outcome differs from an implication. Using the same context, the following table represents "I clean my room, if and only if, I get a reward".

p	q	p ⇔ q	Explanation
T	T	T	I DO clean my room, and I DO get a reward, so the equivalence is true.
T	F	F	I DO clean my room, but I DO NOT get a reward. This cannot be true, since I did clean, but I did not get a reward.
F	T	F	I DO NOT clean my room, and I DO get a reward is also false because equivalence means that getting a reward meant I cleaned my room. I did not clean, so I should not have received a reward.
F	F	T	I DO NOT clean my room, and I DO NOT get a reward is true, since a clean room gets a reward. A room that is not clean will not get a reward.

Notice that equivalence is only **true** when the individual truth values are equivalent (both are true or both are false).

➡ EXAMPLE 3.30

Construct a truth table for the following compound propositions:

1. $p \land \neg q$
2. $\neg(p \lor q)$
3. $\neg q \Rightarrow p$
4. $(p \land q) \Leftrightarrow (\neg p \veebar q)$

Answer Explanations

1.

p	q	¬q	p ∧ ¬q
T	T	F	F
T	F	T	T
F	T	F	F
F	F	T	F

- A column for $\neg q$ was required first. This switches the truth values of q.
- $p \land \neg q$ is only true when **both** p and $\neg q$ are true.

2.

p	q	p ∨ q	¬(p ∨ q)
T	T	T	F
T	F	T	F
F	T	T	F
F	F	F	T

- A column for $p \lor q$ was required first. This is true when either p or q or both are true.
- Next, $\neg(p \lor q)$ is found by switching all the truth values of $p \lor q$.

3.

p	q	¬q	¬q ⇒ p
T	T	F	T
T	F	T	T
F	T	F	T
F	F	T	F

- A column for ¬q was required first. This switches the truth values of q.
- The only way for the implication ¬q ⇒ p to be false is for ¬q to be true but p to be false.

4.

p	q	¬p	p ∧ q	¬p ⊻ q	(p ∧ q) ⇔ (¬p ⊻ q)
T	T	F	T	T	T
T	F	F	F	F	T
F	T	T	F	F	T
F	F	T	F	T	F

- A column for ¬p, p ∧ q, ¬p ⊻ q were required.
- For p ∧ q to be true, both p and q had to be true.
- For ¬p ⊻ q to be true, ¬p or q could be true, but not both.
- For (p ∧ q) ⇔ (¬p ⊻ q) to be true, the columns for p ∧ q and ¬p ⊻ q had to be equal.

p	q	r	p ∧ q ∧ r	p ∨ q ∨ r
T	T	T	T	T
T	T	F	F	T
T	F	T	F	T
T	F	F	F	T
F	T	T	F	T
F	T	F	F	T
F	F	T	F	T
F	F	F	F	F

When there are three propositions in a compound statement, the truth table will be larger. There will be eight rows instead of four. However, the logic connectives all work the same, just on a larger scale.

Conjunction, or "and", is still only true when all propositions are true. Disjunction, or "or", is true when at least ONE of the propositions is true.

➡ **EXAMPLE 3.31**

Construct a truth table for the following compound propositions:

1. $(p \wedge \neg q) \Rightarrow \neg r$
2. $p \Rightarrow \neg(r \wedge q)$

Answer Explanations

1.

p	q	r	¬q	p ∧ ¬q	¬r	(p ∧ ¬q) ⇒ ¬r
T	T	T	F	F	F	T
T	T	F	F	F	T	T
T	F	T	T	T	F	F
T	F	F	T	T	T	T
F	T	T	F	F	F	T
F	T	F	F	F	T	T
F	F	T	T	F	F	T
F	F	F	T	F	T	T

- After p, q, and r, the first column needed is $\neg q$. This switches all the truth values of q.
- The next column needed is $p \wedge \neg q$. This column is true whenever p and $\neg q$ are BOTH true.
- Before determining the truth values for the problem, $\neg r$ must be determined. This switches all the truth values of r.
- Finally, $(p \wedge \neg q) \Rightarrow \neg r$ is only **false** when $p \wedge \neg q$ is TRUE, but $\neg r$ is FALSE.

2.

p	q	r	r ∧ q	¬(r ∧ q)	p ⇒ ¬(r ∧ q)
T	T	T	T	F	F
T	T	F	F	T	T
T	F	T	F	T	T
T	F	F	F	T	T
F	T	T	T	F	T
F	T	F	F	T	T
F	F	T	F	T	T
F	F	F	F	T	T

- After p, q, and r, the first column required is $r \wedge q$. This is true whenever BOTH r and q are true.
- The next column required is $\neg(r \wedge q)$. This switches all the truth values of $r \wedge q$.
- $p \Rightarrow \neg(r \wedge q)$ is only **false** when p is true, but $\neg(r \wedge q)$ is false.

Consider the proposition, p, I passed my test. Then $\neg p$ would be "I did not pass my test". We can use a truth table to determine when the compound statement $p \vee \neg p$ is true.

p	¬p	p ∨ ¬p
T	F	T
T	F	T
F	T	T
F	T	T

The $p \vee \neg p$ column is always true. This is called a **tautology**. The argument $p \vee \neg p$ is **logically valid**, since all outcomes are true. In words, $p \vee \neg p$ is "I passed my test or I did not pass my test". No matter the true outcome of the test, one of these outcomes would have to be true.

If the compound statement was instead $p \wedge \neg p$, the truth table would be different.

Logic/Sets/Probability

p	¬p	p ∧ ¬p
T	F	F
T	F	F
F	T	F
F	T	F

The $p \wedge \neg p$ column is always false. This is called a **contradiction**. Any argument that is a contradiction is **logically invalid**. In words, $p \wedge \neg p$ would be "I passed my test, and I did not pass my test". This can never be true, since you cannot both pass and fail one test.

If an argument has a mixture of true and false, it is also logically invalid.

➡ EXAMPLE 3.32 _____

Determine if the following arguments are logically valid, and then identify as a tautology, contradiction, or neither.

1. $(p \vee q) \vee (\neg p \wedge \neg q)$
2. $(p \vee r) \Leftrightarrow (\neg p \wedge \neg q)$

Answer Explanations

1.

p	q	¬p	¬q	p ∨ q	¬p ∧ ¬q	(p ∨ q) ∨ (¬p ∧ ¬q)
T	T	F	F	T	F	T
T	F	F	T	T	F	T
F	T	T	F	T	F	T
F	F	T	T	F	T	T

This argument is logically valid. All outcomes are true, thus it is a tautology.

- The columns for $\neg p$ and $\neg q$ are the opposite truth values.
- For $p \vee q$ to be true, either p or q or both needed to be true.
- For $\neg p \wedge \neg q$ to be true, both $\neg p$ and $\neg q$ had to be true.
- For $(p \vee q) \vee (\neg p \wedge \neg q)$ to be true, either $p \vee q$, $\neg p \wedge \neg q$, or both needed to be true.

2.

p	q	r	¬p	¬q	p ∨ r	¬p ∧ ¬q	(p ∨ r) ⇔ (¬p ∧ ¬q)
T	T	T	F	F	T	F	F
T	T	F	F	F	T	F	F
T	F	T	F	T	T	F	F
T	F	F	F	T	T	F	F
F	T	T	T	F	T	F	F
F	T	F	T	F	F	F	T
F	F	T	T	T	T	T	T
F	F	F	T	T	F	T	F

This argument is not logically valid. The outcomes are not all true or all false, thus it is neither a tautology nor a contradiction.

- The columns for $\neg p$ and $\neg q$ are the opposite truth values.
- For $p \vee r$ to be true, either p or r or both needed to be true.
- For $\neg p \wedge \neg q$ to be true, both $\neg p$ and $\neg q$ had to be true.
- For $(p \vee r) \Leftrightarrow (\neg p \wedge \neg q)$ to be true, the truth values needed to be the same: either both true or both false.

3.4 CONVERSE, INVERSE, CONTRAPOSITIVE, AND LOGICAL EQUIVALENCE

Each implication, "If p, then q" can be written into three related statements: the converse, inverse, and contrapositive.

	Words	Symbols
Implication	If p, then q	$p \Rightarrow q$
Converse	If q, then p	$q \Rightarrow p$ or $p \Leftarrow q$
Inverse	If not p, then not q	$\neg p \Rightarrow \neg q$
Contrapositive	If not q, then not p	$\neg q \Rightarrow \neg p$ or $\neg p \Leftarrow \neg q$

It might be easier to remember that **I**mplication and **I**nverse are negations, since they both start with **I**. The **C**onverse and **C**ontrapositive are negations, since they both start with **C**.

You must be able to write these related statements both in words and in symbols.

➡ EXAMPLE 3.40

Let p and q be the propositions:

p: My mom is at work.

q: I have to babysit.

(a) Write in words the implication $p \Rightarrow q$.

(b) Write in words the converse of $p \Rightarrow q$.

(c) Given the following statement, write in symbolic form:

"If I do not have to babysit, then my mom is not at work."

(d) Is the proposition in part c the inverse, converse, or contrapositive of $p \Rightarrow q$?

Answer Explanations

(a) **If my mom is at work, then I have to babysit.**

$p \Rightarrow q$ is simply "If, then".

(b) **If I have to babysit, then my mom is at work.**
The converse is the reverse of the implication.

> **QUICK TIP**
>
> An **iMPLICATION** is $p \Rightarrow q$ ⎫ Inverse is the negation
> The **INVERSE** is $\neg p \Rightarrow \neg q$ ⎭ of the Implication
>
> The **CONVERSE** is $q \Rightarrow p$ ⎫ Contrapositive is
> The **CONTRAPOSITIVE** is $\neg q \Rightarrow \neg p$ ⎬ the negation
> ⎭ of the **C**onverse
>
> The **CONVERSE** is the **reverse** of the **IMPLICATION**.

(c) $\neg q \Rightarrow \neg p$

Proposition q is written first and it is negated, so the first symbol is $\neg q$. Proposition p is written second and it is also negated, so it gets the symbol $\neg p$. Since this is an "If, then" compound statement, these are joined with the $\Rightarrow$ symbol.

(d) **Contrapositive**

The contrapositive is the negation of the converse. Since this statement is the negation of part b, it is the contrapositive.

Compound statements can be **logically equivalent**, which means two or more compound statements have the same outcome on the truth table. In order to determine if compound statements are logically equivalent, a truth table for each is constructed and the final truth values are compared. If every truth value in the answer column is the same, then the statements are logically equivalent.

➡️ **EXAMPLE 3.41** _____

Determine which of the following compound statements are logically equivalent.

(a) $(p \Rightarrow q) \wedge (q \Rightarrow p)$
(b) $(p \Rightarrow q) \wedge q \Rightarrow p$
(c) $\neg(p \veebar q)$
(d) $(p \wedge q) \wedge \neg q$

Answer Explanations

Compound statements (a) $(p \Rightarrow q) \wedge (q \Rightarrow p)$ and (c) $\neg(p \veebar q)$ are logically equivalent.
Truth Tables:

				(a)		(b)
p	q	$p \Rightarrow q$	$(q \Rightarrow p)$	$(p \Rightarrow q) \wedge (q \Rightarrow p)$	$(p \Rightarrow q) \wedge q$	$(p \Rightarrow q) \wedge q \Rightarrow p$
T	T	T	T	**T**	T	**T**
T	F	F	T	**F**	F	**T**
F	T	T	F	**F**	T	**F**
F	F	T	T	**T**	T	**F**

			(c)			(d)
p	q	$p \veebar q$	$\neg(p \veebar q)$	$p \wedge q$	$\neg q$	$(p \wedge q) \wedge \neg q$
T	T	F	**T**	T	F	**F**
T	F	T	**F**	F	F	**F**
F	T	T	**F**	F	T	**F**
F	F	F	**T**	F	T	**F**

An important fact to remember is that the IB Math Studies formula packet contains a truth table with a column for p, q, $\neg p$, $p \wedge q$, $p \vee q$, $p \veebar q$, $p \Rightarrow q$, and $p \Leftrightarrow q$.

3.5 SET THEORY AND VENN DIAGRAMS

A set is a group of numbers or objects represented by capital letters and curly brackets.

For example, $A = \{2, 4, 6\}$ is the set named A, which contains the elements 2, 4, and 6. Elements are the individual members of the set. The symbol $\in$ means "element of", so $2 \in A$, but $5 \notin A$.

A subset is a smaller set that fits completely in the larger set.

For example, if $A = \{3, 5, 7, 9\}$, then a possible subset could be $B = \{3, 5\}$. This means $B \subset A$, which states B is a subset of A. The symbol $\not\subset$ means "not a subset".

The empty set, $\varnothing$, is exactly that: a set with no elements. The empty set can be represented as $\varnothing$ or $\{ \}$.

In Topic 1 the number sets were defined (pages 1–3). These are important to review as they are often used in set theory problems.

If	Then
$x \in \mathbb{R}$	$x \in \{$all real numbers$\}$
$x \in \mathbb{Q}$	$x \in \{$all numbers that can be written as a fraction$\}$
$x \in \mathbb{Z}$	$x \in \{$integers$\}$ $x \in \{\ldots -3, -2, -1, 0, 1, 2, 3 \ldots\}$
$x \in \mathbb{Z}^+$	$x \in \{$positive integers$\}$ $x \in \{1, 2, 3 \ldots\}$
$x \in \mathbb{Z}^-$	$x \in \{$negative integers$\}$ $x \in \{\ldots -3, -2, -1\}$
$x \in \mathbb{N}$	$x \in \{$natural numbers$\}$ $x \in \{0, 1, 2, 3 \ldots\}$

The set $A = \{x \mid 0 < x < 5, x \in \mathbb{N}\}$ is written in set notation. Each part has its own translation.

$$A = \{x \mid 0 < x < 5, x \in \mathbb{N}\}$$

- The set of all x
- such that
- x is between 0 and 5
- and an element of the natural numbers

This could also be written as $A = \{x \in \mathbb{N} \mid 0 < x < 5\}$. Either way $A = \{1, 2, 3, 4\}$.

➡ EXAMPLE 3.50

Consider the following sets.

$$A = \{x \mid -5 \leq x < 1, x \in \mathbb{Z}\}$$
$$B = \{x \mid |x| < 3, x \in \mathbb{Z}\}$$
$$C = \{x \mid x^2 - 4 = 0, x \in \mathbb{N}\}$$

1. List the elements of each set.
2. Determine if the following are true or false.
   (a) $A \subset B$
   (b) $B \supset C$
   (c) $C \not\subset A$
   (d) $-2 \notin C$

Answer Explanations

1. $A = \{-5, -4, -3, -2, -1, 0\}$
   - x is between -5 and 1, including -5 but not 1.
   - x must be an integer.

   $B = \{-2, -1, 0, 1, 2\}$
   - You can take the absolute values of both positive and negative numbers, so x can be any value between -3 and 3, exclusively.
   - x must be an integer.

   $C = \{2\}$
   - To solve $x^2 - 4 = 0$, you can factor, use the quadratic formula, or use PLYSMLT on the GDC. The solution is $x = -2$ or 2.
   - x is a natural number, so -2 is not a part of set C.

2. (a) **False**
   Every element of set A is not a part of set B; therefore, A cannot be a subset of B.

   (b) **True**
   The subset symbol is opening in the opposite direction, so the problem reads "C is a subset of B". This is true, since the only element of C is also a member of B.

   (c) **True**
   C is not a subset of A, since 2 is not an element of A.

   (d) **True**
   -2 is not part of set C, so it is true that -2 is not an element of C.

The **universal set** contains all possible elements for a given problem. For example, if $U = \{x \mid x < 10, x \in \mathbb{N}\}$, then the only numbers available for the problem would be natural numbers less than 10. $U = \{0, 1, 2, 3, 4, 5, 6, 7, 8, 9\}$.

Every set has its own **complement**. When combined, a set and its complement create the universal set. The notation for the complement of A is A'. $A + A' = U$

In other words, A' is made up of all the elements in U that are **not** in A.

For example, if $A = \{$prime numbers$\}$, then $A = \{2, 3, 5, 7\}$. Therefore, $A' = \{0, 1, 4, 6, 8, 9\}$, since these are the numbers in the universal set that are **not** in set A.

Two or more sets can be combined using **union** or **intersection**. Given the sets A and B:

- $A \cup B$ (A union B) is the set of all the elements in A <u>or</u> B. Union is like a marriage between sets. When sets "marry", *ALL* parts (or elements) of their "lives" are combined to form a new set.

- $A \cap B$ (A intersect B) is the set of elements in A __and__ B. Continuing with the marriage analogy, intersection is like the reason the sets "married" in the first place; the intersection is what their "lives" have in common.

$$\text{Union } (\cup) = \text{OR} = \text{ALL}$$
$$\text{Intersection } (\cap) = \text{AND} = \text{COMMON}$$

Sets can also be represented using a Venn diagram. The circles represent the individual sets with the overlapping portion showing the intersections of those sets.

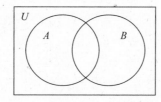

$A \cup B$ is represented by the shading shown. The shading is ALL of A with ALL of B. (Union = ALL)

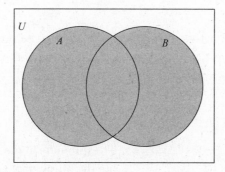

$A \cap B$ is represented by the shading shown. The shading is only what A and B have in COMMON. We can refer to this as "the football", since the shape is similar to an American football. (Intersection = COMMON)

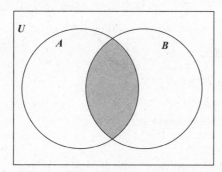

Venn diagrams are a very useful tool when working with two or three sets.

➡ **EXAMPLE 3.51** ──────────────────────────

Consider the following sets.

$$U = \{x \mid x < 9, x \in \mathbb{Z}^+\}$$
$$A = \{\text{even numbers}\}$$
$$B = \{\text{factors of 18}\}$$

1. List out the elements of set:
   (a) A
   (b) B
   (c) A'

2. Represent the sets on a Venn diagram. Place elements in the appropriate location.

3. Hence or otherwise, list the elements of
   (a) $A \cup B$
   (b) $A \cap B$
   (c) $A' \cap B'$

Answer Explanations

1. (a) **2, 4, 6, 8**

   Elements in set A are even, but they must be in the universal set. Therefore, positive integers less than 9, which are also even, are the numbers 2, 4, 6, and 8.

   (b) **1, 2, 3, 6**

   Elements in B are factors of 18, but they also must be positive integers less than 9. The factor pairs are 1 and 18, 2 and 9, 3 and 6. Out of those, only the factors 1, 2, 3, and 6 are in the universal set.

   (c) **1, 3, 5, 7**

   Elements in A' are all the elements from the universal set that are **not** in A.

2.

The numbers in the middle (the "football") are the numbers A and B have in common. The remaining numbers from each set go in the outer part of the circle. In the universal set, 5 and 7 were the only numbers not in A or B, so they must be represented in the rectangle.

3. (a) $A \cup B = \{1, 2, 3, 4, 6, 8\}$

   The union is ALL the elements in A or B.

   (b) $A \cap B = \{2, 6\}$

   The intersection is only the overlapping portion, or the "football" shape. It is what A and B have in COMMON.

   (c) $A' \cap B' = \{5, 7\}$

   The intersection of **not** A and **not** B are the elements that are **not** in both A and B.

Often, locating specific regions on a Venn diagram or shading specific areas is a troublesome point. As long as you can remember,

Logic/Sets/Probability

- ∪ = Union = **ALL**
- ∩ = Intersection = **COMMON**

then you can easily work with Venn diagrams.

For example, the region that corresponds to $A \cup B'$ can be found using the following procedure:

1. Identify the areas on the Venn diagram that correspond to set A. Represent these with the symbol, ∘. Anywhere set A exists in the Venn diagram will get marked with "∘".

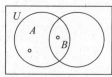

2. Next, identify the areas on the Venn diagram that correspond to set B'. Remember, this is where set B is **not**. We will represent these will the symbol, •. Anywhere set B does **not** exist in the Venn diagram will get marked with "•".

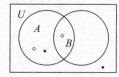

3. Finally, look at the problem. Since the problem is asking for $A \cup B'$, shade ALL the indicated regions. If the problem had asked instead for $A \cap B'$, you would shade the area where BOTH symbols appeared.

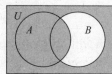

➡ **EXAMPLE 3.52** ——————————————————————————————

Shade the indicated area on the Venn diagram.

1. $A \cap B'$

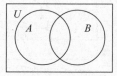

2. $A' \cup (A \cap B)$

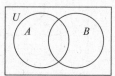

3. $A \cap B \cap C$

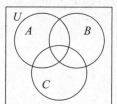

4. $(A \cup B)' \cap C$

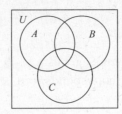

Answer Explanations

1. $A \cap B'$

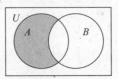

The shaded area is the part A and **not** B have in common. In other words, it is where both symbols, ∘ and •, appear.

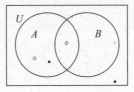

2. $A' \cup (A \cap B)$

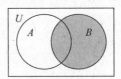

The shaded area is ALL of **not** A with the common part of A and B.

First, note all the regions where **not** A exists (shown with ∘). Then, note the region for $A \cap B$ (the "football"—shown with •). Since this problem asks for union, shade ALL indicated regions.

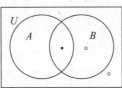

3. $A \cap B \cap C$

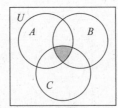

The shaded area is where sets A, B, and C intersect, which is the area all have in common. Set A is represented with ∘, set B with •, and set C with *.

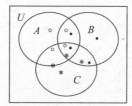

The only place where all three symbols exist is right in the middle.

4. $(A \cup B)' \cap C$

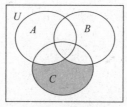

The shaded region represents the area that is **not** in A *or* B with the area in common with C.

The best way to work this problem out is to first find $(A \cup B)'$. Set A is represented with ∘ and set B with •, thus the union is ALL of the indentified regions. However, you want what is **NOT** in that identified region, which is the shading shown below.

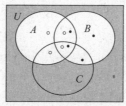

Using this shading, look and see what the shaded portion has in **common** with set C.

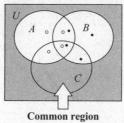

Common region

(a) **Copy and complete** the tree diagram to represent this information.

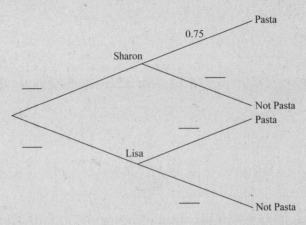

(b) Find the probability that Lisa cooks dinner, and they do not have pasta.

(c) Find the probability that they do not have pasta.

(d) Given that they do not have pasta, find the probability that Lisa cooked dinner.

Part B

A survey was carried out in a year 12 class. The pupils were asked which pop groups they like out of the *Rockers* (R), the *Salseros* (S), and the *Bluers* (B).

The results are shown in the following diagram.

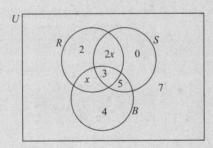

(a) Write down $n(R \cap S \cap B)$.

(b) Find $n(R')$.

(c) Describe which groups the pupils in the set $S \cap B$ like.

(d) Use set notation to describe the group of pupils who like the *Rockers* and the *Bluers* but do not like the *Salseros*.

There are 33 pupils in the class.

(e) (i) Find x.

   (ii) Find the number of pupils who like the *Rockers*.

(See page 440 for solutions)

Logic/Sets/Probability

3.6 SAMPLE SPACE AND BASIC PROBABILITY

The probability of an event A is the number of outcomes in the event out of the total possible outcomes in the sample space, S.

$$P(A) = \frac{\text{number of outcomes in } A}{\text{total number of outcomes}}$$

The sample space contains all the possible outcomes of an event. For example, if a coin is flipped, the sample space would be heads (H) or tails (T). If a random number cube or die is rolled, the sample space would be 1, 2, 3, 4, 5, or 6.

Some important facts about probability:

1. An event that is certain to happen every time has a probability of one. This is the largest probability possible.

    For example, the probability of rolling a die and landing on a number is one. There is no other option when a die is rolled other than landing on a number.

2. An event that will never happen has a probability of 0. This is the smallest probability possible.

    For example, the probability of rolling a number cube and landing on the letter A is 0. It is impossible to roll a number cube and land on a letter.

3. The probability of event A is $0 \leq P(A) \leq 1$, meaning the probability is any number between 0 and 1 inclusive.

4. The complement of any event A would be all the outcomes in the sample space that are not in A. This is very similar to the complement of a set.

    For example, if we flip a coin and look at landing on tails, then the complement of the event is landing on heads. So $P(T) + P(T') = 1$. This says the probability of landing on tails and the probability of **not** landing on tails is one. You know that when you flip a coin you will either land on tails or you will not. Therefore, the probability of an event and its complement always add to one.

    The IB Math Studies formula packet has complementary events as $P(A') = 1 - P(A)$.

5. Probabilities can be written as fractions or as decimals.

➡ EXAMPLE 3.60

Joe has a bag containing 5 red marbles, 7 blue marbles, and 3 yellow marbles. He reaches in and randomly selects one marble.

What is the probability Joe selects:

1. a red marble?
2. a yellow marble?
3. a marble that is not blue?

> **QUICK TIP**
>
> Remember: **probability** is the number of **possible** outcomes over the **total** number of outcomes

1. $P(R) = \dfrac{5}{15} = \dfrac{1}{3}$ **or 0.333**

   - There are 5 red marbles, which is the number of possible outcomes in the event "selecting a red marble".
   - There are a total of 15 marbles (5 red + 7 blue + 3 yellow = 15), which is the total number of outcomes in the sample space.
   - Therefore, the probability is the fraction $\dfrac{5}{15}$. The answer can be written as the reduced fraction or the decimal rounded to 3 significant figures.

2. $P(Y) = \dfrac{3}{15} = \dfrac{1}{5}$ **or 0.2**

   - There are 3 yellow marbles, which is the number of possible outcomes in the event "selecting a yellow marble".
   - There are a total of 15 marbles (5 red + 7 blue + 3 yellow = 15), which is the total number of outcomes in the sample space.
   - Therefore, the probability is the fraction $\dfrac{3}{15}$. The answer can be written as the reduced fraction or the decimal.

3. $P(B') = \dfrac{8}{15}$ **or 0.533**

   There were two ways to work out this probability.

   - The event desired is a marble that is **not** blue, so that means the marble must be red or yellow. The number of marbles in the event would be 5 red + 3 yellow, which is 8 in total.
   - Since the event is the marble that is **not** blue, you know that $P(B') = 1 - P(B)$. Therefore, $P(B') = 1 - \dfrac{7}{15} = \dfrac{8}{15}$.

In a set number of probability trials, the **expected value** represents how many times you would expect a certain event to occur. For example, suppose you flip an unfair coin where landing on tails has a probability of 0.75. If you flip the coin 100 times, you would expect 75 of the flips to be tails. This is found by multiplying the number of trials times the given probability; $100(0.75) = 75$.

➡ **EXAMPLE 3.61** ————————————————————————

Ava-Taylor is interested in what students are eating for lunch in the school cafeteria. For one week she records the choices of 100 students. The results are displayed in the table below.

Selected Food	Number of Students
Pizza	21
Salad	15
Hot Dog	16
French Fries	28
Fresh Fruit	20

1. Use the given information to find the probability that a student chosen at random:
    (a) selected pizza
    (b) selected fresh fruit
    (c) did not select a hot dog

2. If 350 students are randomly sampled, how many students would you expect to select French fries?

Answer Explanations

1. (a) $\dfrac{21}{100}$ or 0.21

    There are 21 students out of 100 that selected pizza.

    (b) $\dfrac{20}{100} = \dfrac{1}{5}$ or 0.2

    Twenty students selected fresh fruit out of 100 students.

    (c) $\dfrac{84}{100} = \dfrac{21}{25}$ or 0.84

    The easiest way to calculate this probability is to realize that since 16 students ordered a hot dog, 84 students did **not** (100 total students – 16 hot dog students = 84).

2. **98 students**

    The probability of a student selecting French fries is $\dfrac{28}{100}$, so the expected number of students who would select French fries out of 350 students is $350\left(\dfrac{28}{100}\right) = 98$.

3.7 MUTUALLY EXCLUSIVE, COMBINED, AND INDEPENDENT EVENTS

Mutually exclusive means two or more probability events have no common outcome. This means $P(A \cap B) = 0$.

For example, roll a die and focus on the events of rolling a number less than 4 or a number greater than 5. This can be expressed as P (number less than 4 or greater than 5). These two outcomes have no common outcome, so they are mutually exclusive. A number cannot be less than 4, while also greater than 5.

On a Venn diagram the event would be shown as follows

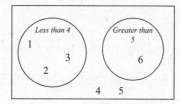

To find the probability of the event, add the individual probabilities together.

P (number less than 4) + P (number greater than 5) $= \dfrac{3}{6} + \dfrac{1}{6} = \dfrac{4}{6}$, so $P(A) = \dfrac{2}{3}$ or 0.667.

➡ EXAMPLE 3.70 _____

Brayden wants to determine the likelihood of the basketball team making a basket from various distances around the court. One afternoon he had a few players from the school team attempt shots from the following distances. The number of successful baskets out of 35 attempts is recorded below.

Distance from goal (in meters)	Number of Successful Baskets
1	9
2	8
3	8
4	6
5	4

What is the probability that a randomly selected attempt will be successful from a distance of:

1. 2 meters?
2. 5 meters?
3. 1 meter or 4 meters?
4. any distance except 5 meters?

Answer Explanations

1. $\dfrac{9}{35}$ or **0.257**

   At the 2-meter distance, there were 9 successful attempts out of a total of 35 attempts.

2. $\dfrac{4}{35}$ or **0.114**

   At the 5-meter distance, there were 4 successful attempts out of a total of 35 attempts.

3. $\dfrac{15}{35} = \dfrac{3}{7}$ or **0.429**

   These are mutually exclusive events, so add the individual probabilities together.

$$P(1\,\text{meter}) + P(4\,\text{meter}) = \frac{9}{35} + \frac{6}{35} = \frac{15}{35}$$

4. $\dfrac{31}{35}$ or **0.886**

   This is the complement of making the shot from 5 meters, so

$$P(\text{not 5 meters}) = 1 - P(\text{5 meters})$$

$$P(\text{not 5 meters}) = 1 - \frac{4}{35} = \frac{31}{35}$$

If two or more events have a common outcome, they are considered **combined events**. Using the example of rolling a die, focus on the events rolling an even number or rolling a number less than 5. This can be expressed as P (an even number or a number less than 5). The events do have a common outcome, since the even numbers on a die are 2, 4, and 6, while the numbers less than 5 are 1, 2, 3, and 4. Both events have the numbers 2 and 4.

On a Venn diagram the event would be shown as

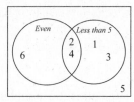

To calculate the probability of the combined event, you cannot simply add the probability of getting an even number with the probability the number is less than 5, since they have a common outcome. If you added the individual probabilities together, the probability for landing on a 2 or 4 would then get counted twice.

The IB Math Studies formula packet has the formula $P(A \cup B) = P(A) + P(B) - P(A \cap B)$ for combined events. This ensures you do not accidentally count the common probability twice.

$$P(\text{even or less than 5}) = P(\text{even}) + P(\text{less than 5}) - P(\text{even AND less than 5})$$

$$P(\text{even or less than 5}) = \frac{3}{6} + \frac{4}{6} - \frac{2}{6} = \frac{5}{6} \text{ or } 0.833.$$

You could also simply look at the Venn diagram and see there are 5 numbers within both circles, so that would mean the probability is $\frac{5}{6}$.

➥ EXAMPLE 3.71

A local dog breeder has 50 small dogs. Twenty of the small dogs have brown hair and 30 have long hair. If 15 of the small dogs have long and brown hair, calculate the probability that a randomly selected small dog has:

1. brown or long hair.
2. long hair.
3. short and brown hair.
4. neither brown hair nor long hair.

Answer Explanations

1. $\frac{35}{50} = \frac{7}{10}$ or 0.7

   These are combined events, since there is a common outcome—15 dogs have long hair and brown hair. Let $P(B)$ = probability of brown hair and $P(L)$ = probability of long hair.

   $$P(B \cup L) = P(B) + P(L) - P(B \cap L)$$
   $$P(B \cup L) = \frac{20}{50} + \frac{30}{50} - \frac{15}{50} = \frac{35}{50}$$

2. $\frac{30}{50} = \frac{3}{5}$ or 0.6

   There are 30 dogs with long hair out of 50.

> **QUICK TIP**
>
> **Combined** events have a shared outcome.
>
> You cannot simply ADD the probabilities together; you must **subtract** the common outcome from the sum.
>
> $P(A \cup B) = P(A) + P(B) - P(A \cap B)$

3. $\dfrac{5}{50} = \dfrac{1}{10}$ **or 0.1**

There are 20 dogs with brown hair, but 15 of those have brown hair **and** long hair. If the selected dog has short hair, then it cannot have long hair. So 20 − 15 = 5 brown and short-haired dogs.

4. $\dfrac{15}{50} = \dfrac{3}{10}$ **or 0.3**

A dog with neither brown nor long hair is the complement of question 1.

$$P((B \cup L)') = 1 - P(B \cup L)$$
$$P((B \cup L)') = 1 - \frac{35}{50} = \frac{15}{30}$$

Independent events are events that happen together but one does not affect the outcome of the other. For example, if a coin is flipped twice, it does not matter if the coin lands on heads the first time because the second flip is not affected by the first flip.

The IB Math Studies formula packet has the following formula for independent events: $P(A \cap B) = P(A)P(B)$.

To calculate the probability of two or more events that occur one after the other, multiply individual event probabilities.

For example, if Taylor has a 0.45 probability of winning a 100m race, then the probability she would win two races is (0.45)(0.45) = 0.2025, assuming the outcomes are independent.

➡ EXAMPLE 3.72

Aamon knows she has a probability of 0.85 of passing her math test and a probability of 0.65 of passing her Spanish test. If the outcomes of the tests are independent, find the probability she passes:

1. math and Spanish,
2. math but not Spanish,
3. neither test.

Answer Explanations

> **QUICK TIP**
>
> **Independent** events do not affect one another.
>
> Simply **MULTIPLY** the probabilities of each event together.
>
> $P(A \cap B) = P(A)P(B)$

1. **0.5525**

   The events are independent, so multiply the probability of passing math and the probability of passing Spanish. (0.85)(0.65) = 0.5525. Since this answer is exact, you do not have to round to 3 significant figures.

2. **0.2975**

   The probability of not passing Spanish is 1 − 0.65 = 0.35.

   Multiply the probability of passing math and the probability of not passing Spanish. (0.85)(0.35) = 0.2975

3. **0.0525**

   The probability of not passing math is 1 − 0.85 = 0.15.

   Multiply the probability of not passing math and the probability of not passing Spanish. (0.15)(0.35) = 0.0525

Venn diagrams are useful tools for probability events as well as set theory. When labeling each region on the Venn diagram, it is best to begin in the middle (the common area). Then, you will most likely subtract your way out to find the remaining areas.

➡ EXAMPLE 3.73

An afterschool club has 80 students. On Friday, the students are offered three snack items: fruit (*F*), crackers (*C*), or granola (*G*).

40 students choose fruit
35 students choose crackers
37 students choose granola
13 students choose fruit and crackers
5 students choose crackers and granola
20 students choose fruit and granola
3 students choose all three

> **QUICK TIP**
>
> Always begin in the **innermost region** of a **Venn diagram**.
>
> Subtract your way out, unless the given information says "only".

1. Draw and label a Venn diagram showing the snack choice of the students.
2. Find the probability that a randomly selected student chooses:
   (a) only fruit.
   (b) fruit and crackers but not granola.
   (c) neither fruit, crackers, nor granola.

Answer Explanations

1.

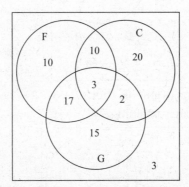

To begin the Venn diagram, the very center of the diagram is always labeled first. This is what all three circles have in common and is the statement "3 students choose all three".

Now label the common regions between each circle. There are three. The Venn diagram below shows the shaded region that corresponds to fruit and crackers. The problem states "13 students choose fruit and crackers", but notice 3 students are already identified in the shaded area. Therefore, 10 students go in the remaining space.

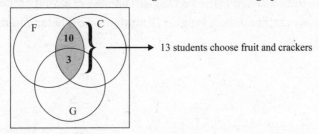

13 students choose fruit and crackers

Repeat for the other two regions.

Logic/Sets/Probability

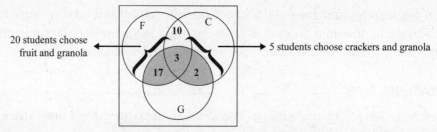

20 students choose fruit and granola

5 students choose crackers and granola

Now, there are only 4 regions without numbers. The problem states 40 students choose fruit, but in the fruit circle, which is darkened, you already have $10 + 3 + 17 = 30$ students identified. Therefore, only 10 students are remaining.

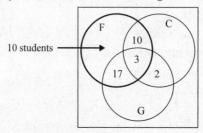

10 students

Thirty-five students choose crackers, but $10 + 3 + 2 = 15$ students have been identified. Therefore, 20 students are remaining.

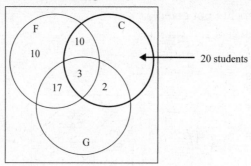

20 students

Finally, 37 students choose granola, but $17 + 3 + 2 = 22$ students have been identified. Therefore, 15 students are remaining.

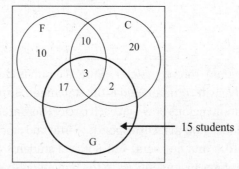

15 students

The last region to be labeled is the rectangle. There were a total of 80 students. Adding up all the numbers inside the circles give 77 students, so 3 belong in the rectangle.

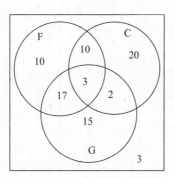

2. (a) $\dfrac{10}{80} = \dfrac{1}{8}$ or 0.125

"Only fruit" corresponds to the number in the fruit circle that does not belong to any other circle.

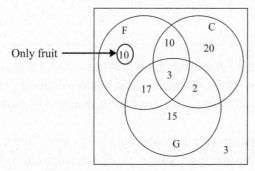

(b) $\dfrac{10}{80} = \dfrac{1}{8}$ or 0.125

A student who chooses fruit and crackers but not granola is in the region between the fruit and cracker circles, but not in the granola circle.

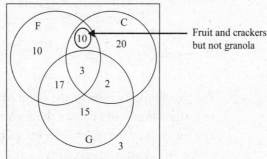

(c) $\dfrac{3}{80}$ or 0.0375

Neither fruit, crackers, nor granola is the region in the rectangle because it cannot be a part of any circle.

Logic/Sets/Probability

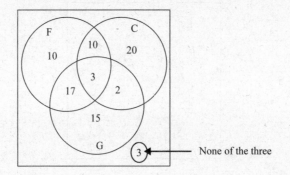

None of the three

Venn diagrams are helpful when dealing with events that share an outcome.

Tree diagrams are helpful when dealing with one event that leads to another. The key to a tree diagram is to map out the pathways of the problem and make sure the "branches" always add to one.

➥ EXAMPLE 3.74

On any given day, there is a 0.35 probability that Justin's alarm clock will not go off. If the alarm clock does not go off, the probability that Justin will be late to school is 0.85. Even if the alarm clock does go off, Justin still has a 0.25 probability of being late to school.

1. Represent the scenario using a tree diagram.
2. What is the probability that Justin's alarm clock did not go off, yet he is on time to school?
3. What is the probability that Justin is late to school?

Answer Explanations

1.

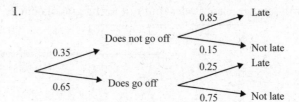

The first branch would be the event of the alarm going off or not, since this event causes the second event of being late. The probability the alarm does not go off is 0.35, so the probability the alarm does go off is $1 - 0.35 = 0.65$.

If the alarm does not go off, then the next event is being late or not. The probability of being late, if the alarm does not go off, is 0.85 so the probability of being on time is 0.15.

If the alarm does go off, the probability of being late is 0.25, and the probability for being on time is 0.75.

2. **0.0525**

His alarm clock does not go off, and he is on time, so following the tree diagram multiply $0.35(0.15) = 0.0525$.

Logic/Sets/Probability

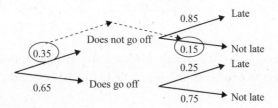

3. **0.46**

There are two ways for Justin to be late: his alarm can either not go off, or it can go off.

Following those two pathways yields:

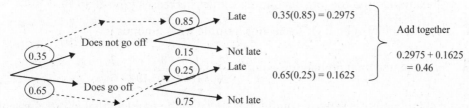

Probability that involves selecting an item, keeping it, and then selecting another is called **without replacement.** This means the probability of the event A changes with each trial.

For example, there is a bag with six red candies and five blue candies. You randomly select a candy, eat it, and then select another. This is without replacement. If you select a red candy first, there will only be five red candies remaining. If you select a blue candy first, there will only be four blue candies left. Regardless of selections, there are only ten candies instead of eleven for the second pick.

➥ **EXAMPLE 3.75** ────────────────────────────────

Chelsey and Coleen have a box of 20 chocolate truffles. Twelve of the truffles are plain and the remaining truffles have almonds. They each take a turn reaching into the box and randomly selecting a truffle to eat. Chelsey selects first, and then Coleen.

1. What is the probability both girls choose a plain truffle?
2. What is the probability that Coleen chooses a truffle with almonds?

Answer Explanations

1. $\frac{33}{95}$ **or 0.347**

   ■ The probability that Chelsey selects a plain truffle is $\frac{12}{20}$.

   ■ When Coleen takes her turn, there is one less plain truffle, so her probability of selecting a plain truffle is $\frac{11}{19}$.

   ■ To find the probability of the event, multiply $\frac{12}{20}\left(\frac{11}{19}\right) = \frac{33}{95}$.

> **QUICK TIP**
>
> • **Without replacement** means the selected item is not put back into the sample space
> • The denominator decreases by one with **each** selection
> • The numerator decreases by one **only** when the desired outcome was selected previously

2. $\dfrac{2}{5}$ or 0.4

- There are two ways that Coleen can select a truffle with almonds: Chelsey gets a plain truffle, and then Coleen gets almonds, OR both Chelsey and Coleen select a truffle with almonds.

- The probability that Chelsey selects a plain truffle and then Coleen selects a truffle with almonds is $\dfrac{12}{20}\left(\dfrac{8}{19}\right)$. Since Chelsey ate a plain truffle, there are still eight truffles with almonds, but the total amount of truffles decreases from 20 to 19.

- The probability of both girls selecting a truffle with almonds is $\dfrac{8}{20}\left(\dfrac{7}{19}\right)$.

- Therefore, the total probability would be $\dfrac{12}{20}\left(\dfrac{8}{19}\right)+\dfrac{8}{20}\left(\dfrac{7}{19}\right)=\dfrac{2}{5}$.

Conditional probability is where a condition is stated in the problem that affects the sample space. Knowing an event has occurred changes the denominator of the probability fraction.

For example, if there is a group of 10 girls and 15 boys, where 6 girls and 7 boys are wearing red, then the probability of selecting a person wearing red is $\dfrac{13}{25}$. However, if you know that the person selected was a girl, the probability the person was wearing red is $\dfrac{6}{10}$. You are no longer using the total amount of girls and boys in the sample space, since you know the person was a girl.

Sometimes a problem involves the use of "common sense", such as in the above example, but other times the use of the conditional probability formula is required.

Conditional Probability: $P(A|B)=\dfrac{P(A\cap B)}{P(B)}$.

$$P(A\mid B)=\dfrac{P(A\cap B)}{P(B)}$$

$P(A\cap B) \longleftarrow$ *Probability of A and B*

$P(B) \longleftarrow$ *Probability of B*

Probability of A   *given*   *B*

➡ EXAMPLE 3.76

Paco has a 0.75 probability of winning his first baseball game. If he wins the first game, the probability he will also win the second game is 0.9. If he does not win the first game, the probability he will win the second game is only 0.2.

1. Draw a tree diagram to represent the given scenario.
2. What is the probability that Paco wins the second baseball game?
3. Given that Paco won the second baseball game, what is the probability that he did not win the first game?

Answer Explanations

1.

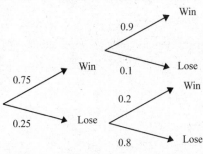

QUICK TIP

Conditional Probability gives additional information affecting the sample space.

The "given" information is always event B.

The desired probability is always event A.

$$P(A \mid B) = \frac{P(A \cap B)}{P(B)}$$

2. **0.725**

There are two ways for Paco to win his second game: win the first and second OR lose the first but win the second.

Following the pathways on the tree diagram gives:

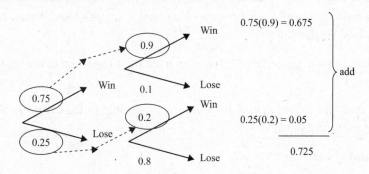

3. **0.0690**

This is a conditional probability problem. You know that Paco won the second game, so that is event B. You want to know if he did not win the first game, so that is event A.

Put the formula $P(A \mid B) = \dfrac{P(A \cap B)}{P(B)}$ into context.

$$P(\text{lost first} \mid \text{won 2nd}) = \frac{P(\text{lost first AND won 2nd})}{P(\text{won 2nd})}$$

Using the tree diagram, the probability that he lost the first game but won the second is $0.25(0.2) = 0.05$. You already found the probability that he won the second game in part 2. Therefore, the probability is $\dfrac{0.05}{0.725} = 0.0690$.

Paper 1 Questions

1. Consider the set $B = \{x \mid -6 < x \leq 2, x \in \mathbb{Z}\}$

   (a) Write down the elements in set B.

   A number is chosen at random from set B.

   What is the probability that the number is

   (b) a positive even integer;

   (c) greater than 0;

   (d) a negative integer?

2. Katherine is researching what type of car to buy when she graduates college. She has narrowed down her search to three car dealerships: Frank's Used Autos, Big City Cars, and Piedmont Dealership.

   She asks two friends, Amanda and Shannon, what their experiences are regarding the car dealerships.

   Shannon said, "If I looked at cars at Piedmont Dealership or Big City Cars, then I did not look at Frank's Used Autos".

   Let p, q, and r be the propositions:

   p = I looked at cars at Frank's Used Autos
   q = I looked at cars at Big City Cars
   r = I looked at cars at Piedmont Dealerships

   (a) Write down Shannon's experience in symbolic logic form.

   Amanda wrote her experience down as $p \wedge \neg q \wedge r$.

   (b) Write Amanda's statement in words.

   Katherine decides that if she looks at Frank's Used Autos, then she will look at Big City Cars.

   (c) Write down the implication using symbolic logic.

   (d) Write in words the converse of the implication.

3. The probability that Ethan brought his lunch money on any given day is 0.6. If he brought his lunch money, the probability he will eat in the cafeteria is 0.8. If he did not bring his money, the probability he will eat in the cafeteria is 0.2.

(a) Complete the tree diagram below.

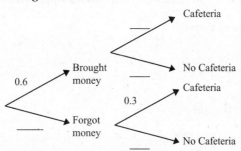

(b) What is the probability Ethan does not eat in the cafeteria?

(c) Knowing that Ethan does not eat in the cafeteria, what is the probability he forgot his lunch money?

4. A group of first graders are going on a field trip to the local zoo. The teacher asked each student which animal the student was most excited to see. The results are displayed in the table below.

Animal	Number of Students
Elephant	10
Lion	14
Giraffe	8
Alligator	3
Monkey	11

(a) What is the probability that a student selected at random is excited to see the alligators?

(b) What is the probability that a student selected at random is excited to see the lions or elephants?

(c) Two students are selected at random. What is the probability both are excited to see the monkeys?

5. Consider the propositions p and q.

(a) Complete the truth table below.

p	q	$\neg p$	$\neg p \wedge q$	$\neg q$	$\neg p \Rightarrow \neg q$
T	T				
T	F				
F	T				
F	F				

(b) Determine if the compound statement $(\neg p \wedge q) \vee (\neg p \Rightarrow \neg q)$ is a tautology, contradiction, or neither. Support your answer with a valid reason.

6. Let $U = \{x \mid x < 7, x \in \mathbb{Z}^+\}$, with subsets A = {even numbers} and B = {multiples of 3}.

    (a) List the members of the set

        (i) B

        (ii) A'

        (iii) $A' \cap B$

    (b) Write down one element in the set $(A \cap B)'$

    (c) Which of the following is false?

        (i) $2 \notin (A \cap B)$

        (ii) $B \subset A$

        (iii) $(A \cup B) \subset B$

SOLUTIONS

1. (a) $B = \{-5, -4, -3, -2, -1, 0, 1, 2\}$

      Set $B = \{x \mid -6 < x \leq 2, x \in \mathbb{Z}\}$, which means x is an integer between -6 and 2, excluding -6 but including 2.

    (b) $\dfrac{1}{8}$ **or 0.125**

      There is only one positive even integer, 2.

    (c) $\dfrac{2}{8} = \dfrac{1}{4}$ **or 0.25**

      There are 2 numbers greater than zero, 1 and 2.

    (d) $\dfrac{5}{8}$ **or 0.625**

      There are 5 negative integers, $-5, -4, -3, -2, -1$.

2. (a) $(r \vee q) \Rightarrow \neg p$

      "If I looked at cars at Piedmont Dealership or Big City Cars, then I did not look at Frank's Used Autos".

      Piedmont Dealership (r) or Big City Cars (q) means $r \vee q$.

      Did not look at Frank's Used Autos (p) means $\neg p$.

      The "If, then" means the symbol $\Rightarrow$ joins the compound statements together.

    (b) **I looked at Frank's Used Autos and Piedmont Dealerships but not Big City Cars.**

      $p \wedge \neg q \wedge r$ joins all three propositions together with "and"; however q is negated. You could have also written "I looked at Frank's Used Autos and not at Big City Cars, and I looked at Piedmont Dealerships", since this was the order of the compound statement.

    (c) $p \Rightarrow q$

      "If she looks at Frank's Used Autos, then she will look at Big City Cars" contains propositions p and q. The "If, then" means the symbol $\Rightarrow$ joins the compound statements together.

(d) **If she looks at Big City Cars, then she will look at Frank's Used Autos.**

The converse is the reverse of the implication so switch the order of the propositions.

3. (a)

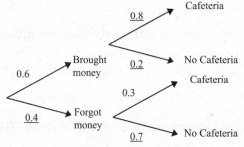

Each "branch" must add to one, so using the given probabilities you can find the missing one. The second "branch" on the top had to be filled in using the given information. You know that if Ethan brought his lunch money, the probability he will eat in the cafeteria is 0.8. Therefore, the probability he does not is $1 - 0.8 = 0.2$.

(b) **0.4**

There are two ways for Ethan to not eat in the cafeteria. Using the tree diagram you find:

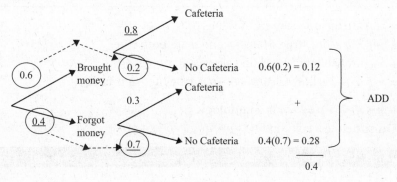

(c) **0.7**

Knowing that Ethan does not eat in the cafeteria means this is a conditional probability.

Using the formula and inserting the context of the problem yields:

$$P(\text{Forgot money}|\text{no cafeteria}) = \frac{P(\text{Forgot money AND no cafeteria})}{P(\text{no cafeteria})}.$$

Looking at the tree diagram, you already have the information you need. The probability Ethan forgot his lunch money and did not eat in the cafeteria is $0.4(0.7) = 0.28$. You just found the probability he did not eat in the cafeteria is 0.4, so the answer would be $\frac{0.28}{0.4} = 0.7$.

4. (a) $\frac{3}{46}$ **or 0.0652**

Adding up the number of students in the chart you find there were a total of 46. Of the 46, three are excited about the alligator.

(b) $\dfrac{24}{46} = \dfrac{12}{23}$ **or 0.522**

These are mutually exclusive events, so you can add the probabilities together.

(c) **0.0531**

Since two students are being selected, this is a without replacement problem. You would not select a student and then allow that student to be chosen again. Since 11 students like the monkeys, the probability the first student is excited about the monkeys is $\dfrac{11}{46}$. When the second student is selected, there are only 10 students remaining that are excited about the monkeys and a total of 45 students. Therefore, $\dfrac{11}{46}\left(\dfrac{10}{45}\right) = 0.0531$.

5. (a)

p	q	$\neg p$	$\neg p \wedge q$	$\neg q$	$\neg p \Rightarrow \neg q$
T	T	F	F	F	T
T	F	F	F	T	T
F	T	T	T	F	F
F	F	T	F	T	T

- $\neg p$ are the opposite truth values of p.
- $\neg p \wedge q$ is only true when $\neg p$ and q are **both** true.
- $\neg q$ are the opposite truth values of q.
- $\neg p \Rightarrow \neg q$ is only false when $\neg p$ is **true** but $\neg q$ is **false**.

(b) $(\neg p \wedge q) \vee (\neg p \Rightarrow \neg q)$ **is a tautology.**

Constructing a column for $(\neg p \wedge q) \vee (\neg p \Rightarrow \neg q)$

p	q	$\neg p \wedge q$	$\neg p \Rightarrow \neg q$	$(\neg p \wedge q) \vee (\neg p \Rightarrow \neg q)$
T	T	F	T	T
T	F	F	T	T
F	T	T	F	T
F	F	F	T	T

- $(\neg p \wedge q) \vee (\neg p \Rightarrow \neg q)$ is true when either $\neg p \wedge q$, $\neg p \Rightarrow \neg q$, or both are true.
- All truth values are true, so it is a tautology.

6. (a) (i) **{3, 6}**

Multiples of three begin with 3 then proceed to 6, 9, 12, 15, …
The universal set are only positive integers less than 7, so the only multiples of 3 in the universal set are 3 and 6.

(ii) **{1, 3, 5}**

The set A are even numbers, which would be 2, 4, and 6. A would **not** be all the other numbers left in the universal set (meaning odd numbers).

(iii) **{3}**

Three is the only number in both B and A'.

(b) **1, 2, 3, 4, or 5 would be accepted.**

$(A \cap B)'$ means the element is **not** in the intersection of A and B. The only number A and B share is 6, so the rest of the universal set could be an answer.

Logic/Sets/Probability

(c) (i) **2 is not an element of the intersection of A and B, so this is true.**

(ii) **B is not a subset of A because 3 is not an element of A. This is false.**

(iii) **The intersection of A and B is a subset of B, since 6 is an element in B. This is true.**

Paper 2 Questions

1. Consider the sets

$U = \{x \in \mathbb{Z} \mid 0 < x \le 12\}$
$P = \{\text{prime numbers}\}$
$Q = \{\text{factors of 36}\}$
$R = \{\text{multiples of 3}\}$

(a) List the elements of:
   (i) P
   (ii) Q
   (iii) R
   (iv) $P \cap Q \cap R$

(b) Represent the given information on a Venn diagram. Be sure to appropriately place each element from U on the Venn diagram.

(c) Describe in words the set $P \cap R'$

(d) Shade the region corresponding to $P \cap R'$ on the Venn diagram.

(e) Let p, q, and r be the statements:

$p = x$ is a prime number
$q = x$ is a factor of 36
$r = x$ is a multiple of 3

   (i)   Write in words $(p \veebar q) \Rightarrow \neg r$.
   (ii) Construct a truth table for $(p \veebar q) \Rightarrow \neg r$.

p	q	r
T	T	T
T	T	F
T	F	T
T	F	F
F	T	T
F	T	F
F	F	T
F	F	F

   (iii) Write down an element for which $(p \veebar q) \Rightarrow \neg r$ is true.

2. At a recent bake sale, 60 people were asked to select their favorite dessert from cake (C), pie (P), or ice cream (I).

10 selected all three
5 selected pie and ice cream *only*
10 selected cake and ice cream *only*
12 selected cake and pie *only*
4 selected *only* cake
1 selected *only* pie
8 selected *only* ice cream

(a) Draw a Venn diagram and clearly indicate the number of people in each region.

(b) How many people did not like any of the three?

(c) A person is chosen at random. Find the probability that the person selected:
   (i)   cake,
   (ii)   cake or ice cream,
   (iii) ice cream or pie but not cake,
   (iv)   pie given the person also selected cake.

(d) Write down in words what $(C \cap I)' \cup P$ represents.

(e) Shade the region corresponding to $(C \cap I)' \cup P$ on the Venn diagram.

(f) Two people are randomly selected. What is the probability neither of the two preferred cake, ice cream, or pie?

SOLUTIONS

1. (a) (i) $P = \{2, 3, 5, 7, 11\}$
      (ii) $Q = \{1, 2, 3, 4, 6, 9, 12\}$
      (iii) $R = \{3, 6, 9, 12\}$
      (iv) $\{3\}$

   The universal set is $\{1, 2, 3, 4, 5, 6, 7, 8, 9, 10, 11, 12\}$.

   Set A is the prime numbers within that range. The first prime number is 2. Remember, a prime number is only divisible by one and itself.

   Set B is the factors of 36. The factor pairs are 1 and 36, 2 and 18, 3 and 12, 4 and 9, and 6 and 6. Of these, you must select from the numbers that are in the universal set.

   Set C is the multiples of 3 that are also within the universal set.

(b)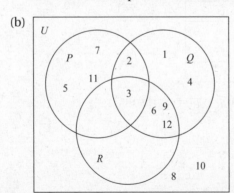

Begin by placing the 3 inside all three circles. Next, identify what else P and Q have in common (2) and place that in their overlap. Repeat for Q and R (6, 9, 12) and also P and R (nothing else).

Then, place the remaining elements in the outer circle.

Finally, write the unused universal set values in the rectangle.

(c) $P \cap R'$ means x is a prime number and not a multiple of 3.

(d)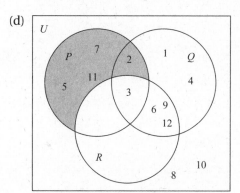
$P \cap R'$ is the common part of P and not R.

(e) (i) **If x is a prime number or a factor of 36, but not both, then x is not a multiple of 3.**
Remember, $\underline{\vee}$ is exclusive disjunction.

(ii)

p	q	r	$p \underline{\vee} q$	$\neg r$	$(p \underline{\vee} q) \Rightarrow \neg r$
T	T	T	F	F	T
T	T	F	F	T	T
T	F	T	T	F	F
T	F	F	T	T	T
F	T	T	T	F	F
F	T	F	T	T	T
F	F	T	F	F	T
F	F	F	F	T	T

- $p \underline{\vee} q$ is true when either p or q is true, but not both.
- $\neg r$ are the opposite truth values of r.
- $(p \underline{\vee} q) \Rightarrow \neg r$ is only false when $p \underline{\vee} q$ is true and $\neg r$ is false.

(iii) **Answers may be 1, 2, 3, 4, 5, 7, 8, 10, or 11.**
The truth table shows where the statement $(p \underline{\vee} q) \Rightarrow \neg r$ is true. The truth value of p, q, and r show where to go on the Venn diagram.
For example, the first line has a true outcome. On this line p, q, and r are all true, so this is where the three circles intersect.

p	q	r	$p \underline{\vee} q$	$\neg r$	$(p \underline{\vee} q) \Rightarrow \neg r$
T	**T**	**T**	F	F	**T**
T	**T**	**F**	F	T	**T**
T	F	T	T	F	F
T	**F**	**F**	T	T	**T**
F	T	T	T	F	F
F	**T**	**F**	T	T	**T**
F	**F**	**T**	F	F	**T**
F	**F**	**F**	F	T	**T**

2. (a)

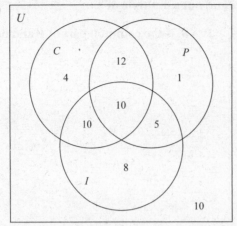

The 10 people that selected cake, pie, and ice cream are placed in the center of the Venn diagram first. Everything else gets placed without subtraction because of the word "*only*". The phrase "5 selected pie and ice cream *only*" means that those 5 people go in the common overlap between pie and ice cream, but **not** in the portion in common with cake. This is true of all the statements.

(b) **10**

There were a total of 60 people. Adding up the numbers within all the circles gives $4 + 10 + 10 + 12 + 1 + 5 + 8 = 50$. Thus, $60 - 50 = 10$ people.

(c) (i) $\dfrac{36}{60} = \dfrac{3}{5}$ **or 0.6**

If the randomly selected person picked cake, then **all** the numbers in the cake circle must be added. $4 + 12 + 10 + 10 = 36$; there were a total of 60 people.

(ii) $\dfrac{49}{60}$ **or 0.817**

If a randomly selected person picked cake or ice cream, this means **all** the numbers in cake and ice cream must be combined. Remember, or = all.

(iii) $\dfrac{14}{60} = \dfrac{7}{30}$ **or 0.233**

If a randomly selected person picked ice cream or pie but not cake, then the numbers in ice cream or pie must be added. However, cake cannot be included, since you know the person did not pick cake. The shading represents those people who chose pie or ice cream but not cake.

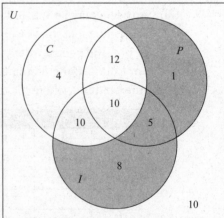

(iv) $\dfrac{22}{36} = \dfrac{11}{18}$ or 0.611

You know the randomly selected person chose cake, so the denominator of the fraction would be the total number of people who chose cake (36). Of those, focus on the ones who also chose pie. On the Venn diagram, this is the overlapping "football" between cake and pie. Adding 12 and 10 means 22 people selected pie.

(d) **A person did not select cake and ice cream or they selected pie.**

If you break down $(C \cap I)' \cup P$ into the two parts, $(C \cap I)'$ means "cake and ice cream **not**" and then $\cup P$ means "or pie".

(e)

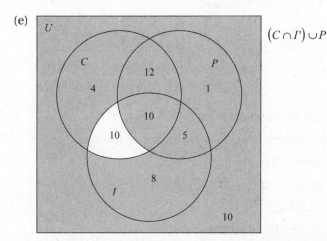

$(C \cap I') \cup P$

The best way to determine the region to be shaded is to go in stages.

$(C \cap I)'$ would be the entire Venn diagram **except** the intersection of cake and ice cream. This means everything but the "football" between sets C and I will be shaded. Next, $\cup P$ means to add on the shading for pie. Remember union = or = all.

(f) $\dfrac{10}{60}\left(\dfrac{9}{59}\right) = 0.0254$

Since two people are randomly selected, you will multiply two probabilities together. Once one person is selected, they will not go back into the sample space because you do not want to pick the same person twice. Thus, the probabilities are without replacement. The probability the first person did not like any of the three is $\dfrac{10}{60}$, but for the second person each total decreases by one, since a person has already been chosen.

CHAPTER OBJECTIVES

Before you move on to the next chapter, you should be able to:

☐ Identify if a statement is a proposition

☐ Given propositions, write a compound statement using words or symbols

☐ Complete a truth table involving two or three propositions

☐ Describe a compound statement as a tautology, contradiction, or neither

☐ Given a proposition, write the converse, inverse, and contrapositive

☐ Identify logically equivalent compound statements

☐ List members in a given set

☐ Determine if elements are a member of a given set

☐ Determine if one set is a subset of another

☐ Find the complement to a set

☐ Find the union or the intersection of two or three sets

☐ Draw a Venn diagram showing the relationship between two or three sets

☐ Shade an area of a Venn diagram corresponding to a given set notation

☐ Write out the sample space of an event

☐ Calculate the probability for one event

☐ Determine the expected value for a probability event

☐ Calculate the probability for mutually exclusive, combined, and independent events

☐ Complete a Venn diagram using given information and use to find probabilities

☐ Draw a tree diagram and use to find the probability of two events

☐ Determine the probability of an event with and without replacement

☐ Determine conditional probability

Statistical Applications

- **NORMAL DISTRIBUTION**: bell-shaped curve centered on mean with the standard deviation the measure of spread; normalcdf (lower, upper, mean, standard deviation) and invNorm (area, mean, standard deviation)

- **SCATTER PLOT**: graph displaying the points of a bivariate data set; shows direction and strength of linear model

- **CORRELATION**: calculates how well a linear regression model fits bivariate data; the closer $|r|$ is to 1, the stronger the correlation

- **REGRESSION LINE**: the line of best fit for bivariate data; (LINREG) in GDC gives $y = ax + b$; can be used to predict

- **CHI-SQUARED TEST OF INDEPENDENCE**: determines if there is a relationship between two categorical factors; must have null and alternate hypotheses; χ^2–Test in GDC

- **EXPECTED VALUE**: determined by multiplying the total of the row and the total of the column for the desired space, and then dividing by the grand total

4.1 NORMAL DISTRIBUTION

Continuous data distributions that are approximately symmetric and follow the bell-shaped curve are said to be normal. The center of the graph is the mean, μ, and the measure of spread is the standard deviation, σ. The line of symmetry for a normal curve is $x = \mu$.

Suppose the mean height of all 16-year-old girls is 163 centimeters, with a standard deviation of 6.5 centimeters. This means the majority of 16-year-old girls will have a height close to the mean, but some girls will be taller and some will be shorter.

The graph below illustrates the hypothetical situation, with the mean of 163 centimeters and standard deviation of 6.5 centimeters.

Heights of 16-year old girls

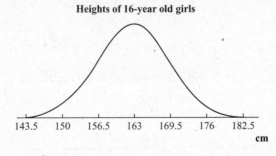

| 143.5 | 150 | 156.5 | 163 | 169.5 | 176 | 182.5 |

cm

An important concept about the normal distribution is that roughly 68% of the data values fall within one standard deviation of the mean, 95% of the data fall within two standard deviations, and 99% of the data fall within three.

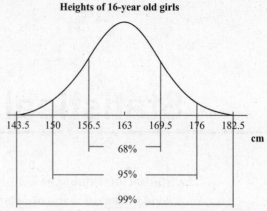

Heights of 16-year old girls

These percentages can be used to find the expected value, or expected number, of a given scenario. Simply multiply the percentage by the number of data in the sample to calculate the expected value.

➡ **EXAMPLE 4.10** ────────────────

The scores on a high school math exit exam are normally distributed with a mean score of 85 and a standard deviation of 6. A total of 450 students took the exit exam.

1. Sketch the normal curve clearly showing the mean and three standard deviations.
2. Find the number of students expected to score between 79 and 91 points.
3. Find the number of students expected to score less than 73 points.

Answer Explanations

> **QUICK TIP**
>
> On a **Normal curve**,
>
> The **mean** is always in the **center**.
>
> **Add** one, two, and three **standard deviations** up.
>
> **Subtract** one, two, and three **standard deviations** back.
>
> The rule is 68–95–99%

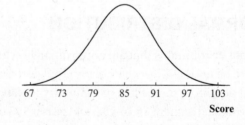

1. The graph is centered on the mean of 85. Add one, two, and three standard deviations to the mean:
   - $85 + 6 = 91$
   - $85 + 2(6) = 97$
   - $85 + 3(6) = 103$

   Subtract one, two, and three standard deviations from the mean:
   - $85 - 6 = 79$
   - $85 - 2(6) = 73$
   - $85 - 3(6) = 67$

2. **306 students**

   The scores 79–91 are one standard deviation below and one standard deviation above the mean. Values within one standard deviation represent 68% of the data. To find the number of students, multiply 450(0.68) = 306.

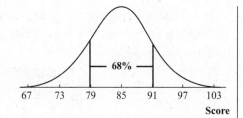

3. **11 students**

   A score of 73 is 2 standard deviations less than the mean. Therefore, scores between 73 and 97 represent 95% of the data. This leaves 5% for scores less than 73 or more than 97. Since the question only asks for scores less than 73, divide 5% in half to get 2.5%. Now multiply 450(0.025) = 11.25. The

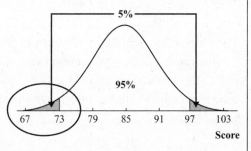

   answer is 11.25 students but we must round for the appropriate context and 11.25 students is not possible.

The normal distribution can also be used to find the probability, or likelihood, of a certain outcome.

Using the previous example of the heights of 16-year-old girls, find the probability of a girl having a height of 173 centimeters or more.

First, sketch the normal curve and shade the area desired. The sketch is very useful and is expected to be drawn on IB examinations.

Heights of 16-year old girls

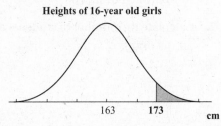

Second, identify the four necessary pieces to calculate the probability using the GDC:

1. Lower bound (where the shading begins)
   - If the shading begins at a data value, like 173, use the data value.
   - If the shading begins at −∞, use −E99 as the value. This is an extremely small negative number used to represent negative infinity.
2. Upper bound (where the shading ends)
   - If the shading ends at a data value, use the data value.
   - If the shading ends at ∞, use E99 as the value. This is an extremely large number used to represent infinity.
3. Mean
4. Standard deviation

To calculate the actual probability, use your GDC. Press [2ND VARS] to get to the [DISTR] (distribution) menu. **Always** select option 2, [NORMALCDF(]. Input the lower bound, upper bound, mean, and standard deviation then press enter.

In the example, the lower bound is 173, the upper bound is E99 to represent infinity since the shading never ends, the mean is 163, and standard deviation 6.5. Now, hit ENTER on PASTE and then ENTER again.

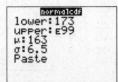 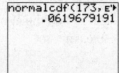

Therefore, the probability a 16-year-old girl is 173 cm or taller is 0.0620.

This can be written mathematically as $P(X \geq 173) = 0.0620$. It is interesting to note that $P(X \geq 173) = 0.0620$ as does $P(X > 173) = 0.0620$. In continuous data, the probability of being exactly a certain value is zero, so the probability that a 16-year-old girl is exactly 173 cm would be zero.

➡ EXAMPLE 4.11

The scores on a high school math exit exam are normally distributed with a mean score of 85 and a standard deviation of 6. A total of 450 students took the exit exam.

> **QUICK TIP**
>
> **Always** sketch a normal curve with the desired area shaded.
>
> **Write down** the calculator command to show your "method".

1. Find the probability that a randomly selected student scored:
   (a) above 95;
   (b) less than 70;
   (c) between 87 and 93.

2. Calculate the number of students expected to score above 95 on the exam.

Answer Explanations

1. (a) **0.0478**

   The lower bound is 95, upper bound is E99 (infinity), the mean is 85, and standard deviation is 6.

    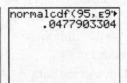

   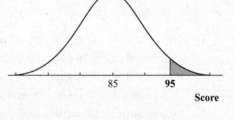

   (b) **0.00621**

   The lower bound is −E99 (negative infinity), upper bound is 70, mean is 85, and standard deviation is 6.

   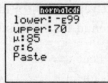 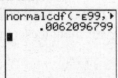

   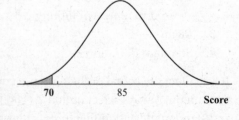

(c) **0.278**

The lower bound is 87, upper bound is 93, mean is 85, and standard deviation is 6.

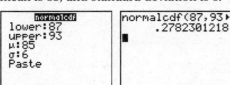

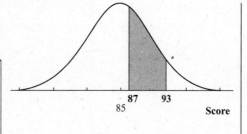

2. **21.5 or 22 students**

The probability a randomly selected student scored above a 95 is 0.0478, which also represents the percent of students who are expected to score above a 95. To find the expected value, multiply the total number of students by the probability: $450(0.0478) = 21.51$.

The normal distribution can also be used to find a data value that corresponds to a given probability or percentage.

For example, the probability that a randomly selected 16-year-old girl has a height of h cm or less is 0.0875. If the mean height of 16-year-old girls is 163 cm, with a standard deviation of 6.5, what is h?

Heights of 16-year old girls

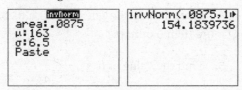

To calculate the value of h, work backwards. You know the probability, which is the value of the shaded region. You are going to determine the height that corresponds to the upper bound. This is done using the inverse normal function on the GDC. In the DISTR menu, select option 3: INVNORM(. The area is 0.0875, the mean is 163, and the standard deviation is 6.5. Press ENTER on PASTE and then ENTER again.

```
   invNorm              invNorm(.0875,16▶
 area:.0875                  154.1839736
 μ:163
 σ:6.5
 Paste
```

Thus, $h = 154$ cm.

Be careful if the probability or percentage given for a data value is **greater** than x. The inverse normal function on the GDC uses the "less than" shading, meaning only shading to the left. If the problem asks for a value greater than x, you must determine the "less than" shading by subtracting the given probability from 1.

For example, the probability a randomly selected 16-year-old girl has a height of h inches or more is 0.15. If the mean height of 16-year-old girls is 163 cm with a standard deviation of 6.5, what is h?

The sketch shows the given probability is the shading to the right, since the probability is for a girl being h cm or **more**. To use the INVNORM function on the calculator, you must convert to a "less than" shading. The entire normal curve represents 100% (or 1), so the two sections of the curve add up to 1.

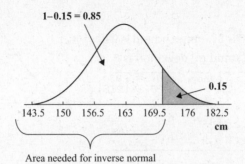

1 − 0.15 = 0.85

0.15

143.5   150   156.5   163   169.5   176   182.5
cm

Area needed for inverse normal

Now the area is 0.85 when calculating inverse normal, which gives $h = 170$ cm (rounded to 3 significant figures).

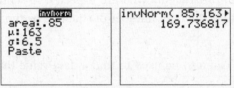

➡ **EXAMPLE 4.13** _____

A local business packages gift boxes of candied almonds. The weights of the boxes are normally distributed with a mean of 16.1 ounces and a standard deviation of 0.25 ounces.

One box is randomly selected from the production line and weighed. Find the probability the box

1. weighs less than 15.7 ounces;

2. weighs more than 16.7 ounces.

The lightest 5% of the gift boxes weigh w ounces and are discarded.

3. Draw a sketch of the normal curve clearly showing w and μ.

4. Calculate the value of w.

A randomly selected gift box has a 0.20 probability of weighing more than b ounces.

5. Find the value of b.

> **QUICK TIP**
>
> When using [INVNORM]:
>
> Sketch a picture showing the given percent or probability.
>
> If the shading is to the **left**, use the given percent as the "area" for [INVNORM].
>
> If the shading is to the **right**, subtract the percent from one. This is the "area" for [INVNORM].

Answer Explanations

1. 0.0548

Lower bound: −E99 (negative infinity)

Upper bound: 15.7

Mean: 16.1

Standard deviation: 0.25

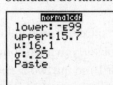 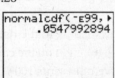

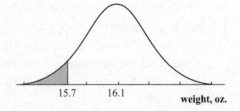

15.7   16.1   **weight, oz.**

2. **0.00820**

   Lower bound: 16.7

   Upper bound: E99 (infinity)

   Mean: 16.1

   Standard deviation: 0.25

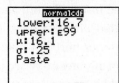 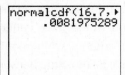

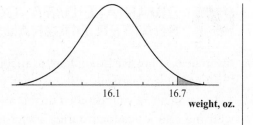

3. The mean, μ, goes in the middle, and since w represents the lightest 5%, the shading is below the mean.

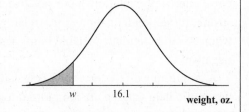

4. **$w = 15.7$ ounces**

   Using the INVNORM option on the GDC, the area is 0.05, the mean is 16.1, and the standard deviation is 0.25.

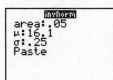 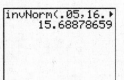

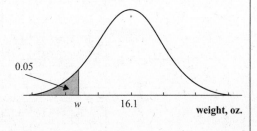

5. **$b = 16.3$**

   Since the probability given is for weighing **more than** b ounces, you must subtract from 1 to find the "less than" shading. Now, the area is 0.80, with a mean of 16.1 and a standard deviation of 0.25.

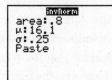 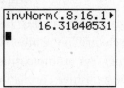

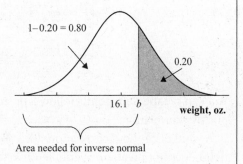

4.2 BIVARIATE DATA: CORRELATION AND SCATTER DIAGRAMS

Bivariate data involve two different variables that may have a relationship. One variable, x, is the independent variable, and the other, y, is the dependent variable.

For example, if you recorded the number of minutes you exercised for one month paired with the calories burned during exercise, you would have generated bivariate data. The minutes exercised would be the independent variable, while the calories burned would be the dependent. There would be a strong, positive relationship between minutes exercised and calories burned. The longer you exercise, the more calories you would burn.

This is called the **correlation**. The closer the data fall to a straight line, the stronger the correlation. If the relationship is positive, meaning when the independent variable increases so does the dependent, then the correlation is positive. If the relationship is negative, meaning when the independent variable increases the dependent decreases, then the correlation is negative.

A graphical way to determine correlation is by a scatter diagram. This is simply a graph where the data value pairs are plotted together. The graph below illustrates the example of minutes exercised and calories burned.

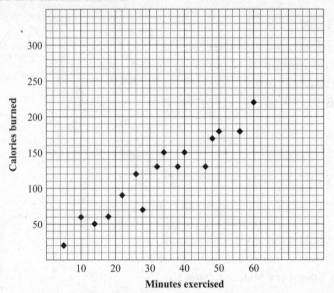

You can see the data points follow a positive, linear trend. Therefore, the data have a strong, positive correlation.

A word of caution, correlation does not mean causation. Just because data follow a linear trend does not mean one variable causes the change in the other. Correlation shows there is a relationship but not that one causes the other. For example, studies show that the taller a person, the larger his shoe size. This would be a positive correlation. However, the height of a person does not *cause* his foot to be a certain size. You cannot prove causation; you can only show correlation.

➡ EXAMPLE 4.20

For each of the following, graph a scatter diagram and describe the correlation.

1. You record the liters remaining in your fuel tank after driving x kilometers.

Kilometers driven	0	60	110	160	210	240	300	390	450	480
Liters remaining	72	64	53	49	38	34	30	19	11	7

2. Ms. Pueller records the minutes each student studied along with the student's score on the math exam.

Minutes studied	5	6	7	10	13	15	18	20	20	24	27	27	30	30
Score	20	60	76	41	68	44	53	20	96	72	48	88	77	100

Answer Explanations

1. **The data have a strong, negative correlation.**

   After plotting each point (kilometers driven, liters remaining), you can see the data form a relatively straight line that is decreasing. This means the data have a strong, negative correlation.

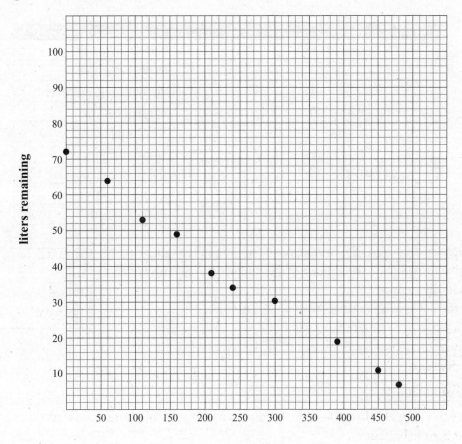

2. **The data have no correlation.**

   After plotting each point (minutes studied, score), you can see the data points are widely scattered and follow no linear pattern. This means the data have no correlation.

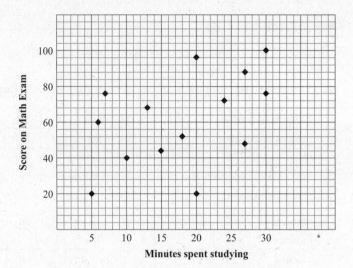

The correlation coefficient, r, can be calculated by hand; although, for the IB exam it is expected the calculation will be done using a GDC. The correlation coefficient, r, is a numerical value associated with the linearity of the data.

Values of r range from -1 to 1, with 0 being no correlation. The chart below gives different ranges of r with their correlation description. There is no exact science of when r is "strong" versus "moderately strong", so use these boundaries only as a guideline.

Value of r	Correlation Description
$r = -1$	Perfect, negative
$-1 < r < -0.7$	Strong, negative
$-0.7 \leq r < -0.3$	Moderately strong, negative
$-0.3 \leq r < 0$	Weak, negative
$r = 0$	No correlation
$0 < r \leq 0.3$	Weak, positive
$0.3 \leq r < 0.7$	Moderately strong, positive
$0.7 < r < 1$	Strong, positive
$r = 1$	Perfect, positive

To calculate the value of r, the diagnostic option on the GDC must be turned on. To do this, press 2ND followed by 0. This is the catalog menu.

Scroll down to DIAGNOSTICON and hit ENTER twice.

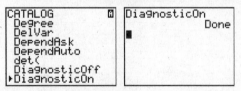

Now, you will be able to calculate r. This option will remain "on" unless you reset the memory or change the batteries.

To actually calculate r, the independent variable is put into L_1 and the dependent variable into L_2.

For example, suppose you recorded the following data while exercising:

Minutes exercised	5	10	14	18	22	26	32	34	40
Calories burned	20	60	50	60	90	120	130	150	150

After inputting the data into the lists, be sure to quit 2ND MODE.

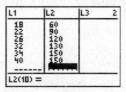

Now, go to the STAT menu, arrow right to the CALC options and select 4: LINREG(AX + B).

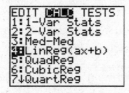

As long as your data have been entered into L_1 and L_2, you can leave XLIST and YLIST as is. If you used any other list, you must change the default settings.

Always leave FREQLIST blank. You can also leave STORE REGEQ blank as well.

Now press ENTER on calculate.

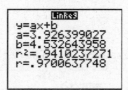

The outputs are the components of the linear regression, which will be discussed in the next section. At the bottom of the screen is the correlation coefficient, $r = .970$. Since r is positive and very close to 1, the data have a strong, positive correlation.

You can check the scatter diagram as well for a second opinion. Press 2ND Y = to get to the statistical plot options.

Press ENTER on PLOT 1 and then ENTER on ON. The first graph option should be darkened since that is the option for a scatter diagram.

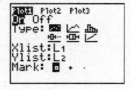

Now, press ZOOM 9 to see the graph.

Notice the points are following a positive trend and form a relatively straight line. This supports the correlation coefficient of 0.970.

➡ **EXAMPLE 4.21** ———————————————————————

QUICK TIP

The closer *r* is to 1 or –1, the stronger the linear relationship.

The sign of *r* is the direction of the linear relationship.

• positive *r* = positive relationship

• negative *r* = negative relationship

Calculate the correlation coefficient for each of the following scenarios. Comment on the value calculated.

1. You record the liters remaining in your fuel tank after driving x kilometers.

Kilometers driven	0	60	110	160	210	240	300	390	450	480
Liters remaining	72	64	53	49	38	34	30	19	11	7

2. You record the height and weight for a group of fourth graders.

Height (cm.)	125	119	132	117	112	127	122	119	102	114	137
Weight (kg.)	28	29	32	26	25	31	27	25	23	26	31

Answer Explanations

1. $r = -0.994$

   Kilometers driven and liters remaining have a strong, negative correlation.

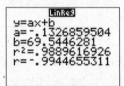

2. $r = 0.901$

   Height and weight have a strong, positive correlation.

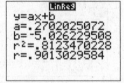

4.3 REGRESSION LINES

If bivariate data follow a linear trend, calculating the equation of the line that best fits the data makes sense. There are many names for this line: line of best fit, linear regression equation, regression line for y on x, and least-squares regression are a few examples.

There are various ways of calculating the regression line, but you will rely on the GDC to generate your equation.

When graphing the regression line on the exam, it **must** pass through the mean point $(\bar{x}, \bar{y})$. Graphing the y-intercept, if the intercept is a reasonable point, is wise. The regression line should roughly cut the data in two sections, with half of the data falling below the line and the other half above the line.

In the example of exercising and calories burned, you can write the regression line by entering the data into L_1 and L_2 as before.

Minutes exercised	5	10	14	18	22	26	32	34	40
Calories burned	20	60	50	60	90	120	130	150	150

Then, just like calculating r, select (LINREG(AX + B)). The output gives all the parts needed to write the equation of the line. The value $a = 3.93$ is the gradient, and $b = 4.53$ is the y-intercept.

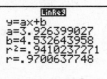

Therefore, the regression line is $y = 3.93x + 4.53$.

You can store this equation in the $Y =$ menu by telling the calculator where to place the equation. When calculating the regression line, move the cursor to (STORE REGEQ), press (VARS), arrow right to (Y-VARS), and press (ENTER) three times.

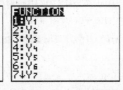
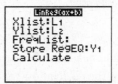

Now, the regression line will be placed into Y_1 for future use.

Understanding what the gradient represents *in context* is an important skill. Since gradient is the change in y over the change in x, our gradient represents the change in calories burned over the change in minutes exercised. Thus, for every one minute exercised, an additional 3.93 calories are burned.

You may be interested in predicting how many calories you would burn if you exercised for 25 minutes. Since 25 minutes falls within the given range data values, using the regression line to predict the number of calories would be appropriate. Substitute 25 into the equation for x to determine y. Thus, $y = 3.93(25) + 4.53 = 103$ calories, rounded to 3 significant figures.

If the regression line is stored in Y_1, the calculator can substitute in a value and simplify. On the home screen, press (VARS), arrow right to (Y-VARS), and press (ENTER) twice. Now, using function notation, tell the calculator to plug in 25 by typing an opening parenthesis, 25, and then a closing parenthesis.

Statistical Applications

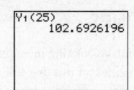

Y₁(25)
        102.6926196

You can also predict how long you must exercise in order to burn 110 calories. Again, 110 falls within the given range of data so using the linear model is appropriate. This time substitute 110 in for y and solve for x. The Y_1 equation cannot be used in this situation.

$$110 = 3.93x + 4.53$$
$$105.47 = 3.93x$$
$$26.8 = x$$

Extrapolation is using the regression line to predict values outside of the data range. If you wanted to predict how many calories you would burn after 60 minutes, this would be extrapolation. This is an unwise decision, since 60 is too far outside the range. You know the data follow a linear trend only for the values recorded. Outside of that range you cannot assume the data still follow the same linear trend.

➡ EXAMPLE 4.30

Amber is running in a 10 kilometer race. She keeps pace by noting her running time at each kilometer of the race.

Time (mins)	6	12	17	25	32	37	44	51	58	63
Distance (km)	1	2	3	4	5	6	7	8	9	10

1. Draw a scatter diagram of the data. Let 1 cm represent 10 minutes on the horizontal axis and 1 cm represent 1 km on the vertical axis.
2. Write down the regression line of y on x.
3. Describe the meaning of the gradient *in context*.
4. Write down the correlation coefficient.
5. Describe the relationship between time and distance run.
6. Calculate the distance Amber will have run after 28 minutes.
7. Using the regression line, predict the time it will take Amber to run 15 kilometers.
8. Explain if using the regression line is appropriate to estimate the time taken to run 15 kilometers.

Statistical Applications

Answer Explanations

1. The scale must be 10 minutes for each cm block on the *x*-axis and 1 kilometer for each cm block on the *y*-axis. Axes must be labeled.

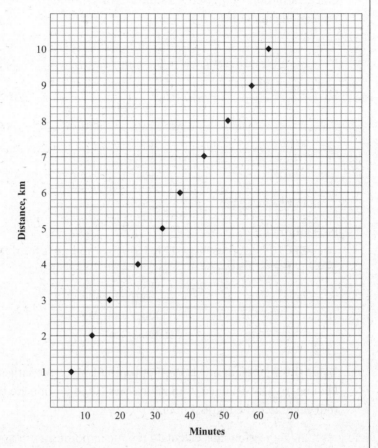

2. $y = 0.154x - 0.172$

   The minutes should be entered into L_1 and distance into L_2. Then, the [LINREG(AX + B)] is calculated.

3. **For every 1 minute, Amber runs an additional 0.154 km.**

   Gradient is the change in *y* over the change in *x*. Therefore, the gradient is $\dfrac{0.154}{1} = \dfrac{\text{kilometers}}{\text{minutes}}$.

4. **$r = 0.999$**

   The correlation coefficient *r* is the last value on the output of [LINREG(AX + B)]. If *r* is not present, turn the diagnostic option back on as discussed earlier.

5. **Time and distance run have a strong, positive correlation.**

   Since $r = 0.999$, which is positive, the data have a strong, positive correlation. Remember, the closer to 1, the stronger the correlation.

Statistical Applications

6. **4.48 km OR 4.50 km**

   Since 28 minutes represents x and is within the data range, substitute 28 for x in the regression equation.

   $$y = 0.154(28) + 0.172$$
   $$y = 4.48$$

   OR

   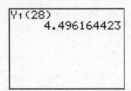

   Therefore $y = 4.50$ km.

7. **96.3 minutes**

   The value 15 km represents y. Plug 15 in for y and solve for x. Be sure to express the answer to 3 significant figures.

   $$15 = 0.154x + 0.172$$
   $$14.828 = 0.154x$$
   $$96.2857 = x$$

8. Using the regression line is not appropriate, since the largest distance in the data is 10 kilometers. The distance of 15 kilometers is too far outside the given range of distances that generated the regression line.

To graph the regression line, calculate the mean point. In the (STAT) menu, arrow right to the (CALC) options. Select option 2: (2-VAR STATS), since this will calculate the mean of the x-values and y-values at the same time. As before, as long as the data is entered into L_1 and L_2, you can leave the default options as is. If not, you must tell the GDC which list represents x and which represents y.

Going back to the exercising example, you calculate the mean minutes exercised and the mean calories burned. After inputting the data and calculating (2-VAR STATS), you find $\bar{x} = 22.3$ and $\bar{y} = 92.2$.

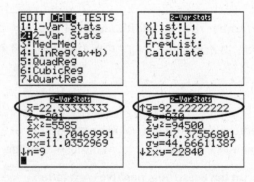

When plotting the data points, clearly plot the mean point (22.3, 92.2). The regression line must pass through this point. Plotting the y-intercept, when appropriate, as a second point is also a wise decision. Next, place a straight edge on the intercept and mean point to create the line.

Roughly half the data points should lie above the line and half below.

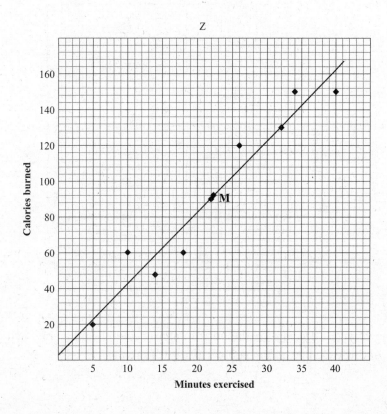

➥ EXAMPLE 4.31

For her IB Math Studies project, Cullen wanted to determine if there is a relationship between a person's age and the time taken to run a mile. She asked 10 people with similar athletic abilities to run 1.6 kilometers. She recorded the results below.

Age	18	22	26	28	30	33	35	19	21	37
Time (minutes)	6.8	7.5	8.3	10	11.1	11.8	11.7	7.8	8.8	12.4

1. Plot the data on a scatter diagram. Let 1 cm represent 5 years on the horizontal axis and 1 cm represent 2 minutes on the vertical axis.
2. Write down the
   (a) mean age;
   (b) mean time to run 1.6 kilometers.
3. Plot the mean point on the scatter diagram. Label the point M.
4. Write down the equation of the regression line of y on x.
5. Draw the line of regression on the scatter diagram.

QUICK TIP

The **regression line must** pass through the **mean point** on a graph.

Half the data should be **above the line** while half is **below**.

Answer Explanations

1.

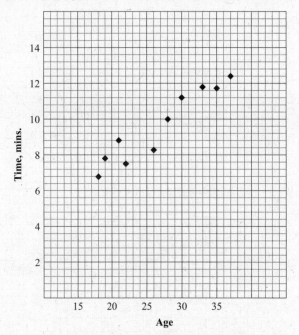

Since the youngest age was 18, the horizontal axis could begin at 15. Each data pair is plotted, noting each line on the vertical axis is 0.4 minutes.

2. (a) **26.9**

   (b) **9.62**

Enter the data into L_1 and L_2 and select 2-VAR STATS .

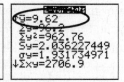

3.

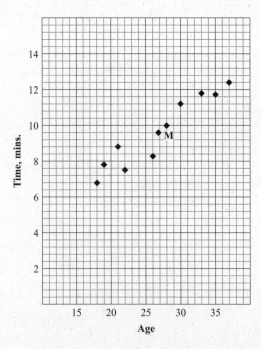

Add the point (26.9, 9.62) to the scatter diagram. Graphing close to (27, 9.6) gives a general idea of the location. Be sure to label the mean point M.

4. $y = 0.286x + 1.93$

    The data are already in the correct lists, so calculate [LINREG(AX + B)].

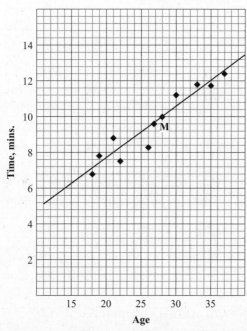

5.

The regression line must pass through the mean point. Roughly half the points are above the line and half are below. Since the scale on the x-axis did not begin with zero, you cannot use the y-intercept as a point on the line.

If the x-axis had begun with zero, then the following could be the scatter diagram and line of best fit.

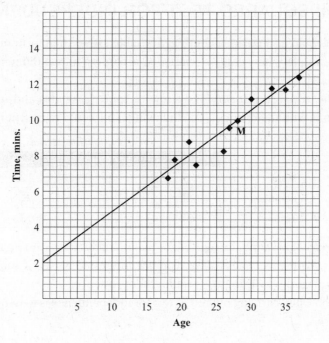

FEATURED QUESTION

May 2011, Paper 2

The heat output in thermal units from burning 1 kg of wood changes according to the wood's percentage moisture content. The moisture content and heat output of 10 blocks of the same type of wood each weighing 1 kg were measured. These are shown in the table.

Moisture content % (x)	8	15	22	30	34	45	50	60	74	82
Heat output (y)	80	77	74	69	68	61	61	55	50	45

(a) Draw a scatter diagram to show the above data. Use a scale of 2 cm to represent 10% on the x-axis and a scale of 2 cm to represent 10 thermal units on the y-axis.

(b) Write down

(i) the mean percentage moisture content, $\bar{x}$;

(ii) the mean heat output, $\bar{y}$.

(c) Plot the point $(\bar{x}, \bar{y})$ on your scatter diagram and label this point M.

(d) Write down the product-moment correlation coefficient, r.

The equation of the regression line y on x is $y = -0.470x + 83.7$.

(e) Draw the regression line y on x on your scatter diagram.

(f) Estimate the heat output in thermal units of a 1 kg block of wood that has 25% moisture content.

(g) State, with a reason, whether it is appropriate to use the regression line y on x to estimate the heat output in part (f).

(See page 442 for solutions)

4.4 CHI-SQUARED TEST FOR INDEPENDENCE

Bivariate data could also create a contingency table rather than form ordered pairs. A contingency table displays how data are classified according to two different factors. The table looks like a matrix.

For example, suppose you surveyed some of your classmates and recorded their lunch preference along with their gender. The following table could represent the results.

Lunch Preference

Gender		Salad	Sandwich	Pizza	Total
	Male	5	9	12	26
	Female	8	6	6	20
	Total	13	15	18	46

The contingency table shows the number of males and females who prefer each of the three lunch options. With this type of data, you are interested to see if there is a relationship between the categories by using a χ^2 test for independence.

The first step when carrying out the χ^2 test for independence is to write the null and alternate hypotheses.

The null hypothesis, or H_o, always follows the pattern of "_____ and _____ are independent." Simply fill in the two categories from the table.

The alternate hypothesis, or H_a, is the exact opposite and follows the pattern "_____ and _____ are dependent."

For this example, the hypotheses would be

H_o: Gender and lunch preference are independent.
H_a: Gender and lunch preference are dependent.

The second step is to calculate the expected values. The contingency table provides the observed values—the values that are collected (or observed). Use the observed data to determine what you would have expected. The IB exam may require a calculation of one or two expected values by hand, but the GDC can also be used.

To find the expected value for each element in the table, multiply the total of the row and the total of the column for the desired space, and then divide by the grand total. Continue this process for each space in the table.

For example, the observed number of males that preferred salad is 5. The expected number of males that preferred salad would be found by multiplying the total number of males, 26, by the total number of students who preferred salad, 13, and dividing by the grand total, 46.

		Lunch Preference		
	Salad	**Sandwich**	**Pizza**	**Total**
Male	5	9	12	26
Female	8	6	6	20
Total	13	15	18	46

Gender

$$\frac{26(13)}{46} = 7.35$$

The expected number of females that preferred pizza is calculated by $\frac{20(18)}{46} = 7.83$.

		Lunch Preference		
	Salad	**Sandwich**	**Pizza**	**Total**
Male	5	9	12	26
Female	8	6	6	20
Total	13	15	18	46

Gender

➡ **EXAMPLE 4.40**

In a local high school, student council officers wanted to determine if there is a relationship between grade level and voting participation. The table below displays the participation of each grade level during the school officer elections.

QUICK TIP

null hypothesis, H_0: _____ and _____ are **independent**

alternate hypothesis, H_a: _____ and _____ are **dependent**

Expected value: multiply row total and column total for the desired cell, and then divide by the grand total

		Participated	Did not Participate	Total
Grade Level	Freshman	121	154	275
	Sophomore	157	105	262
	Junior	175	76	251
	Total	453	335	788

1. State the null and alternate hypotheses for the χ^2 test of independence.

2. Show that the expected number of juniors who did not participate is 107.

Answer Explanations

1. **H_0: Grade level and voting participation are independent.**
   H_a: Grade level and voting participation are dependent.

2. $\dfrac{251(335)}{788} = 106.707$.

   Thus, 107 juniors are expected to not participate.

   When a problem asks for you to "show that," the best choice is to simply work out the problem ignoring the answer given. Once you have calculated the answer, check back to make sure it is the same as the answer given.

An important part of the χ^2 test for independence is determining the degrees of freedom. If the χ^2 critical value was not given on the IB exam, the degrees of freedom would be necessary to find the critical value.

To calculate the degrees of freedom, use the formula $df = (r - 1)(c - 1)$, where r is the number of rows and c is the number of columns.

For the contingency table used previously, the degrees of freedom would be $(2 - 1)(3 - 1) = 2$. There are two rows and three columns. The total row and total column are not considered part of the data.

		Lunch Preference			
		Salad	Sandwich	Pizza	Total
Gender	Male	5	9	12	26
	Female	8	6	6	20
	Total	13	15	18	46

Now, determine the χ^2 calculated value. The GDC will do the work, but you must first enter the data in as a matrix. Press [2ND] and the x^{-1} key to access the [MATRIX] menu. Arrow right two times to [EDIT], and press [ENTER] on [1:[A]]. When entering in data for the χ^2 test for independence, it is best to always use matrix [A].

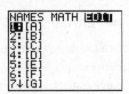

Enter the dimensions of the matrix: *row × column*. For the example, there are 2 rows and 3 columns, so press 2, [ENTER], 3, [ENTER]. The calculator generates a matrix that resembles the contingency table without the total row and total column.

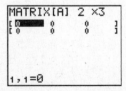

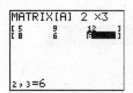

When entering the data, type each element in a row followed by **ENTER**. Remember, the totals are not data values.

Once the matrix is entered, quit (**2ND MODE**). The χ^2 test for independence is located in the **STAT** menu. Arrow right twice to get to the **TESTS** option.

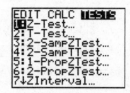

The χ^2 test is at the bottom of the list, so the quickest way to find the test is to arrow UP until you see χ^2–**TEST**. Press **ENTER** on χ^2–**TEST**.

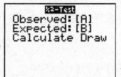

The data were inputted into matrix **A**, so simply hit enter twice, and then again while the cursor is on **CALCULATE**.

The output shows the χ^2 calculated value, the p-value, and the degrees of freedom.

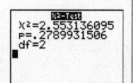

To see all of the expected values, go back to the **MATRIX** menu, arrow right twice to the **EDIT** options, and then down to **2:[B]**. All expected values have been calculated and stored in this matrix.

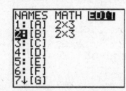
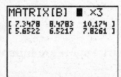

The final step of the χ^2 test for independence is to analyze the results. You have two choices for the end result: you can either reject H_o, which means the categories are dependent, or you can accept H_o, which means the categories are independent.

On the IB exam, you will either use a given χ^2 critical value or the significance level to determine the result.

Analyzing the results using

χ^2 critical value	Significance Level (1%, 5%, or 10%)
Compare your χ^2 calculated value against the critical value.	Compare your p-value against the significance level.
IF: • Calculated < Critical: ACCEPT H_o • Calculated > Critical: REJECT H_o	IF: • p-value > significance level: ACCEPT H_o • p-value < significance level: REJECT H_o

In our example

χ^2 critical value	Significance Level (1%, 5%, or 10%)
The χ^2 critical value at a 5% significance level is 5.991	The p-value was 0.279 and the significance level given was 5%.
2.55 < 5.991 Accept H_o Gender and lunch preference are independent.	0.279 > 0.05 Accept H_o Gender and lunch preference are independent.

No matter which way you analyze the test, the result will be the same.

➡ EXAMPLE 4.41

Jon believes boys and girls prefer different sports, and he is going to carry out a χ^2 test to determine if he is right. He will ask 100 students to select their favourite sport: football, basketball, or hockey.

1. State the null hypothesis for this test.
2. Show that the degrees of freedom would be 2.

Out of the 100 students, he surveyed 65 boys and 35 girls. A total of 48 selected football as their favourite sport.

> **QUICK TIP**
>
> **Degrees of Freedom:** (rows – 1) (columns – 1)
>
> When the **Chi** is **HIGH** or the **p** is **LOW**, reject the H_o

3. Calculate the expected number of girls who prefer football.

Jon used a 10% significance level for the χ^2 test. He calculated a p-value of 0.0678 and rounded to 3 significant figures.

4. State Jon's conclusion to his test. Support your answer with mathematical reasoning.

Answer Explanations

1. **H_o: Gender and favourite sport are independent.**

2. **$(2-1)(3-1) = (1)(2) = 2$**
   There are 2 genders (boy and girl), so there would be 2 rows. There are 3 choices of sports (football, basketball, and hockey), so there would be 3 columns. Degrees of freedom are calculated by (rows – 1)(columns – 1).

3. $\dfrac{35(48)}{100} = 16.8$

   To calculate expected values, multiply the total of the desired row by the total of the desired column, and then divide by the grand total.
   The contingency table for the example could look like:

		Sports Preference			
		Football	Basketball	Hockey	Total
Gender	Male				65
	Female				35
	Total	48			100

4. **$0.0678 < 0.10$, Reject H_o.**

Gender and favourite sport are dependent.

We compared the *p*-value against the significance level. Remember 10% is 0.10. Since the *p*-value was less than the significance level, reject H_o.

➡ EXAMPLE 4.42

Ms. Ghali recorded three classes of her students' grades according to where each student sits in class, as shown in the table below.

		Grade earned			
		A	**B**	**C**	**D or less**
Seating location	**Front**	10	14	7	4
	Middle	6	9	11	9
	Back	7	8	12	8

Ms. Ghali carries out a χ^2 test for independence.

1. State the null and alternate hypotheses.
2. Write down the degrees of freedom.
3. Show that the expected number of students sitting in the middle of the classroom who earned a D or less is 7.
4. Using the GDC, write down the calculated value of χ^2.

The χ^2 critical value at the 5% significance level is 12.59.

5. Determine if the null hypothesis would be rejected. Give a reason to support your answer.

Answer Explanations

1. **H_o:** Seating location and grade earned are independent.
   H_a: Seating location and grade earned are dependent.

2. **6**

   There are three rows and four columns, thus the degrees of freedom would be $(3 - 1)(4 - 1) = (2)(3) = 6$.

3. $\dfrac{35(21)}{105} = 7$

   The contingency table did not include totals, so add those on.

		Grade earned				
		A	**B**	**C**	**D or less**	**Total**
Seating location	**Front**	10	14	7	4	35
	Middle	6	9	11	9	35
	Back	7	8	12	8	35
	Total	23	31	30	21	105

There are a total of 35 students in the middle, 21 students earning a D or less, and a grand total of 105.

4. **6.53**

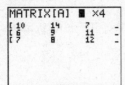

First, enter the data into matrix Ⓐ. The dimensions are 3×4.

Then perform the χ^2 test from the (STAT) menu.

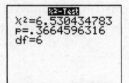

5. **The null hypothesis would be accepted.** Since $6.53 < 12.59$, we accept H_o. Seating location and grade earned are independent.

When determining the results using the χ^2 critical value, the value must be greater in order to reject the null hypothesis.

TOPIC 4 PRACTICE

Paper 1 Questions

1. The marks earned by 10 students on IB Math Studies and Biology final exams are shown below.

IB Math Studies, x	50	55	30	32	39	22	51	56	48	41
Biology, y	62	67	42	44	51	33	64	70	60	53

(a) Write down the correlation coefficient, r.

(b) Describe the correlation between the final exam scores in IB Math Studies and Biology.

(c) Write down the equation of the regression line in the form $y = mx + c$.

(d) If a student scored a 45 on the IB Math Studies final exam, what is the predicted score for Biology? Round the answer to the nearest integer.

(e) Explain why the estimate in part (d) is reasonable.

2. The scatter diagram at right shows the height and arm span in meters for a group of 10 twelfth-grade boys.

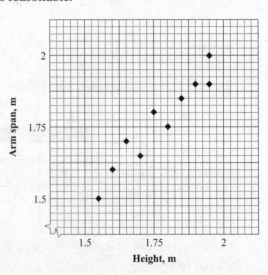

Statistical Applications

(a) If $\bar{x} = 1.77$ and $\bar{y} = 1.77$, draw a line of best fit.

(b) Suppose a student whose arm span is 1.65 m is added. Using the line of best fit, predict the height of the student.

3. On the math portion of the SAT, scores are normally distributed with a mean score of 514 and a standard deviation of 130.

   A randomly selected student scored a 600 on the math portion of the SAT.
   (a) Draw and label a normal curve clearly showing the student's score.
   (b) Determine the probability that a student scores higher than 600 on the math portion of the SAT.

   Suppose the highest 10% of scores earn a scholarship.
   (c) What is the minimum score needed to earn the scholarship?

4. Alyssa wants to determine if the athletes at her school eat healthier breakfasts than non-athletes. She surveys 100 students and asks if they play a sport or not and what they eat for breakfast: eggs, cereal, baked goods, or nothing.

Play Sports?

	Eggs	Cereal	Baked Goods	Nothing	Total
Yes	15	c	10	7	44
No	14	17	b	n	56
Total	29	29	19	23	100

Breakfast

Alyssa will carry out a χ^2 test for independence.

(a) State the null and alternate hypotheses.
(b) Determine the degrees of freedom.
(c) Write down the values of c, b, and n.
(d) Write down the χ^2 calculated value.

At a 5% significance level, the χ^2 critical value is 7.81.

(e) Determine if the null hypothesis would be accepted. Justify your answer.

SOLUTIONS

1. (a) **$r = 0.999$**

   After entering the data into L_1 and L_2, r is found by calculating the linear regression.

   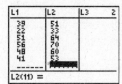 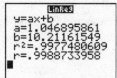

   (b) **There is a strong, positive correlation between the exam scores in IB Math Studies and Biology.**

   Since r is positive and very close to 1, the correlation is strong and positive.

(c) **$y = 1.05x + 10.2$**

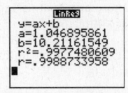

The linear regression output gives the parts to the regression line.

(d) **57**

Substitute 45 in for x and simplify. Round the answer to the nearest integer.

$$y = 1.05(45) + 10.2$$
$$y = 57.45$$
$$y = 57$$

(e) **The estimate is reasonable since a score of 45 is within the given data range. Therefore, using the regression line is appropriate to estimate.**

2. (a)

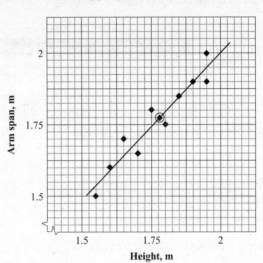

The line of best fit must pass through the mean (1.77, 1.77). The mean is circled on the graph above. Roughly half the data points should be above the line and half below.

(b) **1.65**

Since the arm span given is 1.65 m, draw a horizontal line from 1.65 on the y-axis to the line of best fit. Now, draw down from the line to the x-axis. Answers between 1.625 and 1.675 would be accepted.

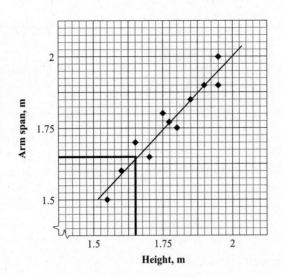

3. (a)

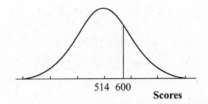

The mean of 514 goes directly in the middle. A score of 600 is slightly less than one standard deviation away. Be sure to label the axis.

(b) **0.254**

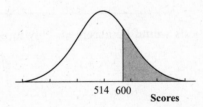

The probability a student scores above a 600 is represented by the shaded region.

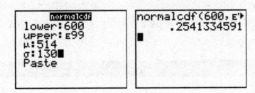

(c) **681**

First sketch a normal curve displaying the highest 10%.

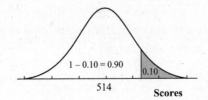

Next, find the "less than" shading in order to use the (INVNORM) function on the GDC.

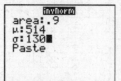 

4. (a) **H_o: Playing sports and breakfast eaten are independent.**
    H_a: Playing sports and breakfast eaten are dependent.

    (b) **3**

    $$(2-1)(4-1) = 1(3) = 3$$

    (c) **$c = 12, b = 9, n = 16$**
    Using the totals, subtract to find the values.

    $$c = 29 - 17 = 12$$
    $$b = 19 - 10 = 9$$
    $$n = 23 - 7 = 16$$

    (d) **3.08**
    After entering the table into matrix Ⓐ, perform the χ^2 test.

    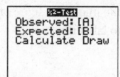 

    (e) **The null hypothesis would be accepted. Playing sports and breakfast eaten are independent.**
    $3.08 < 7.81$, accept H_o.

Paper 2 Questions

1. A local shipping company, Speedy Ship, delivers small packages whose weights are normally distributed with a mean of 2.5 kilograms and a standard deviation of 0.65 kilograms. The shipping charges are determined by the package weight.

Package Weight, w, in kg	Shipping Fee, in USD
$w < 0.45$	1.50
$0.45 \leq w < 1.8$	4.50
$1.8 \leq w < 3.2$	6.75
$3.2 \leq w < 4.5$	8.50

    (a) Sketch a normal curve displaying the distribution of package weights. Clearly identify the boundaries for the shipping fees.

    A package is selected at random.

    (b) Find the probability the shipping fee would be
    (i) $6.75,
    (ii) $4.50 or less.

There is a 0.65 probability that a randomly selected package weighs less than p kg.

(c) Determine the value of p.

(d) Write down the price to ship a package that weighs p kg.

Speedy Ship delivers 350 packages each month.

(e) Calculate the expected number of $6.75 packages shipped in a month.

The regional manager of Speedy Ship offers an incentive to the local store. If the store ships at least 40 packages a month that weigh between 3.2 and 4.5 kg, the store will receive a bonus.

(f) Calculate the probability that a randomly selected package weighs between 3.2 and 4.5 kg.

(g) Is it likely the local Speedy Ship will receive the bonus? Justify your answer.

2. For his IB project, Matt counted how many steps a student could climb in one minute with their heart rate in beats per minute (bpm) immediately after completion. He used 12 male students, age 18, for the experiment. The data is displayed below.

Steps climbed	22	20	18	24	23	21	16	15	25	28	12	22
Heart Rate, bpm	135	130	115	140	141	136	122	112	134	144	110	130

(a) Draw a scatter diagram for the data. Let 2 cm represent 10 steps on the horizontal axis and 2 cm represent 10 bpm on the vertical axis.

(b) Write down the mean
   (i) steps climbed,
   (ii) heart rate in bpm.

(c) Write down the correlation coefficient.

(d) Describe the correlation between steps climbed and heart rate.

(e) Write down the equation of the regression line.

(f) Draw the regression line on the scatter diagram.

A thirteenth student was able to climb 35 steps.

(g) Using the regression line, predict the student's heart rate.

(h) Explain if using the regression line is appropriate to predict the heart rate.

3. Studies have shown that listening to music aids learning. Tanisha performs an experiment with a large group of students of similar math abilities. The students are divided into four groups that listen to a different style of music while being taught the same math lesson. At the end of the lesson, all students take the same quiz. The results of the quiz are shown in the table below.

<table>
<tr><td rowspan="2">Grade Earned
on Quiz</td><td></td><td colspan="5">Type of Music</td></tr>
<tr><td></td><td>Hip Hop</td><td>Classical</td><td>Country</td><td>No Music</td><td>Total</td></tr>
<tr><td></td><td>A or B</td><td>12</td><td>15</td><td>9</td><td>6</td><td>42</td></tr>
<tr><td></td><td>C or D</td><td>8</td><td>7</td><td>14</td><td>12</td><td>41</td></tr>
<tr><td></td><td>F</td><td>10</td><td>8</td><td>7</td><td>12</td><td>37</td></tr>
<tr><td></td><td>Total</td><td>30</td><td>30</td><td>30</td><td>30</td><td>120</td></tr>
</table>

She plans to perform a χ^2 test for independence.

(a) Write the null and alternate hypotheses for the χ^2 test.

(b) Write down the degrees of freedom.

(c) Show that the expected number of students who listened to country music and failed the quiz is 9.25.

(d) Write down the χ^2 calculated value.

Tanisha will use a 5% significance level.

(e) Write down the p-value.

(f) State, with a reason, the conclusion to the test.

SOLUTIONS

1.

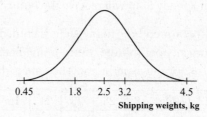

Shipping weights, kg

(a) The mean weight goes in the middle. The standard deviation is 0.65. Using the standard deviation as a guide, the shipping fee boundaries are placed appropriately on the number line. Remember, the normal curve is labeled with the mean in the middle and a distance of one, two, and three standard deviations on both sides.

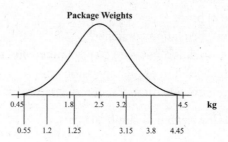

(b) (i) **0.718**

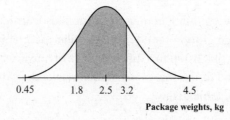

If a package costs \$6.75 to ship, then it weighs between 1.8 and 3.2 kg.

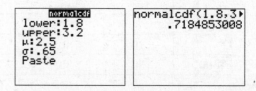

(ii) **0.141**

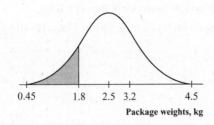

A package that costs \$4.50 or less must weigh less than 1.8 kg.

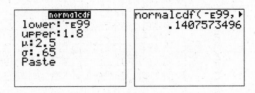

(c) **2.75 kg**

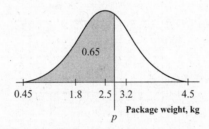

The probability of 0.65 is the area under the curve.

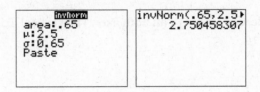

(d) **\$6.75**

A package that weighs 2.75 kg falls in the $1.8 \leq w < 3.2$ range, and thus would cost \$6.75 to ship.

(e) **251**

A package that costs \$6.75 to ship weighs between 1.8 and 3.2 kg. In part (b) we determined this to be 0.718.

To find the expected number, multiply the probability by the number of packages.

350(0.718) = 251.3, which rounds up to 251.

(f) **0.140**

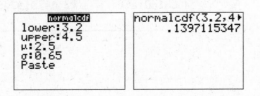

(g) It is likely the store will earn the bonus because they should expect to ship about 49
packages that weigh between 3.2 and 4.5 kg.

350(0.140) = 49, which is a little larger than the 40 packages needed to earn the bonus.

2. (a)

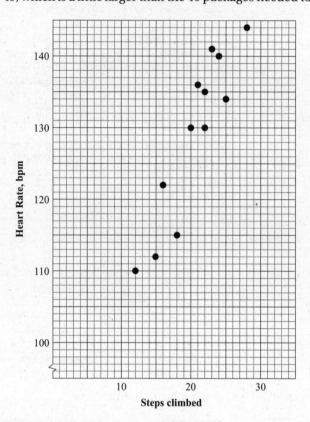

**The scale was 2 cm = 10 on both axes. The vertical axis did not need to begin until
100 or 110.**

(b) (i) **20.5**

(ii) **129**

After entering the data into L_1 and L_2, perform the 2-VAR STATS option.

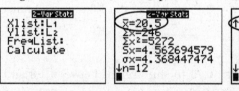

(c) **r = 0.920**

You find the correlation coefficient after doing the linear regression.

(d) **Steps climbed and heart rate have a strong, positive correlation.**

Since $r = 0.920$, the number is positive and close to 1, which yields a strong, positive correlation.

(e) $y = 2.35x + 80.9$

Using the output from the linear regression function, PLOT the regression line.

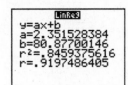

(f)

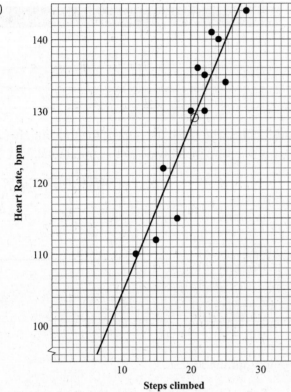

The regression line must pass directly through the mean point, which is designated by the open circle on the line. Since the vertical axis does not start at zero, you cannot use the y-intercept as a point. Draw the line so that roughly half the points are above and half the points are below.

(g) **163**

Plug 35 in for x and simplify.

$$y = 2.35(35) + 80.5$$
$$y = 162.75$$
$$y = 163$$

(h) **It is not appropriate to use the regression line because 35 steps is too far outside the data range.**

3. (a) **H$_o$**: Grade earned and the type of music are independent.
        H$_a$: Grade earned and the type of music are dependent.

    (b) **6**

    $$df = (3 - 1)(4 - 1) = 2(3) = 6$$

    (c) $\dfrac{30(37)}{120} = \mathbf{9.25}$

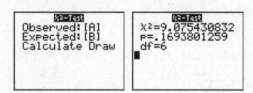

	Type of Music				
	Hip Hop	**Classical**	**Country**	**No Music**	**Total**
A or B	12	15	9	6	42
C or D	8	7	14	12	41
F	10	8	7	12	37
Total	30	30	30	30	120

Grade Earned on Quiz

Multiply the total number of students who failed the quiz by the total number of students who listened to country music, and then divide by the grand total.

    (d) **9.08**

    First, enter the data into matrix Ⓐ.
    Then perform the χ^2-test.

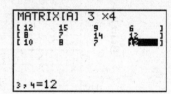

```
χ2-Test
Observed:[A]
Expected:[B]
Calculate Draw
```

```
χ2-Test
χ²=9.075430832
P=.1693801259
df=6
```

    (e) **0.169**

    The p-value is part of the output for the χ^2-test.

    (f) **Accept H$_o$, since 0.169 > 0.05. Grade earned and the type of music are independent.**

CHAPTER OBJECTIVES

Before you move on to the next chapter, you should be able to:

☐ Draw and completely label a normal curve

☐ Find the expected value of a normal distribution for a given scenario

☐ Determine the probability of being less than a certain value in a normal distribution

☐ Determine the probability of being more than a certain value in a normal distribution

☐ Determine the probability of being between two given values in a normal distribution

☐ Draw an appropriate diagram with shading for the above probability problems

☐ Find the data value that corresponds to a certain percent or probability in a normal distribution

☐ Draw an appropriate diagram with shading for given percents or probabilities

☐ Describe the correlation between two variables

☐ Draw a scatter diagram for bivariate data

☐ Find the correlation coefficient, r

☐ Write the regression line of y on x for bivariate data

☐ Calculate the mean point of bivariate data

☐ Draw the regression line on the scatter diagram

☐ Use the regression line to predict values

☐ Identify when the regression line is not appropriate to predict values

☐ Write the null and alternate hypotheses for a chi-squared test of independence

☐ Calculate the expected value of a certain element in a chi-squared test

☐ Determine the number of degrees of freedom in a chi-squared test

☐ Find the chi-squared calculated value using the GDC

☐ Determine the results of a chi-squared test using both the chi-squared value and the p-value

Geometry and Trigonometry

- **LINEAR EQUATION:** $y = mx + c$ or $ax + by + d = 0$, where $a,b,d \in \mathbb{Z}$; parallel and perpendicular lines

- **TRIGONOMETRIC RATIOS:** for a right triangle, $\sin\theta = \dfrac{\text{opposite}}{\text{hypotenuse}}$, $\cos\theta = \dfrac{\text{adjacent}}{\text{hypotenuse}}$, and $\tan\theta = \dfrac{\text{opposite}}{\text{adjacent}}$

- **PYTHAGORAS' THEOREM:** for a right triangle, $a^2 + b^2 = c^2$

- **ANGLE OF ELEVATION AND DEPRESSION:** the angle formed from the line of sight to the horizontal

- **SINE RULE:** for any triangle, $\dfrac{a}{\sin A} = \dfrac{b}{\sin B} = \dfrac{c}{\sin C}$

- **COSINE RULE:** for any triangle, $a^2 = b^2 + c^2 - 2bc\cos A$ and $\cos A = \dfrac{b^2 + c^2 - a^2}{2bc}$

- **AREA OF A TRIANGLE:** for any triangle, $A = \dfrac{1}{2}ab\sin C$

- **THREE-DIMENSIONAL SHAPES:** cuboid, right prism, right pyramid, right cone, cylinder, sphere, and hemisphere

- **SURFACE AREA:** total area of all the two-dimensional faces of the solid

- **VOLUME:** amount of space a solid occupies

5.1 EQUATION OF LINES

There are two ways to write the equation of a line: slope-intercept form and standard form.

Slope-intercept form is $y = mx + c$, where m is the gradient and c is the y-intercept (i.e. constant).

Standard form is $ax + by + d = 0$, where $a,b,d \in \mathbb{Z}$ (i.e. a, b, and d must be integers).

Slope-intercept form is helpful when graphing, while standard form is helpful when finding intercepts.

Gradient measures the steepness of a line. Often gradient is referred to as $\dfrac{\text{rise}}{\text{run}}$. The intercepts are where the graph crosses the axes.

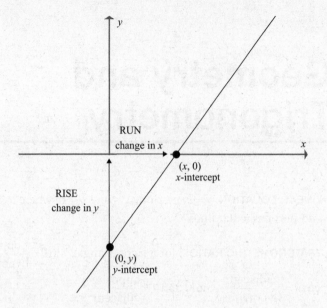

The gradient formula is $m = \dfrac{y_2 - y_1}{x_2 - x_1}$, which is simply the change in y over the change in x.

A more detailed discussion of linear functions can be found in section 6.2.

Recall from section 1.6 that two lines have three possible solutions: no solution (parallel lines), a point (intersecting lines), or the line itself (same lines).

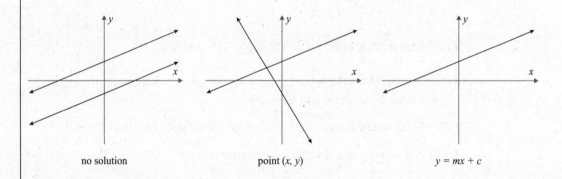

The quickest way to solve a system of linear equations is by graphing the lines using the GDC to find the intersection.

Two lines that have the **same gradient** are parallel. This means $m_1 = m_2$.

Consider the two lines L_1: $y = 2x - 5$ and L_2: $2x - y = -3$.

In order to graph the lines, they should be in slope-intercept form.

Line 1 is already in the correct form.

$$2x - y = -3$$

Line 2: $-y = -2x - 3$

$$y = 2x + 3$$

Since both lines have a gradient of 2, they are parallel.

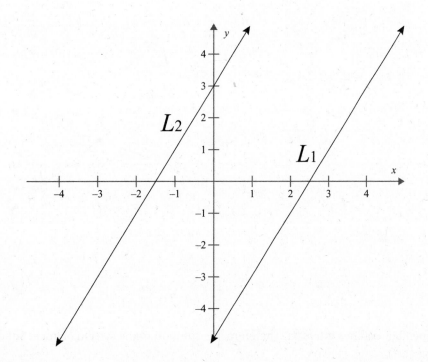

The parallel lines do not intersect. Therefore, there is no solution to the system of linear equations.

Two lines whose gradients are **opposite reciprocals** are perpendicular. This means $m_1 \times m_2 = -1$.

Consider the two lines L_1: $2x - 3y = 3$ and L_2: $3x + 2y = 11$.

First, rewrite both lines in slope-intercept form.

Line 1

$2x - 3y = 3$

$\quad -3y = -2x + 3$

$\quad\quad y = \dfrac{2}{3}x - 1$

Line 2

$3x + 2y = 11$

$\quad\quad 2y = -3x + 11$

$\quad\quad y = -\dfrac{3}{2}x + \dfrac{11}{2}$

The gradients $\dfrac{2}{3}$ and $-\dfrac{3}{2}$ are opposite reciprocals (different signs and flipped). Note $\dfrac{2}{3} \times \left(-\dfrac{3}{2}\right) = -1$.

> **QUICK TIP**
>
> **Parallel Lines: same** gradient
>
> **Perpendicular Lines: opposite reciprocal** gradients
>
> $(m_1 \times m_2 = -1)$

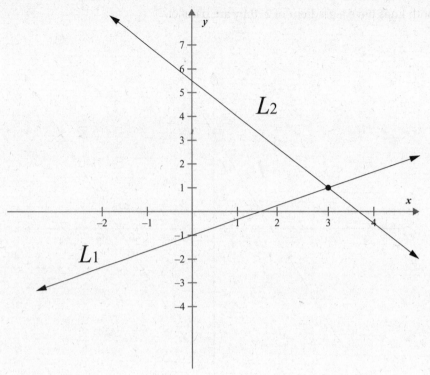

The perpendicular lines intersect. Therefore, the solution to the system of linear equations is the point (3, 1).

➡ EXAMPLE 5.10

The equation of the line L_1 is $2x + y + 3 = 0$. Line L_2 is perpendicular to L_1.

1. Determine the gradient of L_2.

The point of intersection of L_1 and L_2 is $(a, 1)$.

2. Find:
   (a) the value of a,
   (b) the equation of L_2 written in the form $ax + by + d = 0$, where $a, b, d \in \mathbb{Z}$.

Answer Explanations

1. $\dfrac{1}{2}$

   Since L_1 and L_2 are perpendicular, their gradients are opposite reciprocals. First, find the gradient of L_1.

   $$2x + y + 3 = 0$$
   $$y = -2x - 3$$

   L_2 will have a gradient of $\dfrac{1}{2}$, since $-2 \times \left(\dfrac{1}{2}\right) = -1$.

2. (a) $a = -2$

   You cannot graph the two lines in the GDC, since you lack the equation of L_2. All you know is the equation of L_1, so substitute the point $(a, 1)$ into the equation and solve.

QUICK TIP

Standard Form: $ax + by + d = 0$, where $a, b, d \in \mathbb{Z}$

- All terms are on the same side.
- The equation equals zero.
- All numbers must be integers (no fractions or decimals).

$$2a + 1 + 3 = 0$$
$$2a + 4 = 0$$
$$2a = -4$$
$$a = -2$$

(b) $x - 2y + 4 = 0$

The gradient of L_2 is $\frac{1}{2}$, and the line passes through the point $(-2, 1)$.

Plugging what you know into slope-intercept form gives:

$$1 = \frac{1}{2}(-2) + c$$
$$1 = -1 + c$$
$$2 = c$$

Thus, the equation for L_2 is $y = \frac{1}{2}x + 2$.

The problem asks for the line to be written in standard form ($ax + by + d = 0$), so all terms must be on one side and all numbers must be integers.

$$0 = \frac{1}{2}x - y + 2 \qquad \text{(multiply all terms by 2)}$$
$$0 = x - 2y + 4$$

The equivalent answer of $-x + 2y - 4 = 0$ would also be accepted.

➡ EXAMPLE 5.11

The line L_1 has equation $y = -\frac{1}{4}x + 3$.

1. Write down the value of the
   (a) gradient,
   (b) y-intercept.

Line L_2 passes through the points $(-8, 1)$ and $(0, -1)$.

2. Calculate the gradient of L_2.
3. Determine if L_1 and L_2 are parallel, perpendicular, or neither. Support your answer with a valid reason.
4. Write the equation of L_2 in the form $ax + by + d = 0$, where $a, b, d \in \mathbb{Z}$.

Answer Explanations

1. (a) $-\frac{1}{4}$

   The line $y = -\frac{1}{4}x + 3$ is written in slope-intercept form, thus the gradient is the coefficient of x.

   (b) **(0, 3)**

   The y-intercept is the constant of the equation, thus the value of y is 3, and the value of x is 0.

2. $-\dfrac{1}{4}$

When given two points, use the gradient formula.

$$m = \frac{(-1)-1}{0-(-8)} = \frac{-2}{8} = -\frac{1}{4}$$

3. **The lines are parallel since both gradients are $-\dfrac{1}{4}$.**

4. **$0 = -x - 4y - 4$ or $x + 4y + 4 = 0$**

The gradient of L_2 is $-\dfrac{1}{4}$, and the y-intercept is $(0, -1)$. Thus, $m = -\dfrac{1}{4}$ and $c = -1$.

The equation for L_2 is $y = -\dfrac{1}{4}x - 1$.

The problem asks for the line to be written in standard form ($ax + by + d = 0$), so all terms must be on one side and all numbers must be integers.

$$0 = -\frac{1}{4}x - y - 1 \qquad \text{(multiply all terms by 4)}$$

$$0 = -x - 4y - 4$$

The equivalent answer of $x + 4y + 4 = 0$ would also be accepted.

5.2 TRIGONOMETRIC RATIOS

Right triangles have one right angle and two acute angles. Angle BAC in triangle ABC is the right angle.

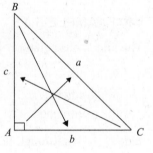

Angles are represented by three capital letters, with the middle letter being the vertex. The side opposite an angle shares the same vertex letter but is lowercase. For example, $B\hat{A}C$ corresponds with side a, $A\hat{B}C$ corresponds with side b, and $A\hat{C}B$ corresponds with side c.

The hypotenuse of triangle ABC is side a. The other two sides, b and c, are called the legs.

All angles in a triangle add up to 180° ($A + B + C = 180°$).

The sides also share a relationship using Pythagoras' theorem: $(\text{leg})^2 + (\text{leg})^2 = (\text{hypotenuse})^2$.

In triangle ABC, Pythagoras' theorem would be $b^2 + c^2 = a^2$.

There are also relationships between the angles and the sides. These relationships generate the six trigonometric ratios, but IB Math Studies focuses only on sine, cosine, and tangent.

The Greek letter θ (Theta) is generally the symbol used to denote an unknown angle measurement.

For an angle θ in any **right** triangle,

$$\sin \theta = \frac{\text{opposite}}{\text{hypotenuse}}$$

Sine

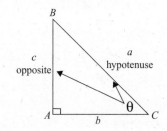

$$\sin \theta = \frac{c}{a}$$

$$\cos \theta = \frac{\text{adjacent}}{\text{hypotenuse}}$$

Cosine

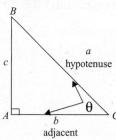

$$\cos \theta = \frac{b}{a}$$

$$\tan \theta = \frac{\text{opposite}}{\text{adjacent}}$$

Tangent

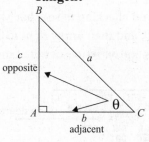

$$\tan \theta = \frac{c}{b}$$

An easy way to remember the three ratios is to remember the phrase SOHCAHTOA (pronounced so–cah–toe–ah). SOHCAHTOA displays each trig ratio with required sides.

SOH = Sine Opposite Hypotenuse   CAH = Cosine Adjacent Hypotenuse
TOA = Tangent Opposite Adjacent

> **Word of caution: these trigonometric ratios only hold true for <u>right</u> triangles.**

Consider triangle BCA, where $B\hat{A}C = 54°$ and $c = 4.7$.

Suppose you need to find $A\hat{B}C$. You know $54° + B = 90°$, since all angles add up to $180°$ and $A\hat{C}B$ is $90°$. Therefore, $A\hat{B}C = 36°$.

To find the measure of side b, the wisest choice is to use the information given. You have $B\hat{A}C$ and the hypotenuse, c. You need to calculate side b, which is adjacent to the given angle. Therefore, the cosine ratio should be used.

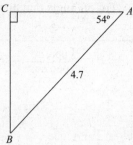

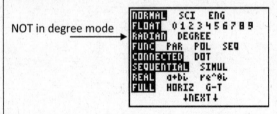 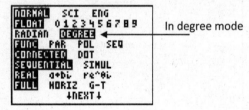

$$\cos 54° = \frac{b}{4.7}$$

To solve, multiply both sides by 4.7.

$$4.7 (\cos 54°) = b$$

Make sure your GDC is in DEGREE MODE by pressing MODE and checking to see that DEGREE is highlighted. If RADIAN is highlighted, arrow down to RADIAN, and then arrow right to DEGREE. Hit ENTER while on DEGREE and the mode will change. Press 2nd MODE to exit out of the mode menu.

NOT in degree mode → 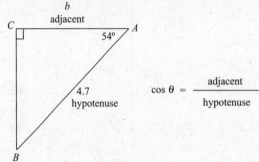 ← In degree mode

Now find the measure of b using the GDC.

Thus, $b = 2.76$. The only missing measure of triangle BCA is side a.

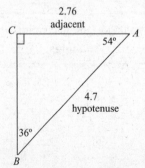

You can calculate side a using Pythagoras' theorem.

$$a^2 + (2.76)^2 = (4.7)^2$$
$$a^2 = (4.7)^2 - (2.76)^2$$
$$a^2 = 14.4724$$
$$a = \sqrt{14.4724}$$
$$a = 3.80$$

You could have also chosen to calculate a using a different trigonometric ratio. Side a is the opposite side of $B\hat{A}C$, and you are given the hypotenuse. Thus, the sine ratio would be used.

$$\sin 54° = \frac{a}{4.7}$$
$$4.7(\sin 54°) = a$$
$$a = 3.80$$

Either way, the answer is the same.

Once the triangle has been solved, checking your answer is always a good idea. You know that the largest side is opposite the largest angle, the medium side is opposite the medium angle, and the smallest side is opposite the smallest angle. If the angles and sides do not follow this rule, part (or all) of your work is incorrect.

In triangle ABC, angle C is the largest since it is the right angle, and its opposite side $c = 4.7$, which is also the largest. Angle A, the medium angle, has the opposite side of $a = 3.80$. The smallest angle, angle B, has the opposite side of $b = 2.76$. Checking the corresponding values does not guarantee your answers are correct, but it is a good indicator.

➡ EXAMPLE 5.20

Given triangle RST below, solve the triangle.

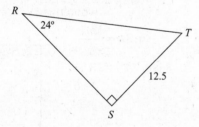

Answer Explanations

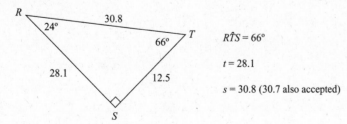

$R\hat{T}S = 66°$

$t = 28.1$

$s = 30.8$ (30.7 also accepted)

The easiest measurement to calculate is $R\hat{T}S$.

$$R + T = 90°$$
$$24° + T = 90°$$

Thus, $R\hat{T}S = 66°$.

Next, calculate side t.

Using $S\hat{R}T$, the given side is opposite and the missing side is adjacent. Tangent is the correct trigonometric ratio to use.

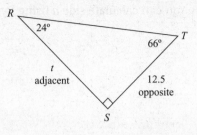

$$\tan\theta = \frac{\text{opposite}}{\text{adjacent}}$$

$$\tan 24° = \frac{12.5}{t}$$

Multiply both sides by t, and then divide by $\tan 24°$.

$$t(\tan 24°) = 12.5$$

$$t = \frac{12.5}{\tan 24°} = 28.1$$

```
12.5/tan(24)
      28.07545967
```

To find the missing hypotenuse, use either Pythagoras' theorem or a different trigonometric ratio. Both methods are shown.

Pythagoras' Theorem	Trigonometric Ratio
$(12.5)^2 + (28.1)^2 = s^2$ $945.86 = s^2$ $\sqrt{945.86} = s$ $30.8 = s$	$\sin\theta = \dfrac{\text{opposite}}{\text{hypotenuse}}$ $\sin 24° = \dfrac{12.5}{s}$ $s(\sin 24°) = 12.5$ $s = \dfrac{12.5}{\sin 24°}$ $s = 30.7$

```
12.5/sin(24)
      30.73241669
```

Both 30.8 and 30.7 would be accepted.

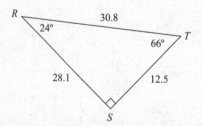

Notice the largest side is opposite the right angle, the medium side is opposite the medium angle, and the smallest side is opposite the smallest angle.

Trigonometric ratios can also be used to find missing angle measurements. However, solving the equations involves the use of the *inverse* trig functions.

For example, set up the ratio $\sin\theta = \dfrac{2}{3}$ based on the given triangle.

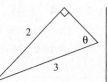

To solve the equation for θ, use the sine inverse function. This is found on the GDC by pressing ⟨2ND⟩ and then ⟨SIN⟩.

$$\sin^{-1}(\sin\theta) = \sin^{-1}\left(\dfrac{2}{3}\right)$$

> A function and its inverse "undo" one another, which is why sine inverse is needed.

$$\theta = \sin^{-1}\left(\dfrac{2}{3}\right) = 41.8°$$

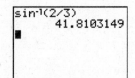

➡ EXAMPLE 5.21

DeVon is 180 meters north of a riverbank. Noah is standing 90 meters due east of DeVon.
1. Draw a diagram to illustrate the information.

Noah is looking at DeVon, and then turns $\theta°$ to look at the riverbank.
2. Show the angle θ on the diagram.
3. Calculate the measure of θ.

DeVon walks directly to the riverbank.
4. Find the shortest distance Noah must walk to meet DeVon at the riverbank.

Answer Explanations

1.

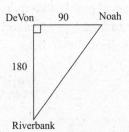

Since DeVon is north of the riverbank, he is directly above the bank at a distance of 180 m. Noah is 90 m east, which is to the right of DeVon. Be sure to clearly show the right angle at DeVon.

2.

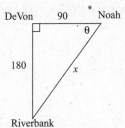

The angle, θ, goes in the vertex of the triangle at Noah. Noah is looking at DeVon, and then turns to look at the riverbank.

3. **63.4°**

   The sides of 180 and 90 are opposite and adjacent to the given angle θ; therefore, tangent is the correct trigonometric ratio.

   $$\tan\theta = \frac{180}{90}$$

   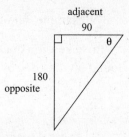

   In order to solve for θ, the tan inverse function is applied to both sides.

   $$\theta = \tan^{-1}\left(\frac{180}{90}\right) = 63.4°$$

   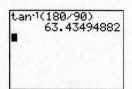

4. **201 meters**

   The shortest distance between two points is a straight line, so you must calculate the length of the hypotenuse. You can either use Pythagoras' theorem or a trigonometric ratio.
   In this situation, Pythagoras' theorem is easiest.

   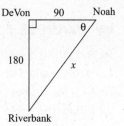

   $$(180)^2 + (90)^2 = x^2$$
   $$40\,500 = x^2$$
   $$\sqrt{40\,500} = x$$
   $$201.246 = x$$

   Rounding the answer to 3 significant figures gives a distance of 201 meters.

Problems involving angles of elevation or angles of depression are commonly found on the IB Math Studies exam. These types of problems require the drawing of a diagram even if it is not explicitly asked for. The key to angles of elevation or depression is to **always** place the angle θ at the bottom of the triangle as shown below.

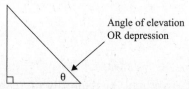

Suppose a boat is 450 km away from the base of a cliff. The angle of elevation from the boat to the top of the cliff is 56°. Determine the height of the cliff.

Assume the cliff is vertical, since the problem did not state otherwise.

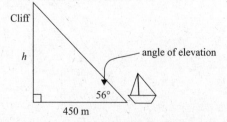

The angle of elevation is from the boat (on the horizontal) up to the top of the cliff; thus, the 56° is placed in the lower corner of the triangle by the boat. You need to find the height of the cliff, h. The given side is adjacent to the angle and the unknown side is opposite. Therefore, tangent is the correct trigonometric ratio.

$$\tan 56° = \frac{h}{450}$$
$$450(\tan 56°) = h$$
$$667 = h$$

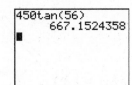

The cliff is 667 meters high.

Next, consider a flagpole that casts a shadow 3.5 meters long when the angle of depression from the sun to the shadow is 50°. How tall is the flagpole?

Angles of depression or elevation are formed from the horizontal to the line of sight.

The diagram illustrates the true angle of depression, which is outside of the right triangle. You need an angle inside the triangle.

The horizontal line from the sun and the base of the right triangle are parallel lines cut by a transversal—the line of sight. When parallel lines are cut by a transversal, the alternate interior angles are congruent. This means the angle of depression is the same as the base angle of the triangle.

You can set up a tangent ratio to solve for the height of the flagpole.

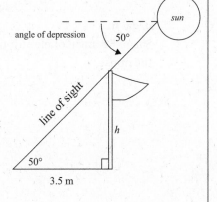

$$\tan 50° = \frac{h}{3.5}$$
$$3.5(\tan 50°) = h$$
$$4.17 = h$$

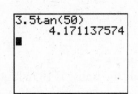

The height of the flagpole is 4.17 meters.

➥ EXAMPLE 5.22

1. Natasha has a 20-foot ladder in order to reach the top of her house.
   (a) Determine the angle of elevation of the ladder if the house is 16 feet tall.
       As Natasha climbed, the ladder slid until the angle of elevation was 49°.
   (b) Calculate the distance the ladder slid from the top of the house.

2. Dmtri is working on the roof of his office building, which is 38 meters tall. He looks down at a 42° angle of depression and sees a car.
   (a) Determine how far the car is from the base of the building.
       The car drives 15 meters closer to the office building.
   (b) Calculate the change in the angle of depression.

> **QUICK TIP**
>
> The angle of elevation or depression **ALWAYS** goes in the bottom corner of the triangle.

Answer Explanations

1. (a) **53.1°**

    First draw a diagram illustrating the given information.

    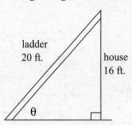

    The height of the house is the side opposite θ, and the length of the ladder is the hypotenuse.

    $$\sin\theta = \frac{16}{20}$$

    $$\theta = \sin^{-1}\left(\frac{16}{20}\right) = 53.1°$$

    (b) **0.906 m**

    Since the angle of elevation has changed, calculate how high the ladder reaches on the house.

    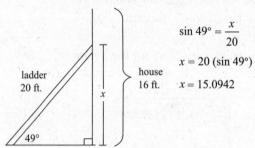

    $$\sin 49° = \frac{x}{20}$$

    $$x = 20\,(\sin 49°)$$

    $$x = 15.0942$$

    The ladder reaches 15.0942 m, while the house is 16 m high. Subtract the new height of the ladder from the height of the house to determine how far the ladder slid.

    $$16 - 15.0942 = 0.906 \text{ m (correct to 3 significant figures)}$$

2. (a) **42.2 m**

    Draw a diagram to represent the information.

    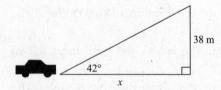

    The height of the building is opposite of the angle, while the distance from the building is adjacent. Therefore, tangent is the correct trigonometric ratio.

    $$\tan 42° = \frac{38}{x}$$

    $$x\,(\tan 42°) = 38$$

    $$x = \frac{38}{\tan 42°} = 42.2$$

(b) **12.4°**

The car has moved 15 meters closer, so the base of the triangle is no longer 42.2 meters but instead 27.2 meters (42.2 − 15 = 27.2).

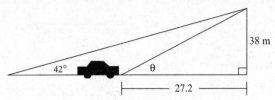

The new angle of depression is calculated by:

$$\tan \theta = \frac{38}{27.2}$$

$$\theta = \tan^{-1}\left(\frac{38}{27.2}\right)$$

$$\theta = 54.4°$$

We need the change in the angle, so subtract the original from the new angle to get 12.4°.

$$54.4° − 42° = 12.4°$$

5.3 SINE AND COSINE RULE; AREA OF A TRIANGLE

The trigonometric ratios sine, cosine, and tangent along with Pythagoras' theorem can only be used with right triangles. When dealing with non-right triangles, there are two options: sine rule and cosine rule.

The formula for the sine rule is $\dfrac{a}{\sin A} = \dfrac{b}{\sin B} = \dfrac{c}{\sin C}$. Only two of the three ratios are used when setting up an equation.

The sine rule is used when the problem gives <u>an angle with its opposite side</u>.

For example, consider the triangle *MNP*. Triangle *MNP* is not a right triangle since there is no angle measure of 90°. Therefore, you cannot use the sine, cosine, or tangent trigonometric ratios to solve for the missing measurements.

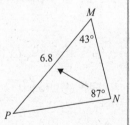

You know $M\hat{N}P$ (angle N) is 87°, and you also know side *n* is 6.8; therefore, use the sine rule to solve the triangle.

The *A*, *B*, and *C* are arbitrary variables in the sine rule formula. The important thing to remember is the side opposite an angle goes in the numerator, and the sine of the angle goes in the denominator. Each ratio is made up of angles with their opposite side. You could easily rewrite the formula as $\dfrac{m}{\sin M} = \dfrac{n}{\sin N} = \dfrac{p}{\sin P}$ for this problem, but that is unnecessary.

Since you know $M\hat{N}P$ (angle N) and side *n*, the ratio of $\dfrac{6.8}{\sin 87°}$ must be used. The only other given information is $N\hat{M}P = 43°$, so solve for the measure of side *m*.

$$\frac{6.8}{\sin 87°} = \frac{m}{\sin 43°}$$

Cross multiply the denominators to the opposite numerator.

$$6.8 \sin 43° = m \sin 87°$$

To solve for m, divide $\sin 87°$ into both sides.

$$\frac{6.8 \sin 43°}{\sin 87°} = m$$

Therefore, $m = 4.64$

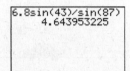

You still need $M\hat{P}N$ and side p. You cannot use Pythagoras' theorem to solve for p, since you do not have a right triangle. You will have to use the sine rule again.

First calculate $M\hat{P}N$. All angles add up to $180°$ in a triangle.

$$P + 43° + 87° = 180°$$
$$P = 50°$$

Looking at triangle MNP again, you need to find side p so use the angle–opposite side relationship of the sine rule.

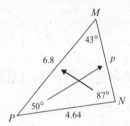

$$\frac{p}{\sin 50°} = \frac{6.8}{\sin 87°}$$

$$p \sin 87° = 6.8 \sin 50°$$

$$p = \frac{6.8 \sin 50°}{\sin 87°} = 5.22$$

Thus, $p = 5.22$

The sine rule can also be used to find missing angle measurements. Suppose you have triangle FGH, where $F\hat{G}H = 102°$, $g = 10.3$, and $f = 9.7$. When a diagram is not provided, the first step should be to draw and label a diagram.

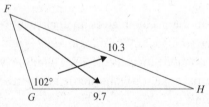

You are given $F\hat{G}H$ (angle G) and its opposite side, so use the sine rule to find $G\hat{F}H$ (angle F) first.

$$\frac{10.3}{\sin 102°} = \frac{9.7}{\sin F}$$ Cross multiply.

$10.3 \sin F = 9.7 \sin 102°$ Divide both sides by 10.3.

$$\sin F = \frac{9.7 \sin 102°}{10.3}$$ Calculate the right side using the GDC.

$\sin F = 0.921168$ To solve for F, sine inverse both sides.

$F = \sin^{-1}(0.921168)$

$F = 67.1°$

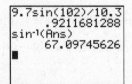

➡ EXAMPLE 5.30

1. A flagpole is leaning 9° away from the vertical. When the
    sun is behind the flagpole and the angle of depression from
    the sun is 20°, the flagpole casts a shadow 12.8 meters in
    length. Determine the height of the flagpole, rounded to
    the nearest meter.

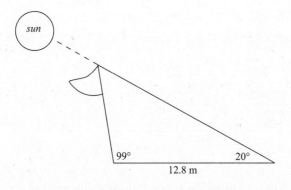

2. A farmer owns the triangular plot of land shown below.

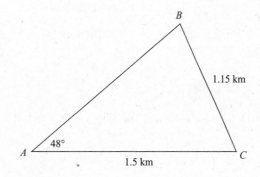

(a) Find the measure of $A\hat{B}C$.

The farmer needs to fence in the perimeter of the land.

(b) Calculate the length of the fence.

Answer Explanations

1. **5 m**

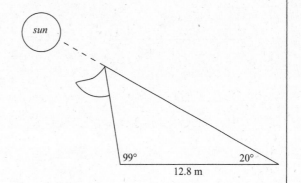

The triangle formed is not a right triangle, so you cannot use the sine, cosine, or tangent trigonometric ratios. In order to use the sine rule, you need an angle with its opposite side. Since only one side is given, you must calculate the angle opposite that side.

$$X + 99° + 20° = 180°$$
$$X = 61°$$

Now, use the sine rule to set up an equation.

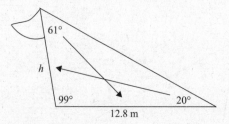

Use the newly calculated angle with its opposite side as one ratio, and the angle of depression with the height of the flagpole as the other ratio.

$$\frac{12.8}{\sin 61°} = \frac{h}{\sin 20°}$$ Cross multiply.

$$12.8 \sin 20° = h \sin 61°$$ Divide both sides by sin 61°.

$$\frac{12.8 \sin 20°}{\sin 61°} = h$$

$$h = 5.0054$$

Now, round the height to the nearest meter. The flagpole is 5 m tall.

2.   (a) **75.8°**

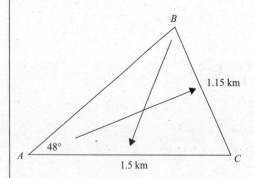

Using the sine rule, solve for $A\hat{B}C$.

$$\frac{1.15}{\sin 48°} = \frac{1.5}{\sin B}$$

$$1.15 \sin B = 1.5 \sin 48°$$

$$\sin B = \frac{1.5 \sin 48°}{1.15}$$

$$\sin B = 0.96932$$

$$B = \sin^{-1}(0.96932)$$

$$B = 75.8°$$

```
1.5sin(48)/1.15
        .9693193376
sin⁻¹(Ans)
        75.77059746
■
```

(b) **3.94 km**

To calculate the perimeter of the triangle you need side c, but first you must find $A\hat{C}B$ (angle C).

$$C + 48° + 75.8° = 180°$$
$$C = 56.2°$$

Next, set up an equation using the sine rule. The wisest choice is to reuse the given information as one ratio.

$$\frac{1.15}{\sin 48°} = \frac{c}{\sin 56.2°}$$
$$1.15 \sin 56.2° = c \sin 48°$$
$$\frac{1.15 \sin 56.2°}{\sin 48°} = c$$

Therefore, $c = 1.29$.

The perimeter of the triangle is the sum of the sides: $1.29 + 1.15 + 1.5 = 3.94$ km.

The other way to solve a problem involving a non-right triangle is to use the cosine rule.

To use the cosine rule, the problem must give either two sides with the included angle **or** all three sides with no angle.

When given two sides and their included angle, find the missing side using the formula $a^2 = b^2 + c^2 - 2bc \cos A$.

Just like in the sine rule, the variables are arbitrary. The important thing to remember is

$$\boxed{a^2} = b^2 + c^2 - 2bc \cos \boxed{A}$$

angle with its opposite side

When given three sides and no angle, calculate any angle using the formula

$$\cos A = \frac{b^2 + c^2 - a^2}{2bc}.$$

Again, the variables are arbitrary, but notice the angle with its opposite side relationship is still present.

angle with its opposite side

$$\cos \boxed{A} = \frac{b^2 + c^2 - \boxed{a^2}}{2bc}$$

Both scenarios are illustrated with examples in the following table.

Two sides and their included angle	Three sides with no angle
$a^2 = b^2 + c^2 - 2bc \cos A$	$\cos A = \dfrac{b^2 + c^2 - a^2}{2bc}$

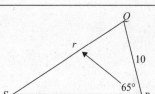

Since $Q\hat{R}S$ (angle R) is given, the formula could be rewritten as $r^2 = q^2 + s^2 - 2qs \cos R$, but it is not necessary.

Plug in the given information:

$$r^2 = 12^2 + 10^2 - 2\,(12)\,(10) \cos 65°$$

Type the entire right side into the GDC

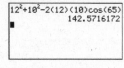

$$r^2 = 142.572$$

Square root both sides to solve for r.

$$r = \sqrt{142.572}$$
$$r = 11.9$$

The triangle now has an angle with its opposite side, so to finish solving, switch to the sine rule.

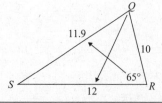

No angle is given, so you must solve for an angle first. If the problem does not specifically ask for an angle, it is best to solve for the largest angle. In this triangle, $R\hat{Q}S$ (angle Q) would be the largest since its opposite side is the largest.

Plug in the given information:

$$\cos Q = \frac{6^2 + 5^2 - 9^2}{2(6)(5)}$$

Type the entire right side into the GDC being careful to put the numerator and denominator in parentheses.

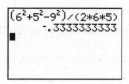

$$\cos Q = -0.33333$$

Cosine inverse both sides to solve for angle Q.

$$Q = \cos^{-1}(-0.33333)$$
$$Q = 109.47°$$

Rounding to 3 significant figures gives you 110°.

The triangle now has an angle with its opposite side, so to finish solving, switch to the sine rule.

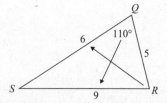

$\dfrac{11.9}{\sin 65°} = \dfrac{12}{\sin Q}$	$\dfrac{9}{\sin 110°} = \dfrac{6}{\sin R}$
$11.9 \sin Q = 12 \sin 65°$	$9 \sin R = 6 \sin 110°$
$\sin Q = \dfrac{12 \sin 65°}{11.9}$	$\sin R = \dfrac{6 \sin 110°}{9}$
$Q = \sin^{-1}(0.91392)$	$R = \sin^{-1}(0.6265)$
$Q = 66.1°$	$R = 38.8°$
Finally, $S + 65° + 66.1° = 180°$	Finally, $S + 110° + 38.8° = 180°$
$S = 48.9°.$	$S = 31.2°.$

The cosine rule is rather straightforward.

When solving for a side, use the formula $a^2 = b^2 + c^2 - 2bc \cos A$, plug the right side into the GDC, and then square root.

When solving for an angle, use the formula $\cos A = \dfrac{b^2 + c^2 - a^2}{2bc}$, plug the right side into the GDC (be sure to include parentheses), and then cosine inverse.

Either way, $\cos A$ and a in the formula are always an angle with its opposite side.

Once the cosine rule is used once, the problem can be finished using the sine rule.

➡ EXAMPLE 5.31 ─────────────────

Olivia jogs in a straight path for 10 minutes at a rate of $\dfrac{1}{5}$ km per minute. The path then turns

at a 115° angle, and she continues jogging a straight path for 15 minutes at a rate of $\dfrac{1}{4}$ km

per minute.

1. Draw a diagram clearly showing the distance Olivia has jogged with the angle of the path.

After jogging 15 minutes along the second straight path, Olivia turns and jogs back to her starting point.

2. Calculate the total distance Olivia has jogged.

Answer Explanations

1.

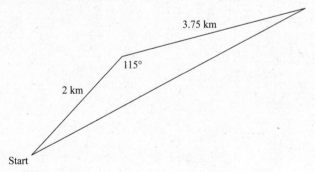

First, calculate the distance of the two legs.

Olivia jogged for 10 minutes at a rate of $\dfrac{1}{5}$ km per minute.

Since distance = rate × time, Olivia has jogged 2 km ($10 \times \dfrac{1}{5} = 2$).

> ### QUICK TIP
>
> Use the **Cosine Rule** when **NOT** given an angle and its opposite side.
>
> When solving for a **side**, $a^2 = b^2 + c^2 - 2bc \cos A$. The side a is opposite of angle A.
>
> When solving for an **angle**, $\cos A = \dfrac{b^2 + c^2 - a^2}{2bc}$. The side a is opposite of angle A. Be sure to use cosine inverse as the last step.

She also jogged 15 minutes at a rate of $\frac{1}{4}$ km per minute. Thus, she has jogged 3.75 km $(15 \times \frac{1}{4} = 3.75)$.

The path turned at a 115° angle, so this is the angle between the two paths.

2. **10.7 km**

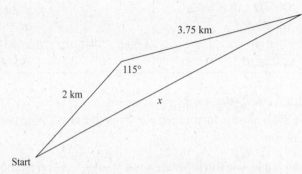

To find the total distance jogged, you need the final leg of her journey. The triangle has two sides with their included angle, so use the cosine rule.

$$x^2 = 2^2 + (3.75)^2 - 2(2)(3.75)\cos 115°$$
$$x^2 = 24.40177$$
$$x = 4.94$$

Add the three legs of her journey for the total distance.

$$2 + 3.75 + 4.94 = 10.7 \text{ km (correct to 3 significant figures)}$$

➡ **EXAMPLE 5.32** _____

Taylor is making the small garden flag shown.
    Calculate the measurements of all the angles.

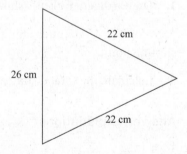

Answer Explanation

53.8°, 53.8°, and 72.4°

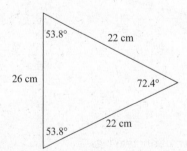

Since the triangle has three sides with no angle, the cosine rule must be used. No angle was specifically mentioned, so finding the largest angle is best. Therefore, $a = 26$.

$$\cos A = \frac{22^2 + 22^2 - 26^2}{2(22)(22)}$$

$$\cos A = 0.30165$$

$$A = \cos^{-1}(0.30165)$$

$$A = 72.4°$$

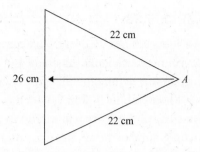

The flag is an isosceles triangle, so instead of using the sine rule to find the remaining angles, you can use algebra.

$$X + X + 72.4° = 180°$$

$$2X = 107.6°$$

$$X = 53.8°$$

Thus, the remaining two angles both have measures of 53.8°.

The angles could have been found using the sine rule.

$$\frac{26}{\sin 72.4°} = \frac{22}{\sin X}$$

$$26 \sin X = 22 \sin 72.4°$$

$$\sin X = \frac{22 \sin 72.4°}{26}$$

$$\sin X = 0.80655$$

$$X = \sin^{-1}(0.80655)$$

$$X = 53.8°$$

To find the area of a right triangle, most are familiar with $A = \frac{1}{2}bh$, where b = base and h = height. However, as with the sine, cosine, and tangent trigonometric ratios, this formula only works with right triangles.

When trying to find the area of a non-right triangle, the quickest way is to use the formula $A = \frac{1}{2}ab\sin C$, where a and b are adjacent sides and C is the included angle.

Looking back at example 5.32, you can find the amount of fabric needed to make the small garden flag, which would be the area of the triangle. You know an angle and its included sides, so plug these measurements into the formula.

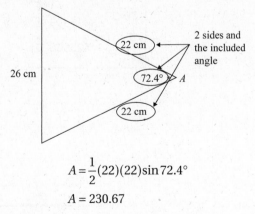

$$A = \frac{1}{2}(22)(22)\sin 72.4°$$

$$A = 230.67$$

231 cm² of fabric will be needed.

➥ **EXAMPLE 5.33** —————————————————————————

PlayRight Builders has installed a new playground for a local preschool. The area of land used for the playground is triangular in shape and covers 130 square meters.

Safety regulations require the playground area to be enclosed by a fence. The vertical side of the area is the preschool building, which is 12 meters in length. From the top corner of the building, one side of the fence will extend 24 meters southeast.

1. Draw a diagram illustrating the playground with the dimensions clearly labeled.
2. Calculate the size of the angle between the building and the known fence length.
3. Determine the amount of fencing needed to enclose the playground. Round to the nearest meter.

Answer Explanations

1.

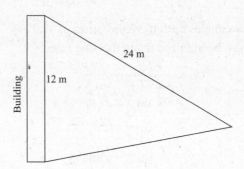

The building makes the vertical edge of the triangle and is 12 meters in length. From the top of the building, extend a segment southeast (down and to the right). This measures 24 meters.

2. **64.5°**

The problem states that the playground covers 130 square meters. This is the area of the triangle. You know two sides of the triangle and want to find the included angle.

$$130 = \frac{1}{2}(12)(24)\sin A \quad \text{Multiply the coefficients together.}$$

$$130 = 144 \sin A \qquad \text{To solve for } \sin A, \text{ divide both sides by 144.}$$

$$0.902778 = \sin A \qquad \text{To solve for } A, \text{ sine inverse both sides.}$$

$$\sin^{-1}(0.902778) = A$$

$$A = 64.5°$$

3. **46 meters**

Now that you have an angle, you can solve for the remaining side.

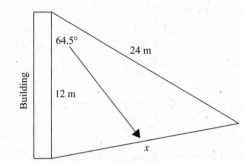

You do not have an angle with the measure of its opposite side, so you cannot use the sine rule. You do have two sides with their included angle, so you can use the cosine rule.

$$x^2 = 12^2 + 24^2 - 2\,(12)\,(24)\cos 64.5°$$
$$x^2 = 472.0256$$
$$x = 21.7$$

Since the playground has one side as the building, only two sides will need fencing. Therefore, the amount of fencing needed is $24 + 21.7 = 45.7$ meters. Now, round to the nearest meter, which means the solution is 46 meters.

➡ EXAMPLE 5.34

Calculate the area of the following triangles.

1. Triangle XYZ, where $x = 3.4$, $y = 2.3$, and $X\hat{Z}Y = 71°$.

2.

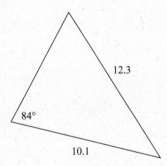

Answer Explanations

1. **3.70**

First, draw a fully labeled diagram to represent triangle XYZ.

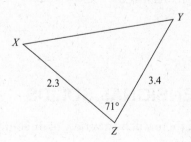

To find the area of a triangle using the formula $A = \frac{1}{2}ab\sin C$, you need two sides with their included angle. Triangle *XYZ* provides those measurements.

Plugging the values into the formula yields

$$A = \frac{1}{2}(2.3)(3.4)\sin 71°.$$

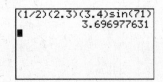

Now, type the right side into the GDC to calculate the answer of 3.70 (rounded to 3 significant figures).

2. **41.0**

The area of the given triangle cannot be calculated from the given information. You do not have two sides with their included angle.

You do have an angle with its opposite side, so you can use the sine rule to find the angle at the top. This is denoted by *A*.

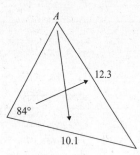

$$\frac{12.3}{\sin 84°} = \frac{10.1}{\sin A}$$

$$12.3 \sin A = 10.1 \sin 84°$$

$$\sin A = \frac{10.1\sin 84°}{12.3}$$

$$A = \sin^{-1}(0.81664)$$

$$A = 54.7°$$

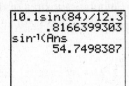

You still do not have two sides with their included angle, but you can calculate the missing angle between the given sides.

$$B + 54.7° + 84° = 180°$$
$$B = 41.3°$$

Now, you have two sides with their included angle.

$$A = \frac{1}{2}(10.1)(12.3)\sin 41.3°$$
$$A = 41.0$$

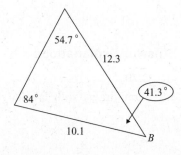

5.4 THREE-DIMENSIONAL SOLIDS

On the IB Math Studies exam there will be a variety of three-dimensional shapes that could be tested.

Required 3-D Shapes	
Cuboid Constructed from 6 rectangular faces at right angles to one another. If the 6 faces are squares, the shape can be called a cube.	
Right Prism Constructed from two bases that are the same shape. All lateral faces are rectangles.	
Right Pyramid Constructed from one base with the height directly above the center of the base. All lateral faces are triangles.	
Right Cone Constructed from a circular base with the height directly above the center of the base.	
Cylinder Constructed from two circular bases directly above one another.	
Sphere Every point on the surface is the same distance from the center.	
Hemisphere Half of a sphere.	

When working with three-dimensional shapes, you will be required to find the distance between two points, whether those two points are two vertices, two midpoints, or a combination of the two.

For example, consider the following rectangular prism.

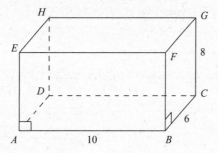

On the IB Math Studies exam, it is common to be asked to calculate the length of *AC* or *AG*.

Since *AC* is the diagonal of rectangle *ABCD*, you can draw a separate diagram of the two-dimensional base.

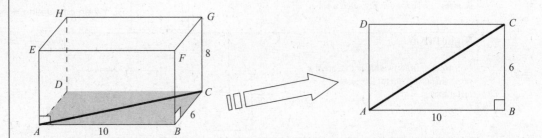

The diagonal *AC* divides the base into two right triangles. Use Pythagoras' theorem to find the length *AC*.

$$10^2 + 6^2 = AC^2$$
$$136 = AC^2$$
$$AC = \sqrt{136} \text{ or } 11.7$$

Now that you have the length of *AC*, you can find the length of *AG*. The length *AG* travels from the base of the prism to the opposite corner. This creates the right triangle *ACG*. The base of the triangle is *AC*, which you just found the length of to be 11.7. The height is given as 8.

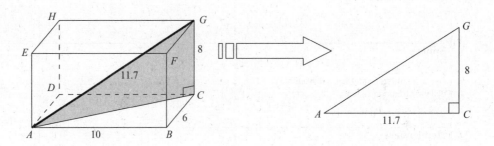

Again, use Pythagoras' theorem to find the hypotenuse (*AG*) of the triangle.

$11.7^2 + 8^2 = AG^2$	$\left(\sqrt{136}\right)^2 + 8^2 = AG^2$
$200.89 = AG^2$ **OR**	$200 = AG^2$
$AG = 14.2$	$AG = \sqrt{200} \text{ or } 14.1$

Since answers can be expressed exactly ($AC = \sqrt{136}$), or rounded to 3 significant figures ($AC = 11.7$), all three answers of *AG* would be accepted ($AG = 14.1$, 14.2, or $\sqrt{200}$).

➡ EXAMPLE 5.40

The following diagram is the square based right pyramid *VABCD*.

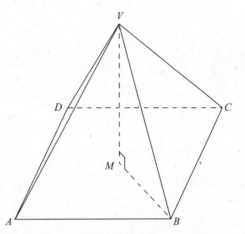

QUICK TIP

- When calculating the length of a diagonal, Pythagoras' theorem **usually** is appropriate.

- When calculating a length involving a midpoint, remember to **halve** the given lengths.

- To find an angle between two faces, form a triangle by drawing a line to represent each face. Then, use **sine, cosine, or tangent** to solve.

Given $AB = 16$ cm and $VM = 20$ cm, find the length of

1. BD,
2. VB.

Answer Explanations

1. **22.6 cm or $\sqrt{512}$ cm**

   First, label the diagram with the given measurements. Since the figure is a square based right pyramid, all sides of the base have a length of 16 cm. You have been asked asked to find BD, which is the diagonal of the square $ABCD$.

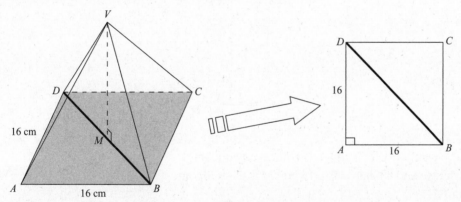

   The diagonal BD is the hypotenuse of right triangle ABD (or BCD), so use Pythagoras' theorem.

   $$16^2 + 16^2 = BD^2$$
   $$512 = BD^2$$
   $$BD = 22.6 \text{ or } \sqrt{512}$$

2. **23.0 cm**

   In order to find the length of VB, identify of which shape VB is an edge.

   VB is the hypotenuse of right triangle BMV. You are given $VM = 20$ cm and have calculated $BD = 22.6$ cm. Since M is the center of the base, MB is half the length of BD. Thus, $MB = 11.3$ cm.

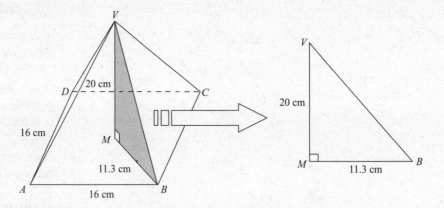

Pythagoras' theorem allows us to calculate *VB*.

$$20^2 + (11.3)^2 = VB^2$$
$$527.69 = VB^2$$
$$VB = 23.0 \text{ cm}$$

➡ **EXAMPLE 5.41**

1. Calculate the slant height of the cone shown below.

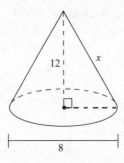

2. The pyramid *VABC* has a base that is an equilateral triangle and a height *VN* = 15 cm. *M* is the midpoint of *BC*.
   (a) Calculate the height of the base, *AM*.
       AN has a length of 11.53 cm.
   (b) Calculate the slant height of the pyramid, *VM*.

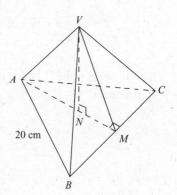

Answer Explanations

1. **12.6**

   The diameter of the base is 8. Thus, the radius is 4.
   To find *x*, use Pythagoras' theorem.

   $$12^2 + 4^2 = x^2$$
   $$160 = x^2$$
   $$x = 12.6$$

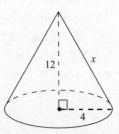

2. (a) **17.3 cm**

Since *M* is the midpoint of the base, you can use the right triangle *ABM* to find the height of the base.

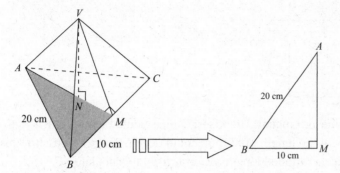

$$10^2 + AM^2 = 20^2$$
$$AM^2 = 300$$
$$AM = 17.3$$

(b) **16.1 cm**

To calculate the slant height *VM*, use right triangle *MNV*. You are given *AN* = 11.53 cm and must find *NM*.

$$AN + NM = AM$$
$$11.53 + NM = 17.3$$
$$NM = 5.77 \text{ cm}$$

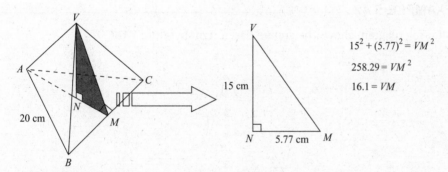

$$15^2 + (5.77)^2 = VM^2$$
$$258.29 = VM^2$$
$$16.1 = VM$$

Calculating the distance between two points in three-dimensional shapes leads into calculating the size of angles in those same shapes.

All exam questions will involve right-angled figures; therefore, you can use sine, cosine, and tangent trigonometric ratios.

For example, consider the previous rectangular prism.

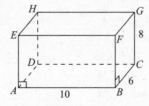

Since you already calculated *AC* = 11.7 and *AG* = 14.1, you can find the size of the angle made by *AG* and the base.

In order to calculate the size of the angle, you need the sides that generate the angle. Obviously *AG* is one side, but you need to draw a line representing the base. The only line that will create the correct angle is *AC*.

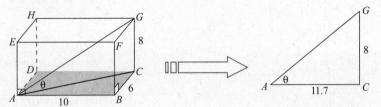

To find $C\hat{A}G$, use sine, cosine, or tangent. The wisest choice would be to use the length of *CG*, since it was given along with the length of *AC*. You could use the length of *AG*, but this is the least accurate answer, since a rounded length (*AC*) was used to calculate *AG*.

CG is opposite $C\hat{A}G$, while *AC* is adjacent. Thus, a tangent ratio is appropriate.

$$\tan\theta = \frac{8}{11.7}$$

$$\theta = \tan^{-1}\left(\frac{8}{11.7}\right)$$

$$\theta = 34.4°$$

➡ EXAMPLE 5.42

The following diagram shows the square based right pyramid *VABCD*.

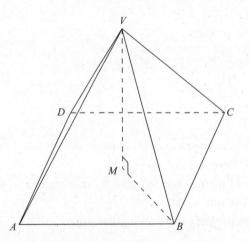

Given *BC* = 10 cm and *VM* = 12 cm, find

1. the length of *MB*,
2. the size of $V\hat{B}M$.

Point *N* is the midpoint on side *BC*. Find

3. the length of *VN*,
4. the size of the angle formed between *VN* and the base.

Answer Explanations

1. **7.07 cm**

   In order to calculate MB, first calculate BD, the hypotenuse of the right triangle ABD.

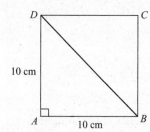

   $10^2 + 10^2 = BD^2$

   $200 = BD^2$

   $14.142 = BD$

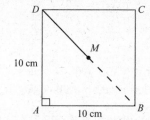

   MB is half of BD;

   $MB = \dfrac{14.142}{2} = 7.07$

2. **59.5°**

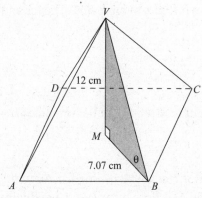

   MB is adjacent to θ, while VM is opposite.

   $\tan \theta = \dfrac{12}{7.07}$

   $\theta = \tan^{-1}\left(\dfrac{12}{7.07}\right)$

   $\theta = 59.5°$

3. **13 cm**

   Since point N is the midpoint of BC, you know the length MN is half of AB.

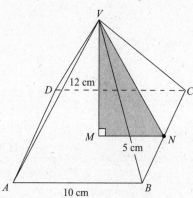

   Using Pythagoras' theorem:

   $5^2 + 12^2 = MN^2$

   $169 = MN^2$

   $13 = MN$

4. **67.3°**

   The angle formed by VN and the base is the angle VNM.

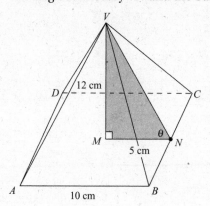

   MN is adjacent to θ while VM is opposite

   $\tan \theta = \dfrac{12}{5}$

   $\theta = \tan^{-1}\left(\dfrac{12}{5}\right)$

   $\theta = 67.3$

FEATURED QUESTION

May 2008, Paper 2

Part A

Mal is shopping for a school trip. He buys 50 tins of beans and 20 packets of cereal. The total cost is 260 Australian dollars (AUD).

(a) Write down an equation showing this information, taking b to be the cost of one tin of beans and c to be the cost of one packet of cereal in AUD.

Stephen thinks that Mal has not bought enough, so he buys 12 more tins of beans and 6 more packets of cereal. He pays 66 AUD.

(b) Write down another equation to represent this information.
(c) Find the cost of one tin of beans.
(d) (i)   Sketch the graphs of these two equations.
     (ii) Write down the coordinates of the point of intersection of the two graphs.

Part B

The triangular faces of a square based pyramid, *ABCDE*, are all inclined at 70° to the base. The edges of the base *ABCD* are all 10 cm and *M* is the centre. *G* is the midpoint of *CD*.

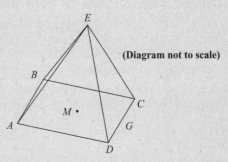

(Diagram not to scale)

(a) Using the letters on the diagram, draw a triangle showing the position of a 70° angle.
(b) Show that the height of the pyramid is 13.7 cm to 3 significant figures.
(c) Calculate
    (i)   the length of EG,
    (ii) the size of angle DÊC.
(d) Find the total surface area of the pyramid.
(e) Find the volume of the pyramid.

(See page 445 for solutions)

5.5 VOLUME AND SURFACE AREA OF THREE-DIMENSIONAL SOLIDS

Surface area is the total area of all the two-dimensional faces of the solid.

Many of the area formulas are located in the "Prior Learning" section of the formula packet.

Shape	Area
Parallelogram (includes squares and rectangles)	$A = b \times h$, where b is the base, and h is the height
Triangle	$A = \frac{1}{2}(b \times h)$, where b is the base, and h is the height
Trapezium (i.e. trapezoid)	$A = \frac{1}{2}(a+b)h$, where a and b are the parallel sides, and h is the height
Circle	$A = \pi r^2$, where r is the radius

For example, consider the cuboid from the last section. To find the surface area, calculate the area of each face, and then add the areas together.

This solid is made up of rectangular faces.

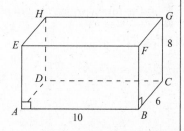

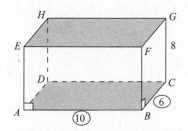

The two bases $ABCD$ and $EFGH$ have a length of 10 and width of 6.

$$A = b \times h$$
$$A = 10 \times 6$$

These two faces each have an area of 60.

The sides $ADEH$ and $BCFG$ have a length of 6 and width of 8.

$$A = b \times h$$
$$A = 6 \times 8$$

These two faces each have an area of 48.

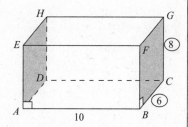

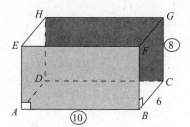

Finally, the front face $ABFE$ and back face $DCGH$ have a length of 10 and width of 8.

$$A = b \times h$$
$$A = 10 \times 8$$

These two faces each have an area of 80.

$$SA = 60 + 60 + 48 + 48 + 80 + 80$$

The surface area is calculated as follows:

$$SA = 376$$

For curved surfaces, such as cones, cylinders, and spheres, the area of the curved surface formula is given in the formula packet.

Area of the curved surface of a cylinder	$A = 2\pi rh$, where r is the radius, and h is the height
Surface area of a sphere	$A = 4\pi r^2$, where r is the radius
Area of the curved surface of a cone	$A = \pi rl$, where r is the radius, and l is the slant height

➡ **EXAMPLE 5.50**

Find the total surface area of each solid.

> **QUICK TIP**
>
> **Surface area:** total area of each face
>
> Area of a circle: πr^2
>
> Area of a triangle: $A = \dfrac{1}{2}bh$
>
> Area of a parallelogram: $b \times h$
>
> The formula packet includes the curved surface formulas for a cone, cylinder, and sphere.

1.

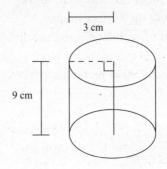

3 cm

9 cm

2.

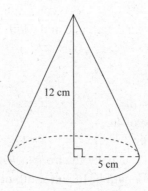

12 cm

5 cm

3.

300 mm

Answer Explanations

1. **226 cm²**

   The surface area of a cylinder is comprised of the two circular bases and the curved surface.

   The top and bottom of the cylinder are identical circles, so $A = \pi r^2$.

   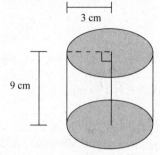

   The radius is 3 cm.

   $$A = \pi (3)^2$$
   $$A = 28.27 \text{ cm}^2$$

   For both circles the total area is 56.5 cm².

   To find the area of the curved surface, use the formula $A = 2\pi rh$, where r is the radius, and h is the height.

   Both pieces are given, so $A = 2\pi (3)(9)$.

   The area of the curved surface is 169.6 cm².

   The surface area for the cylinder is 56.5 cm² + 169.6 cm² = 266 cm² (rounded to 3 significant figures).

2. **283 cm²**

   The surface area of a cone is comprised of one circle and the curved surface.

   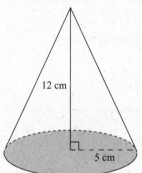

   The radius is 5 cm.

   $$A = \pi (5)^2$$
   $$A = 78.54 \text{ cm}^2$$

   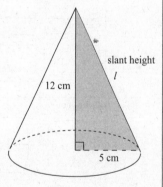

   In order to calculate the area of the curved surface $A = \pi rl$, you must find the slant height of l.

   The slant height is the hypotenuse of the triangle formed from the radius and the height of the cone.

   $$12^2 + 5^2 = l^2$$
   $$169 = l^2$$
   $$13 = l$$

   The area of the curved surface is

   $$A = \pi rl$$
   $$A = \pi (5)(13)$$
   $$A = 204.2 \text{ cm}^2$$

   The surface area for the cone is 78.54 cm² + 204.2 cm² = 283 cm² (rounded to 3 significant figures).

3. **212 000 mm²**

A hemisphere is half a sphere. The area of the curved surface for a sphere can be found using $A = 4\pi r^2$. Thus, a hemisphere is half.

$$A = \frac{4\pi r^2}{2} = 2\pi r^2$$

The diameter is 300 mm, so the radius is 150 mm.

$$A = 2\pi (150)^2$$
$$A = 141\,371.7 \text{ mm}^2$$

The area of the base is simply the area of a circle whose radius is 150 mm.

$$A = \pi (150)^2 = 70\,685.83 \text{ mm}^2$$

The surface area is $141\,371.7 + 70\,685.83 = 212\,057.53$ mm², which rounds to 212 000 (3 significant figures).

The volume is the amount of space a solid occupies.

Consider a gift box in the shape of a cuboid. The surface area would be the amount of wrapping paper needed to cover the entire box. Volume would be the amount of space inside the box.

The volume formulas needed for the exam will be included in the formula packet.

Shape	Volume
Pyramid 	$V = \frac{1}{3} Ah$, where A is the area of the base, and h is the vertical height
Cuboid 	$V = l \times w \times h$, where l is the length, w is the width, and h is the height
Cylinder 	$V = \pi r^2 h$, where r is the radius, and h is the height
Sphere 	$V = \frac{4}{3}\pi r^3$, where r is the radius
Cone 	$V = \frac{1}{3}\pi r^2 h$, where r is the radius, and h is the vertical height
Prism 	$V = Ah$, where A is the area of cross section, and h is the height

➡ EXAMPLE 5.51 ────────────────────────────

Paolo is designing a one-person tent for an upcoming camping trip. His idea is for the front and back of the tent to be equilateral triangles and the lateral sides, which includes the bottom of the tent, to be rectangles.

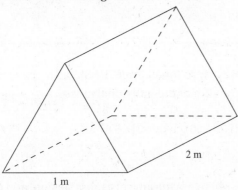

2 m

1 m

> **QUICK TIP**
>
> **Volume:** space occupied by a solid
>
> Formulas for volume are included in the formula packet.
>
> Formulas involving "A" require the area of the base or cross section.

1. Determine the vertical height of the tent.
2. Show that the amount of fabric needed for the tent is 6.87 m².
3. Calculate the volume of the tent.

Answer Explanations

1. **0.866 cm**

   The vertical height can be found using the equilateral triangle.

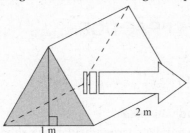 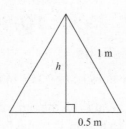

   Using Pythagoras' theorem:

   $$h^2 + (0.5)^2 = 1^2$$
   $$h^2 = 0.75$$
   $$h = 0.866 \text{ m}$$

2. $$2\left[\frac{1}{2}(1 \times 0.866)\right] + 3[2 \times 1] = 6.87$$

   The amount of fabric needed is the surface area.

   The triangular prism is constructed by two equilateral triangles and three rectangles. Since the triangles are equilateral, the three rectangles have the same dimensions.

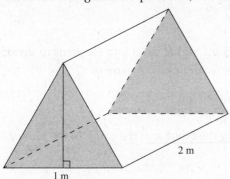

   The triangles have a base of 1 m and a height of 0.866 m.

   The area of one triangle is $A = \frac{1}{2}(1 \times 0.866)$, so the area for both triangles would be $A = 2\left[\frac{1}{2}(1 \times 0.866)\right]$.

   The three rectangular faces all have dimensions 2 m × 1 m.

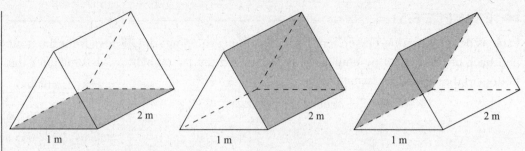

The area of each rectangle is $A = 2 \times 1$, so for all three $A = 3[2 \times 1]$.

The surface area for the tent is the sum of the individual areas:

$$A = 2\left[\frac{1}{2}(1 \times 0.866)\right] + 3[2 \times 1] = 6.87 \text{ m}^2.$$

3. **$V = 0.866$ m^3**

    The volume for a prism is $V = Ah$, where A is the area of a cross section, and h is the height. The cross section is the base shape, which is an equilateral triangle. The area of the triangle is $A = \frac{1}{2}(1 \times 0.866) = 0.433$ m^2.

    Now, multiply this area by the height of the prism, which is 2 m.

    $$V = 0.433(2) = 0.866 \text{ m}^3$$

TOPIC 5 PRACTICE

Paper 1 Questions

1. Consider the lines L_1: $2y - 3x = 14$ and L_2: $y + \frac{1}{2}x - 3 = 0$.

    (a) Graph L_1 and L_2 on the same coordinate plane.
    (b) State, with reason, if L_1 and L_2 are parallel, perpendicular, or neither.
    (c) Find the point of intersection for L_1 and L_2.

2. The equation of the straight line R_1 is $y - \frac{2}{3}x = 4$.

    (a) Write down the gradient of all lines perpendicular to R_1.

    R_2 passes through the point $(-2, -6)$ and is perpendicular to R_1.

    (b) Write the equation of R_2 in the form $ax + by + d = 0$, where $a, b, d \in \mathbb{Z}$.

    R_1 and R_2 intersect at point P.

    (c) Write down the coordinates of point P.

3. A farmer owns a rectangular plot of land with a width of 2.1 km. He plans to divide the land in half by putting up a 3.5 km fence between opposite corners.

    (a) Draw a diagram illustrating the given information.
    (b) Calculate the length of the rectangular plot.
    (c) Determine the size of the angle made by the fence and the width of the plot of land.

4. A hot air balloon is flying 460 meters directly above the ground. When the pilot of the balloon looks at an 18° angle of depression, she sees the local high school.

   (a) Represent the information in a clearly labeled diagram.
   (b) Determine the horizontal distance between the high school and hot air balloon rounded to the nearest meter.

5. The figure below shows two triangular stones, ABC and BCD, to be used in a mosaic at the local park. $AC = 36$ cm, $CD = 32$ cm, $BD = 28$ cm, $A\hat{C}B = 46°$, and $B\hat{C}D = 25°$.

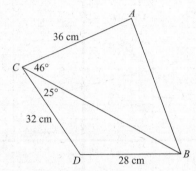

   (a) Using triangle BCD, find the length BC.

   The two stones are to be covered in a protective paint. Each bottle of the protective paint can cover an area of 450 cm².

   (b) Calculate the total area to be covered in the protective paint, rounded to the nearest square centimeter.
   (c) Find how many bottles need to be purchased to ensure there is enough paint.

6. Hasif rides his bicycle along a triangular path. On the first leg of the journey, he rides 30 minutes at a pace of 18 km per hour. On the second leg of the journey, he rides 40 minutes at a pace of 15 km per hour. On the final leg of the journey, he rides 15 minutes at a pace of 20 km per hour.

   (a) Draw a diagram to represent the distance Hasif went on each leg of the journey.
   (b) Calculate the size of the angle between the first and last leg of his journey.

7. A water storage tank is constructed by a cylinder topped with a hemisphere. The height of the cylinder is 3 m. The unit can hold 46.1 cubic meters of water when full.

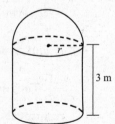

   (a) Calculate the length of the radius, r, rounded to the nearest meter.
   (b) Determine the surface area of the entire storage tank.

8. In the cube shown in the diagram, all edges are 5 cm long, and M is the midpoint of AB.

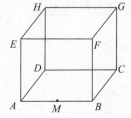

(a) Find the length *CM*.

(b) Find the length *GM*.

(c) Calculate the angle that *GM* makes with the base of the cube.

SOLUTIONS

1. (a)

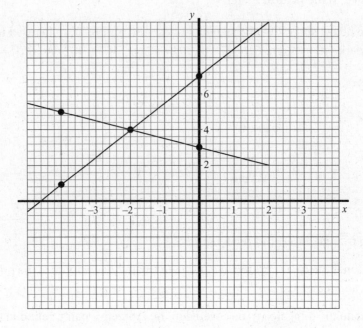

In order to graph L_1 and L_2, the lines should be rewritten in slope-intercept form.

L_1	L_2

$$2y - 3x = 14 \qquad\qquad\qquad\qquad y + \frac{1}{2}x - 3 = 0$$

$$2y = 3x + 14$$

$$y = \frac{3}{2}x + 7 \qquad\qquad\qquad\qquad y = -\frac{1}{2}x + 3$$

The *y*-intercept is (0, 7). The *y*-intercept is (0, 3).

The gradient is $\frac{3}{2}$. The gradient is $-\frac{1}{2}$.

(b) **L_1 and L_2 are not parallel, since the gradient of L_1 is $\dfrac{3}{2}$ and the gradient of L_2 is $-\dfrac{1}{2}$. They are not perpendicular, since $\dfrac{3}{2} \times -\dfrac{1}{2} \neq -1$, nor are the gradients opposite reciprocals. Therefore, L_1 and L_2 are neither.**

(c) **(–2, 4)**

The lines intersect at the point (–2, 4).

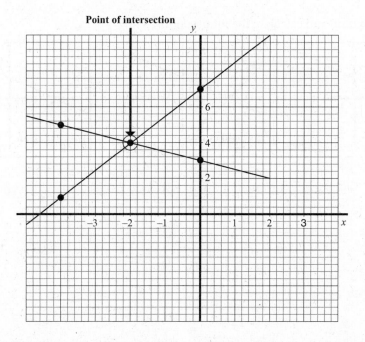

Point of intersection

2. (a) $-\dfrac{3}{2}$

The gradient of R_1 is $\dfrac{2}{3}$ $\left(\begin{array}{c} y - \dfrac{2}{3}x = 4 \\ y = \dfrac{2}{3}x + 4 \end{array} \right)$.

Perpendicular lines have opposite reciprocal gradients, which gives a gradient of $-\dfrac{3}{2}$.

Also, $-\dfrac{3}{2} \times \dfrac{2}{3} = -1$.

(b) $\mathbf{0 = -3x - 2y - 18}$ **or** $\mathbf{3x + 2y + 18 = 0}$

First, find the equation of R_2 by plugging in the given point with the gradient into slope-intercept form.

$-6 = -\dfrac{3}{2}(-2) + c$

$-6 = 3 + c$ Therefore, $y = -\dfrac{3}{2}x - 9$.

$-9 = c$

Now, rewrite the line in standard form.

$0 = -\dfrac{3}{2}x - y - 9$ (multiply all terms by 2)

$0 = -3x - 2y - 18$

(c) $\mathbf{(-6, 0)}$

Using the GDC, enter the lines into Y_1 and Y_2.

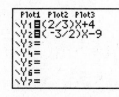

Now graph and calculate the intersection.

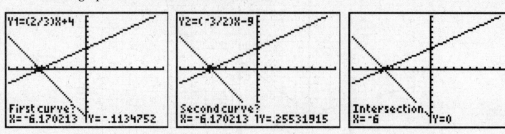

3. (a)

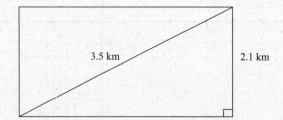

The width given is 2.1 km. The fence creates the diagonal of the rectangle.

(b) **2.8 km**

The measurements given create a right triangle. To solve for the missing length, use Pythagoras' theorem.

$$(2.1)^2 + l^2 = (3.5)^2$$
$$l^2 = 7.84$$
$$l = 2.8$$

(c) **53.1°**

The desired angle is between the fence (the hypotenuse) and the width (adjacent). Therefore, set up a cosine trigonometric ratio.

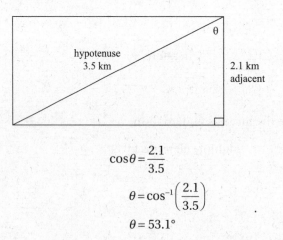

$$\cos\theta = \frac{2.1}{3.5}$$
$$\theta = \cos^{-1}\left(\frac{2.1}{3.5}\right)$$
$$\theta = 53.1°$$

The length could have also been used to arrive at the same answer.

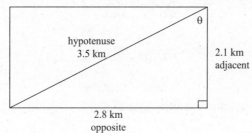

$$\sin\theta=\frac{2.8}{3.5} \qquad\qquad \tan\theta=\frac{2.8}{2.1}$$

$$\theta=\sin^{-1}\left(\frac{2.8}{3.5}\right) \qquad \textbf{OR} \qquad \theta=\tan^{-1}\left(\frac{2.8}{2.1}\right)$$

$$\theta=53.1° \qquad\qquad \theta=53.1°$$

4. (a)

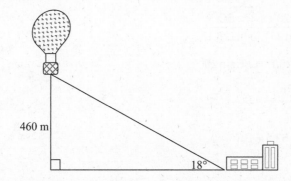

The balloon is directly above the ground, so clearly identify the right angle. The angle of depression goes in the bottom corner by the high school.

(b) **1 416 m**

The height of the balloon is opposite the given angle. You are asked to find the adjacent side, so set up an equation using the tangent ratio.

$$\tan18°=\frac{460}{x}$$

$$x\tan18°=460$$

$$x=\frac{460}{\tan18°}=1\,415.73$$

Rounding to the nearest meter gives an answer of 1 416 m.

5. (a) **BC = 53.5 cm**

The length *BC* cannot be determined until angle *BDC* is known.
In order to find angle *BDC*, you must calculate angle *CBD*.
Since you have an angle (*BĈD*) and its opposite side (*BD*), you can use the sine rule.

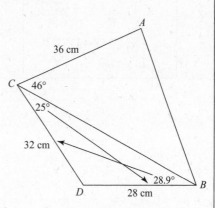

$$\frac{28}{\sin 25°} = \frac{32}{\sin C}$$

$$28 \sin C = 32 \sin 25°$$

$$\sin C = \frac{32 \sin 25°}{28}$$

$$\sin C = 0.482992$$

$$C = \sin^{-1}(0.482992) = 28.9°$$

Now, find angle BDC.

$$25° + 28.9° + B\hat{D}C = 180°$$

$$B\hat{D}C = 126.1°$$

With angle BDC, use the sine rule one more time to find the length BC.

$$\frac{28}{\sin 25°} = \frac{BC}{\sin 126.1°}$$

$$28 \sin 126.1° = BC \sin 25°$$

$$\frac{28 \sin 126.1°}{\sin 25°} = BC$$

$$53.5 = BC$$

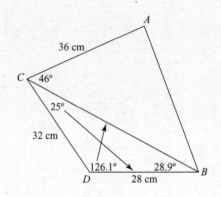

(b) **1 055 cm²**

To find the total area, you must find the area of each triangle.

The area of triangle BCD can be found using any two sides and their included angle.

Option 1	Option 2	Option 3
$A = \dfrac{1}{2}(32 \times 53.5)\sin 25°$	$A = \dfrac{1}{2}(32 \times 28)\sin 126.1°$	$A = \dfrac{1}{2}(28 \times 53.5)\sin 28.9°$
$A = 362$ cm²	$A = 362$ cm²	$A = 362$ cm²

The area of triangle ABC can be found using side AC and BC with their included angle $A\hat{C}B$.

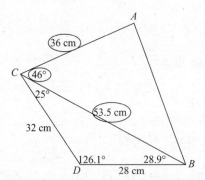

$$A = \frac{1}{2}(36 \times 53.5)\sin 46°$$

$$A = 693 \text{ cm}^2$$

The total area is $362 + 693 = 1\,055 \text{ cm}^2$.

(c) **3 bottles**

If each bottle contains enough paint to cover an area of 450 cm^2, divide the area of $1\,055 \text{ cm}^2$ by this amount.

$$1055 \div 450 = 2.3\overline{4}$$

Since the quotient is larger than 2, 3 bottles will be needed to ensure there is enough paint.

6. (a)

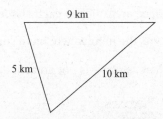

To determine the distance traveled on each leg of the journey, convert the minutes to hours because the pace given is kilometers per **hour**.

Leg 1	Leg 2	Leg 3
$\dfrac{30 \text{ mins}}{x \text{ hour}} = \dfrac{60 \text{ mins}}{1 \text{ hour}}$	$\dfrac{40 \text{ mins}}{x \text{ hour}} = \dfrac{60 \text{ mins}}{1 \text{ hour}}$	$\dfrac{15 \text{ mins}}{x \text{ hour}} = \dfrac{60 \text{ mins}}{1 \text{ hour}}$
$30 = 60x$	$40 = 60x$	$15 = 60x$
$x = \dfrac{1}{2} \text{ hour}$	$x = \dfrac{2}{3} \text{ hour}$	$x = \dfrac{1}{4} \text{ hour}$
$\text{Distance} = \dfrac{1}{2}(18) = 9 \text{ km}$	$\text{Distance} = \dfrac{2}{3}(15) = 10 \text{ km}$	$\text{Distance} = \dfrac{1}{4}(20) = 5 \text{ km}$

(b) **86.2°**

The angle between the starting and ending legs is denoted by θ. You do not have an angle, so you must use the cosine rule.

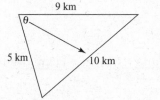

$$\cos\theta = \frac{5^2 + 9^2 - 10^2}{2(5)(9)}$$

$$\cos\theta = 0.0667$$

$$\theta = \cos^{-1}(0.0667) = 86.2°$$

7. (a) **2 m**

The volume of the tank is 46.1 cubic meters and is comprised of the hemisphere and the cylinder.

$$46.1 = \overbrace{\frac{1}{2}\left(\frac{4}{3}\pi r^3\right)}^{Hemisphere} + \overbrace{\pi r^2(3)}^{Cylinder}$$

$$46.1 = \frac{2}{3}\pi r^3 + 3\pi r^2$$

$$0 = \frac{2}{3}\pi r^3 + 3\pi r^2 - 46.1$$

Now solve using [PLYSMLT]. Enter the coefficients exactly even though the calculator will convert to decimals.

The problem tells you to round to the nearest meters; therefore, $r = 2$ m.

(b) **113 m²**

To find the total surface area, look at each shape individually.

Since it is half a sphere, the hemisphere has a surface area of $A = \frac{1}{2}(4\pi r^2)$. Ignore the base of the hemisphere, since it is not an actual part of the tank.

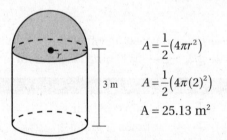

$$A = \frac{1}{2}\left(4\pi r^2\right)$$

$$A = \frac{1}{2}\left(4\pi (2)^2\right)$$

$$A = 25.13 \text{ m}^2$$

Next, calculate the surface area of the cylinder. The surface area of the cylinder would be the area of the curved surface plus the area of the base. Again, do not include the area of the top circle because it is not a solid surface.

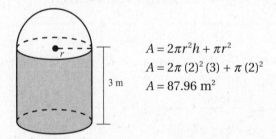

$$A = 2\pi r^2 h + \pi r^2$$

$$A = 2\pi (2)^2 (3) + \pi (2)^2$$

$$A = 87.96 \text{ m}^2$$

The total surface area would be 25.13 m² + 87.96 m² = 113 m² (rounded to 3 significant figures).

8. (a) **5.59 cm**

Since *M* is the midpoint of *AB*, *MB* has a length of 2.5 cm.

The segment *CM* creates the right triangle *BCM*. Pythagoras' theorem will solve for the hypotenuse, *CM*.

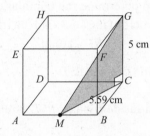

$$(5)^2 + (2.5)^2 = CM^2$$
$$31.25 = CM^2$$
$$5.59 = CM$$

(b) **7.50 cm**

GM is the hypotenuse of right triangle *CGM*. Again, Pythagoras' theorem should be used.

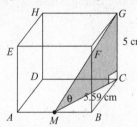

$$(5)^2 + (5.59)^2 = GM^2$$
$$56.25 = GM^2$$
$$7.50 = GM$$

(c) **41.8°**

The angle that *GM* makes with the base is denoted by θ. The side *CG* is opposite, while *CM* is adjacent, so a tangent trigonometric ratio should be used.

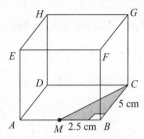

$$\tan\theta = \frac{5}{5.59}$$
$$\theta = \tan^{-1}\left(\frac{5}{5.59}\right)$$
$$\theta = 41.8°$$

Paper 2 Questions

1. The lines L_1 ($y - x - 5 = 0$) and L_2 are shown on the graph below.

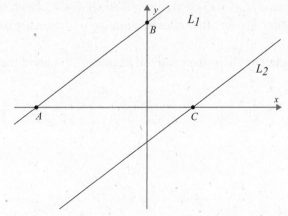

(a) Write down the coordinates of point A and point B.

The line L_2 is parallel to L_1 and passes through the point $(2, 0)$.

(b) Write down the equation of L_2 in the form $y = mx + c$.

The three points A, B, and C are connected to form triangle ABC.

(c) Calculate the length of side
   (i) AB
   (ii) AC

The length of side BC is $\sqrt{29}$.

(d) Determine the size of angle ABC.
(e) Find the area of triangle ABC.

2. A surveyor company sends a team out to determine the boundaries of a piece of land. The team sets up its equipment at the corners of the land, and the triangle RST is formed. The team determines $RS = 85$ m, $ST = 96$ m, and $RT = 110$ m.

(a) Draw and label a diagram representing triangle RST.
(b) Determine the size of $R\hat{S}T$.
(c) Calculate the area of land contained inside triangle RST. Round the answer to the nearest integer.
(d) Calculate the size of $R\hat{T}S$.

The owner of the land wants to subdivide the triangle into two equal pieces. To do so, point Y is placed on side RT so that SY cuts the triangle into two equal pieces.

(e) Write down the area of triangle RSY.
(f) Find the length of RY.

A fence is to be put around the perimeter of triangle RSY.

(g) Determine the length of fencing needed.

The fencing materials cost $12.90 per meter.

(h) Find the total cost of the fence rounded to the nearest dollar.

3. A popular gift option at a local candy store is to choose a plastic container to fill with various colored gumballs. An often-chosen container is the square based pyramid shown below.

The vertex (V) is directly above the center of the base. All sides of the base are 16 cm and $VX = 20$ cm.

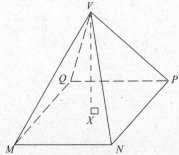

(a) Determine the length of *MX*.

(b) Calculate the size of the angle that *VM* makes with the base.

Consider the midpoint *W* of side *MN*.

(c) Calculate the length of the slant height, *VW*.

(d) Determine the volume of the container.

The gumballs are small spheres with a diameter of 2.54 cm.

(e) Find the volume of one gumball.

(f) Calculate how many gumballs will fit into the container.

Customers can choose to have the container gift wrapped once filled.

(g) Determine how many square centimeters of gift wrap will be needed.

This particular container costs $15 plus $0.05 per gumball. Gift wrapping adds an additional $0.01 per square centimeter.

(h) Calculate the total cost for the container filled with gumballs and gift wrapped.

SOLUTIONS

1. (a) **A: (−5, 0) and B: (0, 5)**

   Points *A* and *B* are the *x*- and *y*-intercepts of the line $y - x - 5 = 0$.

   Point *A* is the *x*-intercept, so let $y = 0$.

   $$0 - x - 5 = 0$$
   $$-x = 5$$
   $$x = -5$$

   Therefore, point *A* is (−5, 0).

   Point *B* is the *y*-intercept, so let $x = 0$.

   $$y - 0 - 5 = 0$$
   $$y = 5$$

   Therefore, point *B* is (0, 5).

   (b) **$y = x - 2$**

   Since the lines are parallel, you need the gradient of L_1 because the gradient will be the same for L_2.

   $$y - x - 5 = 0$$
   $$y = x + 5$$

   The gradient is 1.

   Plugging in the given point with the gradient into slope-intercept form gives:

   $$0 = 1(2) + c$$
   $$-2 = c$$

   Therefore, the equation of L_2 is $y = x - 2$.

(c) (i) **7.07 or $\sqrt{50}$**

The use of the distance formula is considered prior knowledge, meaning students are expected to know how to use the distance formula before taking IB Math Studies. The formula can be found in the "Prior Learning" section of the formula packet.

$$d=\sqrt{(x_1-x_2)^2+(y_1-y_2)^2}$$

To find the length of AB, let point A $(-5, 0)$ be (x_1, y_1) and point B $(0, 5)$ be (x_2, y_2).

$$d=\sqrt{(-5-0)^2+(0-5)^2}$$
$$d=\sqrt{(5)^2+(5)^2}$$
$$d=\sqrt{25+25}=\sqrt{50}$$

(ii) **7**

To find the length of AC, let point A $(-5, 0)$ be (x_1, y_1) and point C $(2, 0)$ be (x_2, y_2).

$$d=\sqrt{(-5-2)^2+(0-0)^2}$$
$$d=\sqrt{(-7)^2+(0)^2}$$
$$d=\sqrt{49}=7$$

(d) **66.8°**

To find the angle ABC, you must use the cosine rule. Since side AC is opposite of angle ABC, $a = 7$ in the cosine rule formula.

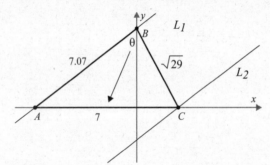

$$\cos\theta=\frac{(7.07)^2+\left(\sqrt{29}\right)^2-7^2}{2(7.07)\left(\sqrt{29}\right)}$$
$$\cos\theta=0.3938$$
$$\theta=\cos^{-1}(0.3938)$$
$$\theta=66.8°$$

(e) **17.5**

To find the area of the triangle ABC, you need two sides and their included angle. Since you know the measure of angle ABC, the two side measures are 7.07 and $\sqrt{29}$.

$$A=\frac{1}{2}(7.07)\left(\sqrt{29}\right)\sin 66.8°$$
$$A=17.5$$

A different way to calculate area is to use the traditional $A=\frac{1}{2}bh$ formula. The base of the triangle is $AC=7$, and the height is the vertical distance from the origin to point B, which is 5.

$$A=\frac{1}{2}(7)(5)=17.5$$

2. (a)

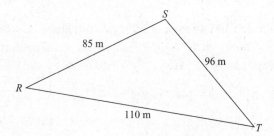

The triangle can be oriented in any direction, but the triangle must be clearly labeled with R, S, and T along with the given measurements.

(b) **74.6°**

Angle RST is denoted by θ, which means $a = 110$ in the cosine rule formula.

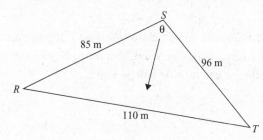

$$\cos\theta = \frac{85^2 + 96^2 - 110^2}{2(85)(96)}$$

$$\cos\theta = 0.26599$$

$$\theta = \cos^{-1}(0.26599) = 74.6°$$

(c) **3 934 m²**

The two sides with their included angle would be sides RS and ST along with $R\hat{S}T$.

$$A = \frac{1}{2}(85)(96)\sin 74.6°$$

$$A = 3933.50$$

Rounding the answer to the nearest integer you get $3\,934$ m².

(d) **48.2°**

Now that you know $R\hat{S}T$, you can use the sine rule to find $R\hat{T}S$.

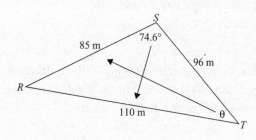

$$\frac{110}{\sin 74.6°} = \frac{85}{\sin T}$$

$$110\sin T = 85\sin 74.6°$$

$$\sin T = \frac{85\sin 74.6°}{110}$$

$$\sin T = 0.74498$$

$$T = \sin^{-1}(0.74498) = 48.2°$$

(e) **1 967 m²**

The area of the entire triangle is $3\,934$ m², so the two smaller triangles have half that area.

$$3\,934 \div 2 = 1\,967$$

(f) **55.1 m**

The triangle *RSY* has an area of 1967 m². You know *RS* = 85 m and *SR̂Y* = 57.2°, since *SR̂Y* = 180° − 74.6° − 48.2°. Side *RY* will be the variable in the area formula.

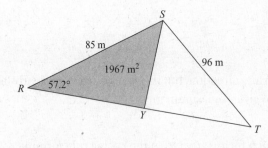

$$1967 = \frac{1}{2}(85)RY\sin 57.2°$$
$$1967 = 35.7241RY$$
$$55.1 = RY$$

(g) **211 m**

To find the perimeter of triangle *RSY*, side *SY* is needed. You do not have an angle with its opposite side, so the sine rule cannot be used. You do have two sides with their included angle, which means the cosine rule will be used.

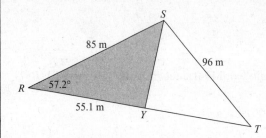

$$SY^2 = (85)^2 + (55.1)^2 − 2(85)(55.1)\cos 57.2°$$
$$SY^2 = 5186.8$$
$$SY = 72.0 \text{ m}$$

The perimeter is the sum of all three sides: 85 + 55.1 + 72.6 = 212.1.
When rounded to 3 significant figures, the perimeter is 212 m.

(h) **$2735**

Multiply the cost per meter by the total number of meters needed.
212 × $12.90 = $2734.8, which rounds to $2735.

3. (a) **11.3 cm**

MX is the hypotenuse of the triangle created when the perpendicular is dropped from point *X* to the midpoint of side *MN*. Point *X* is the center of the square, so it cuts side *MN* in half. The length of the perpendicular is also half of the side.

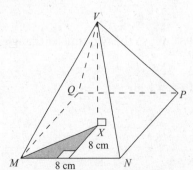

$$8^2 + 8^2 = MX^2$$
$$128 = MX^2$$
$$MX = \sqrt{128} = 11.3 \text{ cm}$$

(b) **60.5°**

The angle *VM* along with the base creates the angle *VMX*.

VX is opposite the angle, while *MX* is adjacent, so set up a tangent ratio.

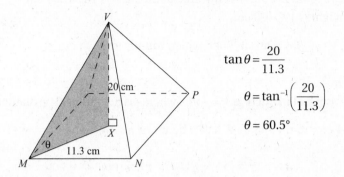

$$\tan\theta = \frac{20}{11.3}$$

$$\theta = \tan^{-1}\left(\frac{20}{11.3}\right)$$

$$\theta = 60.5°$$

(c) **21.6 cm**

Before the length of VW can be found, you must find the length of MV.

Using triangle *MVX*, you can use Pythagoras' theorem to find *VM*.

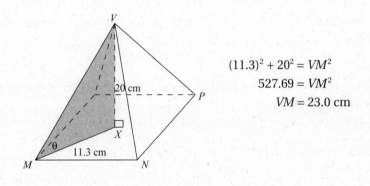

$$(11.3)^2 + 20^2 = VM^2$$

$$527.69 = VM^2$$

$$VM = 23.0 \text{ cm}$$

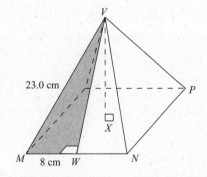

Now, shift your focus to triangle *VMW*.

The hypotenuse of the triangle is *VM* = 23.0, and one leg is *MW* = 8.

$$8^2 + VW^2 = (23.0)^2$$

Using Pythagoras' theorem:

$$VW^2 = 465$$

$$VW = 21.6 \text{ cm}$$

(d) **1 710 cm³**

The volume of a pyramid is $V = \frac{1}{3}Ah$, where A is the volume of the base, and h is the height.

Since the base is a square, the area of the base is $16^2 = 256$ cm².

The height is $VX = 20$ cm.

Thus, $V = \frac{1}{3}(256)(20) = 1706.\overline{6}$, which rounds to 1 710 cm³.

(e) **8.58 cm³**

The volume of a sphere is $V = \frac{4}{3}\pi r^3$, where r is the radius. The diameter of the gumball is 2.54 cm, so the radius is half. Therefore, $r = 1.27$ cm.

$$V = \frac{4}{3}\pi(1.27)^3 = 8.58\,\text{cm}^3$$

(f) **199 gumballs will fit in the container**

The total volume of the container is 1 710 cm³, and each gumball takes up 8.58 cm³ of space.

$$1\,710 \div 8.58 = 199.30$$

The container can only be filled with whole gumballs, so the answer is 199 gumballs.

(g) **947 cm²**

Gift wrap covers the entire surface of the container, so you need to calculate the surface area.

The pyramid is made up of one square and four triangles.

The area of the square is $16^2 = 256$ cm².

The area of triangle MNV is $\frac{1}{2}(16)(21.6) = 172.8\,\text{cm}^2$, since the base of the triangle is 16 cm, and you found the height to be 21.6 cm.

Find the total surface area by adding the area of the square with four times the area of the triangle.

$$256\,\text{cm}^2 + 4(172.8\,\text{cm}^2) = 947.2\,\text{cm}^2$$

Rounding to 3 significant figures gives the surface area as 947 cm².

(h) **$34.42**

The container costs a flat fee of $15.

199 gumballs are needed to fill the container. The cost of the gumballs is $0.05 each. Therefore, the cost of the gumballs is 199(0.05) = $9.95.

To gift wrap the container, 947 cm² of paper is needed, the cost of the gift wrap is $0.01 per cm². The cost to wrap the container is 947(0.01) = $9.47.

The total cost would be $15 + $9.95 + $9.47 = $34.42.

Answers in the form of currency can be left as two decimal places, or the answer of $34.4 would be accepted.

CHAPTER OBJECTIVES

Before you move on to the next chapter, you should be able to:

- ☐ Write a line in the form $y = mx + c$
- ☐ Find the gradient of a given line
- ☐ Calculate the gradient between two points
- ☐ Determine the gradient of all lines parallel or perpendicular to a given line
- ☐ Determine if two lines are parallel, perpendicular, or neither
- ☐ Write the equation of a line in the form $y = mx + c$ or $ax + by + d = 0$, where $a,b,d \in \mathbb{Z}$ given:
  - ○ a point and a gradient
  - ○ two points
  - ○ a point and a parallel line
  - ○ a point and a perpendicular line
- ☐ Calculate the length of a side in a right triangle using Pythagoras' theorem
- ☐ Find the length of an unknown side in a right triangle using a sine, cosine, or tangent ratio
- ☐ Find the measure of an unknown angle in a right triangle using a sine, cosine, or tangent ratio
- ☐ Given a right triangle word problem, draw a diagram representing the described scenario
- ☐ Solve for unknown measurements in a problem involving angles of elevation or depression
- ☐ Determine the length of an unknown side in a non-right triangle using the sine rule or cosine rule
- ☐ Find the length of an unknown side in a non-right triangle using the sine rule or cosine rule
- ☐ Find the measure of an unknown angle in a non-right triangle using the sine rule or cosine rule
- ☐ Calculate the area of a right triangle or a non-right triangle
- ☐ Calculate the length of an unknown distance in a three-dimensional solid
- ☐ Calculate the angle formed between two lines in a three-dimensional solid
- ☐ Calculate the angle formed between two faces in a three-dimensional solid
- ☐ Determine the surface area and volume of a three-dimensional solid
- ☐ Find an unknown measurement for a three-dimensional solid given the surface area or the volume

Geometry/Trig

Mathematical Models

- **DOMAIN:** the set of all possible x-values for a particular function $f(x)$

- **RANGE:** the set of all possible y-values for a particular function $f(x)$

- **LINEAR MODEL:** $f(x) = mx + c$; real-world problems involving a constant rate of change

- **QUADRATIC MODEL:** $f(x) = ax^2 + bx + c$; axis of symmetry; vertex; x-intercepts

- **EXPONENTIAL MODEL:** $f(x) = ka^x + c$; growth or decay; horizontal asymptote at $y = c$

- **POLYNOMIAL MODEL:** $f(x) = ax^m + bx^n + \ldots + m$, $m, n \in \mathbb{N}^+$; axial intercepts; local maxima/minima; x-intercepts

- **RATIONAL MODEL:** $f(x) = ax^m + bx^n + \ldots + m$, $m, n \in \mathbb{Z}$; vertical and horizontal asymptotes; axial intercepts; local maxima/minima

- **DRAWING ACCURATE GRAPHS:** label and scale; axial intercepts; asymptote; local maxima/minima; general shape

- **SOLVING AN EQUATION:** graph each side of equation; find intersection

6.1 FUNCTION, DOMAIN, AND RANGE

y is considered to be a function of x if each value of x only pairs with one y value. Function notation for this relationship is expressed as $f(x) = y$. Each x that is plugged into the equation f yields only one y.

Some examples of functions are

- $f(x) = 3x - 5$ (line)
- $g(x) = 6 - 2x - x^2$ (quadratic)
- $h(x) = 6x^8 - 3x^4 + 2x^3 - 10$ (polynomial)
- $k(x) = \dfrac{2}{x-1} + 3$ (rational)

Functional notation does not always use the variables y and x. If an equation models the height of a rocket after a certain time, the function might be $H(t)$. An equation modeling the cost of selling a certain number of bracelets might be $C(n)$.

➡ **EXAMPLE 6.10**

Adam launches a rocket at 50 meters per second from a table that is one meter tall. The height, h, in meters of the rocket after t seconds can be determined by the function $h(t) = -4.9t^2 + 50t + 1$.

Determine the height of the rocket after

1. 5 seconds,
2. 10 seconds,
3. b seconds.

Answer Explanations

1. **128.5 meters (exact) or 129 meters (3 significant figures)**
   Since $t = 5$, then $h(5) = -4.9(5)^2 + 50(5) + 1$.
   Thus, $h(5) = 128.5$

2. **11 meters**
   Since $t = 10$, then $h(10) = -4.9(10)^2 + 50(10) + 1$.
   Thus, $h(10) = 11$

3. **$-4.9b^2 + 50b + 1$ meter**
   Since $t = b$, replace each t with the variable b. Therefore, $h(b) = -4.9b^2 + 50b + 1$, which cannot be simplified.

Two important parts of a function are its domain and range.

> **Domain: the set of all possible *x*-values for a particular function *f(x)***
> **Range: the set of all possible *y*-values for a particular function *f(x)***

Domain and range can be expressed using inequalities or interval notation.

In interval notation, the inequality $-2 < x \le 5$ would be written as $(-2, 5]$. The parenthesis represents "not equal to," while the bracket represents "equal to."

In interval notation, the inequality $y \le 5$ would be written as $(-\infty, 5]$. Numbers less than or equal to 5 would include all numbers between negative infinity and 5. Since negative infinity is a concept and not an actual number, a parenthesis is used to show a number cannot equal infinity.

For example, consider the function $f(x) = x^2 - 6x + 1$. Since domain is all the x-values of the function, consider values of x that could be inputted into the function. All real numbers can be squared, multiplied, and added or subtracted, so x can be any real number. The domain is all real numbers.

The domain can be written as $x \in \mathbb{R}$ or $(-\infty, \infty)$.

The quickest way to determine range is graphically. Using the GDC, graph the function. Since the y-axis runs from the bottom of the coordinate plane to the top, the range is determined by finding the bottom of the graph and then the top.

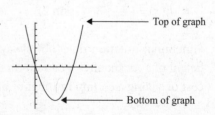

Top of graph

Bottom of graph

The bottom of this particular graph is the minimum. Use the `CALC` graph menu to find the minimum.

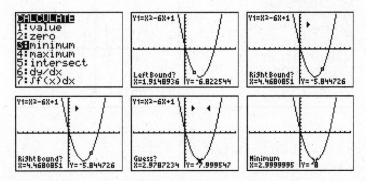

Since the minimum is (3, –8), the range begins at –8 and continues towards infinity.

Therefore, the range is $y \geq -8$ or $[-8, \infty)$.

➡ EXAMPLE 6.11

For each of the following, determine the domain and range.

1. $g(x) = \dfrac{1}{3}x + 2$

2. $h(x) = 7 - x^4$

3. $f(x) = \dfrac{2}{x-1}$

Answer Explanations

1. **Domain: all real numbers OR $(-\infty, \infty)$**

   Range: all real numbers OR $(-\infty, \infty)$

   All real numbers can be multiplied by one-third and added to 2. Thus, x is all real numbers.

   The range is also all real numbers. The graph is a linear function, and all lines travel from negative infinity to infinity.

2. **Domain: all real numbers OR $(-\infty, \infty)$**

   Range: $y \leq 7$ OR $(-\infty, 7]$

   All real numbers can be raised to the fourth power and subtracted from 7. Thus, x is all real numbers.

   Graphically, the range begins at negative infinity and stops at the maximum.

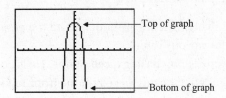

Top of graph

Bottom of graph

Using the GDC the maximum occurs at (0, 7).

 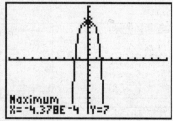

3. **Domain:** $x < 1$ or $x > 1$ **OR** $(-\infty, 1) \cup (1, \infty)$

   Range: $y < 0$ or $y > 0$ **OR** $(-\infty, 0) \cup (0, \infty)$

   The denominator of a fraction can never be zero. Therefore, x cannot be all real numbers. Since the denominator is $x - 1$, $x - 1 \neq 0$. Therefore, $x \neq 1$. This means x can be any real number less than one or greater than one. The domain is in two parts (numbers less than one or numbers greater than one). You need two inequalities or two interval notations.

   Graphically, the range is split when traveling from the bottom of the graph to the top.

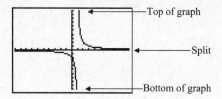

   The graph is split by the line $y = 0$, which is a horizontal asymptote. For more information about asymptotes, see section 6.5.

   Therefore, the range is all y-values less than zero or all y-values greater than zero.

6.2 LINEAR MODELS

Oftentimes real-world scenarios can be modeled using linear functions. Problems involving a constant rate of change would be linear since the rate of change creates a gradient.

All linear functions have a domain and a range of $(-\infty, \infty)$, which also means all real numbers.

Recall from section 5.1 that linear functions can be written in the form $y = mx + c$ and the gradient can be found using the formula $m = \dfrac{y_2 - y_1}{x_2 - x_1}$.

When given a problem involving a linear model, write the equation of the line, and then use the equation to solve for various values.

For example, suppose Kody is purchasing a cell phone. He is considering two companies, TalkForLess and Data Unlimited. If he buys a cell phone from TalkForLess, he must pay a $60 activation fee and $35 each month. Data Unlimited only has a $15 activation fee, but the fee is $60 each month.

In each situation, the activation fee is the *y*-intercept or the constant. The monthly fee is the gradient.

TalkForLess: $T(x) = 35x + 60$ Data Unlimited: $D(x) = 60x + 15$

Now these linear functions can be used to solve various problems, such as the cost after *m* months or when the plans equal in cost.

➡️ **EXAMPLE 6.20**

Kasey needs to rent a large room to host a business conference. He has two options to choose from.

Option 1: A local hotel charges $200 to rent the conference room, along with $7.50 for each person who attends the conference.

Option B: The community center charges $30 to rent the building, along with $10 for each person who attends the conference.

QUICK TIP

Gradient: constant rate of change (fee per month, cost per person, etc.)

y-intercept: flat fee (i.e. one-time amount)

1. Write an equation to represent the total cost for renting the local hotel. Let $H(x)$ represent the cost of *x* people attending.
2. Write an equation to represent the total cost for renting the community center. Let $C(x)$ represent the cost of *x* people attending.
3. If 50 people attend the conference, determine the cost of renting the
   (a) local hotel,
   (b) community center.
4. Determine which option is better for Kasey depending on the number of people who attend.

Answer Explanations

1. $H(x) = 7.50x + 200$
   The hotel charges a flat fee of $200, which is the constant. The per person charge of $7.50 is the gradient.

2. $C(x) = 10x + 30$
   The community center charges a flat fee of $30, which is the constant. The per person charge of $10 is the gradient.

3. (a) **$575**
   (b) **$530**

   Let $x = 50$ and simplify each equation.

   $H(50) = 7.50(50) + 200$ $C(50) = 10(50) + 30$
   $H(50) = 575$ $C(50) = 530$

4. **Option 1 is best if more than 68 people attend. Option 2 is best if 68 people or less attend.**
   To determine which option is cheapest depending upon the number of attendees, you must find the intersection of the two lines.
   Graph both lines. Set the window so that the intersection can be seen. Next, find the intersection.

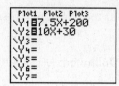  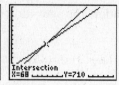

The community center is the cheaper option up until the intersection. If 68 people attend the conference, the rental costs would be the same. Therefore, one additional person would make the hotel cheaper and the community center more expensive.

Looking at the table, the community center (Y_2) is cheapest until 68 people.

X	Y₁	Y₂
63	672.5	660
64	680	670
65	687.5	680
66	695	690
67	702.5	700
68	710	710
69	717.5	720

Press + for △Tbl

➡ **EXAMPLE 6.21**

QUICK TIP

When calculating the **gradient** given two points, be sure to subtract in the same order for both the numerator and denominator.

A cup of tea is slowly cooled so that after 5 minutes of cooling, the tea's temperature, T, is 75°C, and after 15 minutes of cooling, the temperature is 25°C.

The temperature after m minutes of cooling can be expressed by the linear function $T = am + c$.

1. Calculate the gradient of the linear function, a.
2. Determine the value of the constant, c.
3. Write down the temperature of the tea after 8 minutes of cooling.

Answer Explanations

1. $a = -5$

   The given information generates two points: (5, 75) and (15, 25). The gradient formula gives $m = \dfrac{25 - 75}{15 - 5} = \dfrac{-50}{10} = -5$.

2. $c = 100$

   Using one of the points, substitute into $y = mx + c$.

$$75 = -5(5) + c$$
$$75 = -25 + c$$
$$100 = c$$

3. **60°C**

   Substituting 8 in for x yields

$$y = -5(8) + 100$$
$$y = 60$$

6.3 QUADRATIC MODELS

Quadratic functions are in the form $f(x) = ax^2 + bx + c$. However, the value of a cannot be zero, otherwise the function would be linear. The graphs of quadratics are often called parabolas. There are many key features to a parabola.

Consider the quadratic $f(x) = -3x^2 + 18x - 20$. The graph of the quadratic, or the parabola, would look like the following.

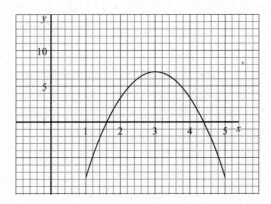

All quadratics are symmetric about the line $x = -\dfrac{b}{2a}$. This is called the axis of symmetry, which passes directly through the vertex. Recall that the vertex of a quadratic is either the maximum or minimum. The equation for the axis of symmetry can be found in the formula packet.

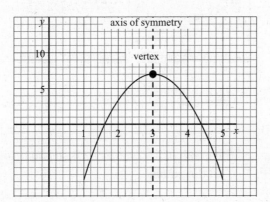

The axis of symmetry divides the graph into two perfectly equal halves. For each point on the left there is another point at the same height on the right.

The domain for all quadratics is $(-\infty, \infty)$, or all real numbers. The range, however, is based on the vertex. For quadratics that open upward, the range begins at the y-value of the vertex and continues to infinity. For quadratics that open downward, the range begins at negative infinity and stops at the y-value of the vertex.

For the quadratic $f(x) = -3x^2 + 18x - 20$, the axis of symmetry is $x = -\dfrac{18}{2(-3)}$, which means $x = 3$.

To find the coordinates of the vertex, substitute in the axis of symmetry since it is the x-coordinate of the vertex. $f(3) = -3(3)^2 + 18(3) - 10$, which means $f(3) = 7$.

The domain would be $(-\infty, \infty)$ and the range would be $(-\infty, 7]$ or $y \le 7$.

The vertex can also be found using the GDC. After graphing the function, decide if the vertex is a maximum (graphs opens downward) or minimum (graph opens upward). Then, find the appropriate point using the [CALC] menu. For the example, the vertex is the maximum of the graph.

 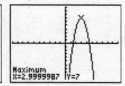

Thus, the vertex is (3, 7).

The two remaining key features of a quadratic are the axial intercepts. The y-intercept is when $x = 0$, and the x-intercepts are when $y = 0$. Using the table, the y-intercept can be quickly located at $x = 0$.

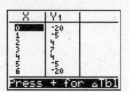

Therefore, the y-intercept is (0, –20).

The x-intercepts are also called the roots or zeros of the function. Section 1.6 covered how to find these values. The easiest way is using the [PLYSMLT] application.

There are two x-intercepts: (4.53, 0) and (1.47, 0).

➡ **EXAMPLE 6.30**

The graph of $y = 2x^2 + bx - 3$ is shown below with points A, B, and C on the graph.

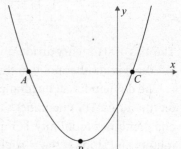

1. If the axis of symmetry for the quadratic is $x = -\dfrac{5}{4}$, find the value of b.

2. Hence, find the coordinates of point
   (a) A,
   (b) B,
   (c) C.

Mathematical Models

3. Draw a point clearly identifying the *y*-intercept of the graph. Label the point *Y*.

4. Draw a second point on the graph that has the same *y*-coordinate as *Y*.

Answer Explanations

1. **$b = 5$**

   The equation for the axis of symmetry is $x = -\dfrac{b}{2a}$, so substitute in what you know: $x = -\dfrac{5}{4}$ and $a = 2$.

   $$-\frac{5}{4} = -\frac{b}{2(2)}$$
   $$-\frac{5}{4} = -\frac{b}{4}$$

   Therefore, *b* must equal 5.

2. (a) **(−3, 0)**

   (b) $\left(-\dfrac{5}{4}, -\dfrac{49}{8}\right)$ **OR (−1.25, −6.125)**

   (c) $\left(\dfrac{1}{2}, 0\right)$ **OR (0.5, 0)**

   Now that you know $b = 5$, use your GDC to find the coordinates. Point *A* and point *C* are the *x*-intercepts, which are found by using [PLYSMLT].

   Looking at the given graph, point *A* is on the negative side of the *x*-axis, so the coordinates for *A* are (−3, 0). The coordinates for *B* would be $\left(\dfrac{1}{2}, 0\right)$.

   Point *B* is the vertex, which is the minimum of the graph.

   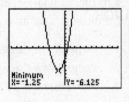

   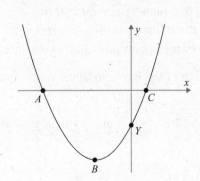

3. The *y*-intercept is where the graph crosses the *y*-axis.

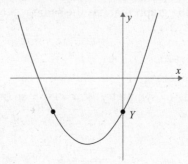

4. All quadratics are symmetric about the axis of symmetry, so the point with the same *y*-coordinate will be equidistant from the axis of symmetry as the *y*-intercept.

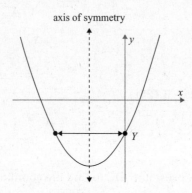

Quadratic functions are also used to model real-world situations, such as cost functions, motion of a projectile, or area functions.

Words such as maximum, minimum, greatest, or least indicate you should find the vertex of the quadratic.

Questions asking for ground level, zero profit, or a specific function value indicate you will find the *x*-intercepts.

➡ EXAMPLE 6.31 ───────────────────────────────

Alexa owns a custom framing business. A customer has ordered a frame for a large painting with dimensions 1.4 m by 1 m. The frame is to have an equal width, *x*, around the painting.

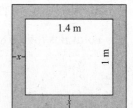

1. Write down the length of the framed painting in terms of *x*.
2. Write down the width of the framed painting in terms of *x*.
3. Show that the area of the framed painting is $A(x) = 4x^2 + 4.8x + 1.4$.

The customer requests the area of the frame to be no more than one-fourth the area of the painting.

4. Find the largest possible width of the frame. Express your answer in **centimeters**.

Answer Explanations

1. **$1.4 + 2x$**

   The framed length is the length of the painting, 1.4 m, plus the width of the frame. Since both sides of the length are increased by x, the new length is $1.4 + 2x$.

   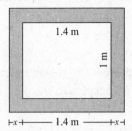

2. **$1 + 2x$**

   As with the length, the framed width is the width of the painting, 1 m, plus the width of the frame. Since both sides of the width are increased by x, the new width is $1 + 2x$.

3. **$(1.4 + 2x)(1 + 2x) = 1.4 + 2.8x + 2x + 4x^2$**
   Thus, $A(x) = 4x^2 + 4.8x + 1.4$.

   The area of the framed painting is found by multiplying the length by the width.

4. **6.90 cm**

   The area of the painting is $1.4(1) = 1.4 \text{ m}^2$. The frame is to be one-fourth this area or less, which gives an area of 0.35 m^2 for the frame.

   To use the quadratic equation generated in question 3, you need the total area of the painting and frame: $1.4 + 0.35 = 1.75 \text{ m}^2$.

   Therefore, $4x^2 + 4.8x + 1.4 = 1.75$.

   You need the value of x that yields an area of 1.75 m^2, so solve the equation for x.

   In order to use $\boxed{\text{PLYSMLT}}$, the equation must equal zero.

   $$4x^2 + 4.8x - 0.35 = 0$$

   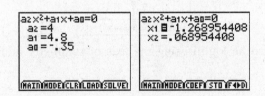

   There are two possible values for x, but the width of a frame cannot be negative. The only plausible answer is $x = 0.0690$ m.

   The question asks for the width to be expressed in centimeters, so move the decimal two places to the right.

kilo-	hecto-	deca-	Base Unit	deci-	centi-	milli-
k	h	da		d	c	m

   $0.0690 \text{ m} = 6.90 \text{ cm}$

EXAMPLE 6.33

QUICK TIP

Initial amount = **y-intercept**

Maximum or Minimum = **VERTEX**

A rocket is fired upward with an initial velocity of 25 meters per second. The height of the rocket above the ground after t seconds is represented by $h(t) = -4.9t^2 + 25t + 3$.

When the rocket is launched, it is on a platform x meters above the ground.

1. Find the height of the platform.
2. Determine the maximum height that the rocket reaches.
3. Calculate the total time that the rocket is in the air.

Answer Explanations

1. **3 meters**

   To determine the height of the platform, let $t = 0$ because the rocket is on the platform when time is zero.

   $$h(0) = -4.9(0)^2 + 25(0) + 3$$
   $$h(0) = 3$$

   This can also be seen on the table when the equation is typed in for Y_1.

   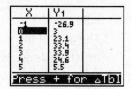

2. **34.9 meters**

   The maximum height occurs at the vertex of the quadratic. Utilizing the GDC will quickly locate the maximum.

   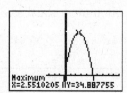

   The vertex can also be found using the axis of symmetry and substituting into the quadratic.

   $$x = -\frac{25}{2(-4.9)} = -\frac{25}{-9.8} = 2.55 \text{ seconds}$$
   $$h(2.55) = -4.9(2.55)^2 + 25(2.55) + 3$$
   $$h(2.55) = 34.9 \text{ meters}$$

3. **5.22 seconds**

   The total time that the rocket is in the air will be found once the rocket hits the ground. Graphically, this is the x-intercept or zero of the function.

   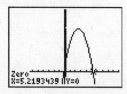

Alternatively, the time can be found using (PLYSMLT).

6.4 EXPONENTIAL MODELS

An exponential function constantly grows or decays by the same multiplier. Geometric sequences are a special type of exponential functions. The variable x is in the exponent for this type of function.

Exponential models that represent growth are in the form $f(x) = ka^x + c$. The graph looks like the one shown below. The coefficient k can be any real number other than zero. The base a must be a positive rational number. The constant c forms a horizontal asymptote at $y = c$.

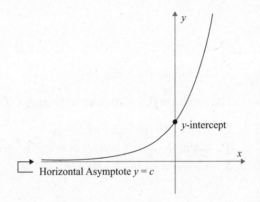

Models that represent decay are in the form $f(x) = ka^{-x} + c$. The variables are the same as with growth, but the negative in the exponent creates decay.

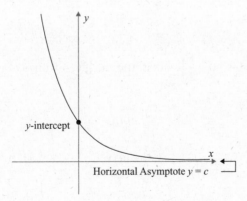

The domain of all exponential functions is $(-\infty, \infty)$, or all real numbers. The range either begins or ends at the horizontal asymptote, but will never equal the asymptote.

➡ **EXAMPLE 6.40**

The graph below shows the function $f(x) = b(a)^{-x} - 1$. The y-intercept of $f(x)$ is $\left(0, -\dfrac{1}{2}\right)$.

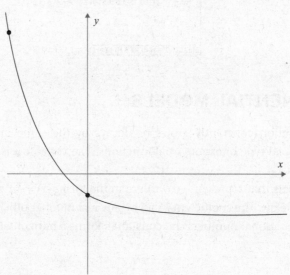

QUICK TIP

Growth: $f(x) = ab^x + c$

Decay: $f(x) = ab^{-x} + c$

Any value raised to the zero power is **ONE**.

The **horizontal asymptote** is **always** $y = c$.

1. Calculate the value of b.

The function $f(x)$ also passes through $(-2, 3.5)$.

2. Calculate the value of a.
3. Write down the equation of the horizontal asymptote.
4. Write down the range of $f(x)$.

Answer Explanations

1. $b = \dfrac{1}{2}$ **OR** $b = 0.5$

   Since the y-intercept is $\left(0, -\dfrac{1}{2}\right)$, substitute the given values of y and x into the function and solve for b.

   $$-\frac{1}{2} = b(a)^0 - 1$$

   $$\frac{1}{2} = b(1)$$

   $$\frac{1}{2} = b$$

   > Any number raised to the zero power equals one.

2. **$a = 3$**

   You know the value of b and are given another point. Substituting this given information into the function yields:

   $$3.5 = 0.5(a)^{-(-2)} - 1$$
   $$3.5 = 0.5(a)^2 - 1$$
   $$4.5 = 0.5a^2$$
   $$9 = a^2$$
   $$3 = a$$

3. **$y = -1$**

   The horizontal asymptote is always the constant of the exponential function. Since $f(x) = 0.5(3)^{-x} - 1$, the constant is -1. Thus, $y = -1$.

4. **$(-1, \infty)$ or $y > -1$**

   The graph begins at the horizontal asymptote of $y = -1$ and continues to infinity.

Real-world data often follows an exponential model. Population growth, appreciation or depreciation, and half-life all are exponential functions. Use of the GDC is expected when solving these types of functions.

➡️ **EXAMPLE 6.41** _____

During a biology experiment, bacteria are grown in a Petri dish. The number of bacteria, $B(t)$, after t hours can be represented by the function $B(t) = 50e^{0.56t}$.

   Some values of B, rounded to the nearest integer, are shown in the following table.

t	0	2	4	6	8
$B(t)$	m	153	n	1 440	4 412

1. To the nearest integer, write down the value of
   (a) m
   (b) n

2. Using the values in the table, draw a graph of B for $0 \le t \le 8$. Let 1 cm on the horizontal axis represent 1 minute and 1 cm on the vertical axis represent 500 bacteria.

3. Using the graph, estimate the number of bacteria in the Petri dish after seven hours.

> **QUICK TIP**
>
> **Always** use the scale provided.
>
> **Always** label the axes with context.
>
> **Solve** exponential equations using the graph or the graphing calculator.

Answer Explanations

1. (a) **50**

    (b) **470**

    Using the exponential function, substitute the given values in for t.

    $$B(0) = 50e^{0.56(0)} = 50$$
    $$B(4) = 50e^{0.56(4)} = 469.67$$

    469.67 will round to 470.

2.

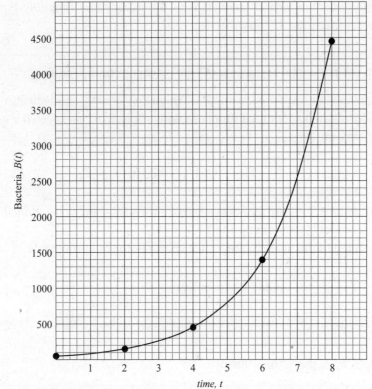

Be sure to label both axes and to scale as instructed. All points from the table should be plotted.

3. **2 600 bacteria (±100 bacteria)**

Draw a line from seven hours up to meet the graph. From that intersection, draw a horizontal line to the *y*-axis. This is the amount of bacteria present after seven hours. Due to graphing variations, answers between 2 500 and 2 700 would be accepted.

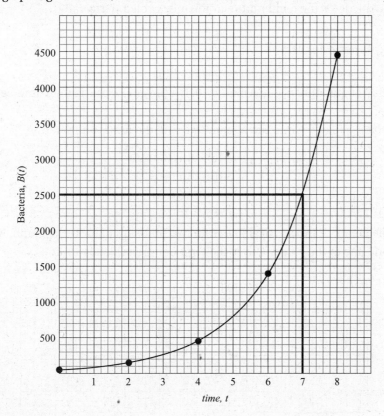

EXAMPLE 6.42

The following graph displays the temperature of a small refrigerator h hours after being turned on. The equation of the graph is $T(h) = 5 + 16(0.65)^t$, where $T(h)$ is the temperature of the refrigerator in degree Celsius h hours after being turned on.

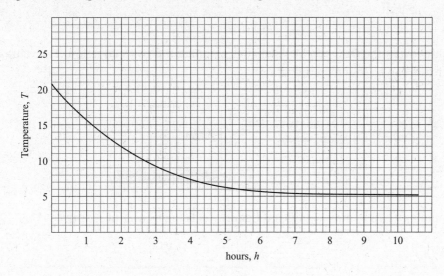

1. Write down the initial temperature of the refrigerator.
2. Find the temperature of the refrigerator after two hours.
3. Determine the number of hours taken to cool the refrigerator to 10°C.
4. Write the equation of the horizontal asymptote.
5. Hence, write down the temperature of the fully cooled refrigerator.

Answer Explanations

1. **21°C**

   The initial temperature occurs when $t = 0$. On the graph, this is the y-intercept, which is at (0, 21).

   Alternatively, plug $t = 0$ into the equation to find the initial temperature.

   $$T(0) = 5 + 16(0.65)^0$$
   $$T(0) = 5 + 16 = 21$$

2. **11.8°C or 12°C**

   Using the graph, draw a vertical line from two hours up to meet the graph. Next, draw a horizontal line to the y-axis. This is the temperature, which is 12°C.

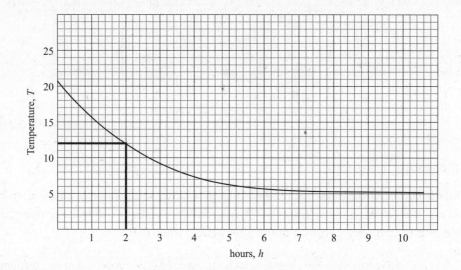

   Alternatively, the GDC can be used after graphing the given equation. The value of the function can be found on the table.

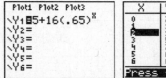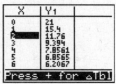

   Rounding 11.76 to three significant figures yields the answer of 11.8°C.

3. **2.7 hours (±0.2)**

   Using the graph, draw a horizontal line from 10°C over to meet the graph. Next, draw a vertical line to the x-axis. This is the number of hours, which could be between 2.5 and 2.9 hours.

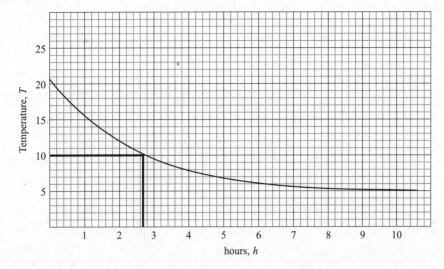

Alternatively, the GDC can be used after graphing the given equation along with the second equation of $y = 10$. The intersection of the two graphs gives the solution.

 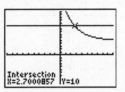

The x-coordinate is the number of hours, 2.70.

4. **$y = 5$**

   The horizontal asymptote is the value of the constant. In the equation $T(h) = 5 + 16(0.65)^t$, the constant is 5.

5. **5°C**

   The graph will follow the horizontal asymptote of $y = 5$. Therefore, when the refrigerator is fully cooled, the temperature is 5°C.

6.5 POLYNOMIAL AND RATIONAL MODELS

The two remaining function families that are part of the IB Math Studies curriculum are polynomial and rational functions.

A polynomial function is in the form $f(x) = ax^m + bx^n + \ldots$, when m and n are counting numbers.

Examples of common polynomial functions are

Function	General Form	Example	Graph
Line	$f(x) = ax + c$	$f(x) = \dfrac{x}{3} - 2$	
Quadratic	$g(x) = ax^2 + bx + c$	$g(x) = 3x^2 - x + 2$	
Cubic	$h(x) = ax^3 + bx^2 + cx + d$	$h(x) = x^3 - x + 1$	
Quartic	$k(x) = ax^4 + bx^3 + cx^2 + dx + e$	$k(x) = 16 - x^4 + 3x^2$	

For all polynomial functions, the domain is all real numbers, or $(-\infty, \infty)$. The range, however, depends on the type of polynomial.

Polynomials whose highest exponent is an even number (quadratic, quartic, etc.) have a range that starts or stops at the maximum or minimum. Refer back to section 6.3 for review.

Polynomials whose highest exponent is an odd number (line, cubic, etc.) have a range of all real numbers, or $(-\infty, \infty)$.

As with quadratics, when graphing you must accurately display the axial intercepts along with any local maxima or minima. The GDC is to be utilized when graphing and locating key features.

➡ EXAMPLE 6.50

Given $f(x) = x^3 - 2x^2 - 4x + 2$,

1. Write down the coordinates of the y-intercept for $f(x)$.
2. Write down the solution to $0 = x^3 - 2x^2 - 4x + 2$ if $x \in \mathbb{R}^-$.
3. Determine the value of x for which $f(x) = -1$ if $x \in \mathbb{N}$.

Answer Explanations

1. **(0, 2)**

   The y-intercept occurs when $x = 0$.

   $$f(0) = 0^3 - 2(0)^2 - 4(0) + 2 = 2.$$

   Alternatively, use the GDC to find the y-intercept.

    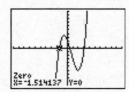

2. **−1.51**

   The question is asking for the x-intercept, root, or zero of the function, but only negative values. There are two options:

   (a) Use the GDC to find the negative zero.

   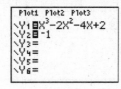

   (b) Use PLYSMLT to solve the equation. Since the function is a cubic, the order (or degree) is 3.

    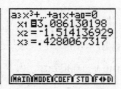 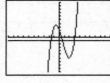 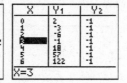

   Again, the question states $x \in \mathbb{R}^-$, which means x must be a negative real number.

3. **$x = 3$**

   Let $Y_2 = -1$ and graph with $f(x)$. The curves intersect three times, but our answer must be a natural number. The table displays x as a natural number, so the intersection can be found, or the table used to locate, when the functions are equal.

QUICK TIP

Solve **polynomial functions** using the PLYSMLT app. These are the zeros (x-intercepts).

Most problems can be solved by graphing on the GDC or using the PLYSMLT app.

Mathematical Models

In section 5.5, the volume of 3-dimensional figures was explored. Volume can be modeled using a cubic function.

A simple example would be a cube with side length x. The volume of the cube is represented by $V = x^3$.

➡ EXAMPLE 6.51

The rectangular box below has a width of x cm, a height that is three times the width, and a length that is three centimeters more than the height.

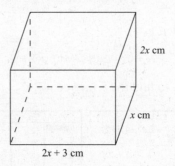

2x cm

x cm

2x + 3 cm

1. Show that the volume is $V = 4x^3 + 6x^2$ cm^3.
2. Calculate the volume of the box when the width is 4 cm.

The box can hold 4 600 cubic centimeters.

3. Determine the width of the box.
4. Write down the dimensions of the box.

Answer Explanations

1. $V = x(2x + 3)(2x)$

   Volume of a cuboid is $V = lwh$, so substitute each of the given dimensions.

2. **352 cm^3**

   Using the volume equation, substitute 4 in for x.

   $$V = 4(4)^3 + 6(4)^2 = 352$$

3. **10 cm**

   Since volume is 4 600 cm^3, let $V = 4\,600$ and solve for x.

   $$4\,600 = 4x^3 + 6x^2$$

   Option 1: Set the equation equal to zero and find the x-intercept.

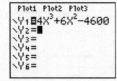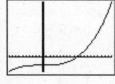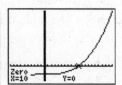

Option 2: Let $Y_2 = 4\,600$ and find the intersection.

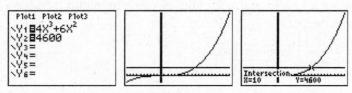

Option 3: Look at the table for when $Y_1 = 4\,600$.

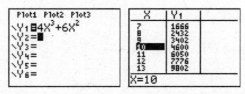

4.   **10 cm $\times$ 20 cm $\times$ 23 cm**

Since the width, x, is 10 cm, find the remaining measures using 10 cm.

$$\text{Height} = 2(10) = 20$$
$$\text{Length} = 2(10) + 3 = 23$$

Rational functions are created when two polynomial functions are divided.

The functions $f(x) = 3x^2 + \dfrac{2}{x}$, $g(x) = \dfrac{x}{x+1}$, and $h(x) = \dfrac{x-1}{x^2-9}$ are all examples of rational functions. Since dividing by zero is not possible, all rational functions have vertical and horizontal asymptotes. Asymptotes are boundary lines that the graph follows.

Vertical asymptotes block the x value, which would turn the denominator of the rational function into zero. Rational function can never cross or touch vertical asymptotes.

Horizontal asymptotes show the "end behavior" of the graph and x goes to infinity. Horizontal asymptotes can be crossed by the graph, but this does not happen frequently. Not all rational functions have horizontal asymptotes, but all rational functions have vertical asymptotes.

The graph of $f(x) = x^2 - \dfrac{2}{x}$ is shown below.

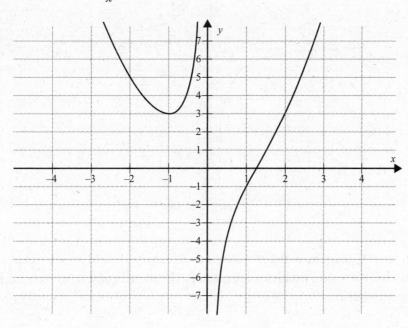

Notice the graph is split about the y-axis. This means the vertical asymptote is $x = 0$. This graph is not following a horizontal asymptote as x goes to infinity, which means there is no horizontal asymptote.

The graph of $g(x) = \dfrac{2}{x}$ is shown below.

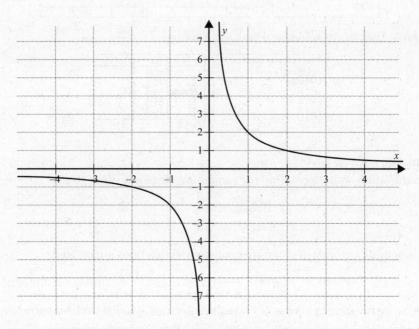

This graph has both a vertical asymptote of $x = 0$ and a horizontal asymptote of $y = 0$. The graph is split about the y-axis, which means the vertical asymptote is $x = 0$. The ends of the graph are also following the x-axis, which means $y = 0$ is the horizontal asymptote.

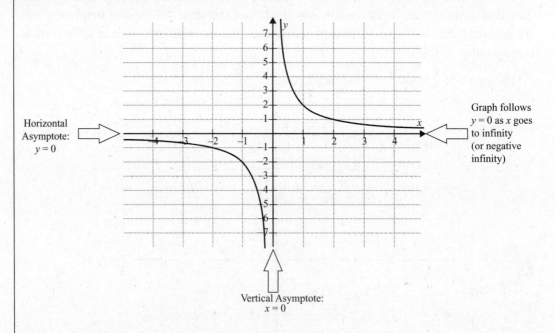

In IB Math Studies, all rational functions tested will have a vertical asymptote of $x = 0$, and the use of the GDC to solve problems is expected. The domain and range vary based on the rational function given.

➡ **EXAMPLE 6.52**

Consider the function $g(x) = \dfrac{3}{x^2} - 2x$.

1. Write down the x-intercept of $g(x)$, correct to three decimal places.
2. Determine the minimum value of $g(x)$, when $x < 0$.
3. Write down the equation of the vertical asymptote.

Diamond believes the graph has a y-intercept larger than 100.

4. Explain if Diamond is correct in her statement.

Answer Explanations

1. **(1.145, 0)**

   Graph the function using the GDC, and then find the zero, which is the x-intercept.

   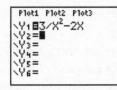 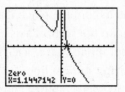

   Be sure to round the x-coordinate correctly to three decimal places.

2. **4.33**

   When $x < 0$, the graph takes on a quadratic-like shape. The minimum is easily located using the GDC. The question asks for the minimum value, which is the value of y.

   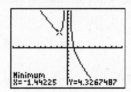

3. **$x = 0$**

   The graph is split at the y-axis, which means the vertical asymptote is $x = 0$.

4. **Diamond is not correct**. There is no y-intercept since the vertical asymptote at $x = 0$ means the graph will not cross the y-axis.

   The y-intercept occurs when $x = 0$, but x can never equal zero. Thus, the graph will never cross the y-axis.

6.6 ACCURATE GRAPH DRAWING

Being able to draw an accurate graph is an important mathematical skill. In each section of this chapter, a function family has been explored with its appropriate graph including key features such as axial intercepts and asymptotes. You must be able to read, interpret, and use the graph to make predictions.

The GDC is a very useful tool; however, you must be able to take the graph from the calculator and draw it on paper.

On the IB exam, you will be asked to either sketch or draw a graph. There is a difference between these.

Draw means you must have a labeled, accurate graph. Lines are to be drawn using a ruler, and the scale must be appropriate. Given points should be correctly plotted and connected in a straight line or smooth curve.

Sketch means you have a labeled graph, but the sketch gives a general idea of the shape. Key features such as asymptotes, maximum or minimum values, and/or axial intercepts should be displayed.

Both your drawing and your sketch must be accurate, labeled, and scaled. However, the command "draw" requires much more precision. When asked to "sketch," you must show the key features, but only the general shape of the graph.

Regardless of the command term, be sure to always:

- Label the axes with the appropriate variable. If graphing $y = f(x)$, then label with x and y. If the problem is in context, place the independent variable on the horizontal axis and the dependent on the vertical axis.
- Be sure to scale appropriately. If a scale is given, you must graph accordingly.
- Use a ruler for drawing straight lines for the axes and the function.
- When a problem involves a table of coordinate pairs, plot each point correctly.
- When given a restricted domain, do not graph beyond the stated values of x.

➡ **EXAMPLE 6.60**

QUICK TIP

When graphing, marks are earned for:

- Correct scale
- Labeled axes
- Axial intercepts
- Given points plotted
- Asymptotes clearly shown (if applicable)
- Local maxima or minima

For each of the following, sketch an appropriate graph given the domain.

1. $f(x) = \dfrac{1}{x^2} + 4x$, when $-5 \le x \le 5$, $x \ne 0$

2. $g(x) = \dfrac{x^3}{4} - 2^x$, when $-4 \le x \le 6$

3. $h(x) = \dfrac{1}{2}x^4 - 3x^2 + \dfrac{2}{x}$, when $-3 \le x \le 3$, $x \ne 0$

Answer Explanations

1.

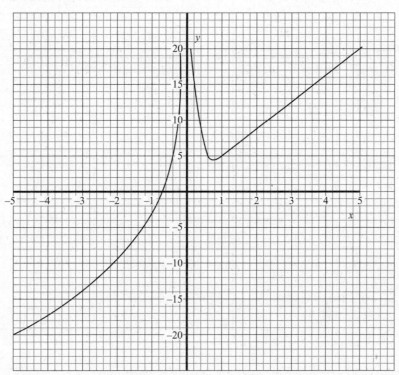

Using the GDC as a guide, locate the following key features:

- The *x*-intercept (−0.63, 0)

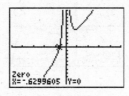

- Vertical asymptote at *x* = 0
- A local minimum at (0.79, 4.8)

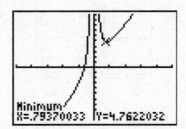

- A few points off the table, including the beginning and ending of the domain

X	Y₁		X	Y₁
-5	-19.96		-1	-3
-4	-15.94		0	ERROR
-3	-11.89		1	5
-2	-7.75		2	8.25
-1	-3		3	12.111
0	ERROR		4	16.063
1	5		5	20.04
X= -5			X=5	

- Then, use the GDC to sketch the general shape of the curve on fully labeled and scaled axes.

2.

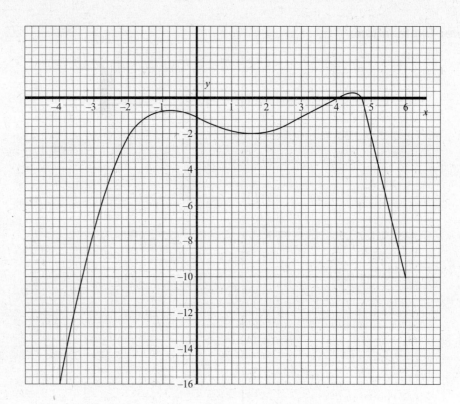

Using the GDC as a guide, locate the following key features:

- A local maximum at (–0.74, –0.70)

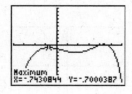

- The *y*-intercept (0, –1)
- A local minimum at (1.78, –2.02)

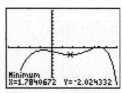

- The *x*-intercept (4.67, 0)

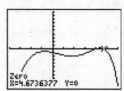

- Another local maximum at (4.37, 0.19)

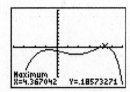

- A few points off the table, including the beginning and ending of the domain

Then, use the GDC to sketch the general shape of the curve on fully labeled and scaled axes.

3.

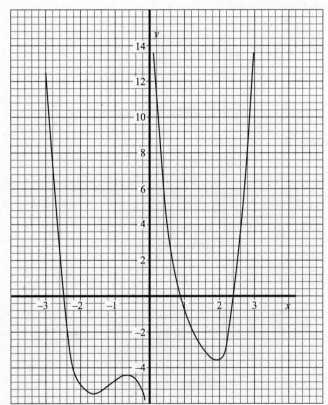

Using the GDC as a guide, locate the following key features:

- Vertical asymptote at $x = 0$
- The x-intercept (−2.50, 0)

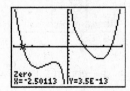

- A local minimum at (–0.74, –0.70)

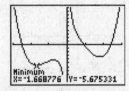

- A local maximum at (–0.74, –4.20)

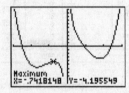

- Second *x*-intercept (0, –1)

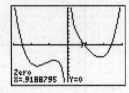

- Local minimum at (1.78, –2.02)

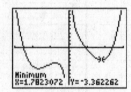

- Final *x*-intercept (2.39, 0)

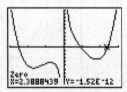

- A few points off the table, including the beginning and ending of the domain

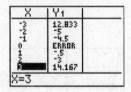

- Then, use the GDC to sketch the general shape of the curve on fully labeled and scaled axes.

6.7 SOLVING EQUATIONS

There are times when you must solve an equation that is considered "unfamiliar," meaning the equation is made up of the different function models. In the IB Math Studies course, solutions cannot be found algebraically, and you must rely on the GDC.

Any equation can be solved using the GDC by graphing. You can either:

- Graph each side of the equation as its own function and find the intersection(s), or
- set the equation equal to zero and find the zeros.

The most direct route is to graph each side of the equation and find the intersection.

➡ EXAMPLE 6.70

Solve for x in each of the following functions.

1. $2x^3 - 5x^2 = 4x - 3$

2. $\dfrac{2}{x^2} = 3^{-x} + 4$

3. $6 - x^2 = e^{0.5x}$

> **QUICK TIP**
>
> Any equation can be solved by graphing each side and finding the **intersection**.
>
> The **solution** is always the x-value.

Answer Explanations

1. $x = -1, 0.5, 3$

    Enter each side of the equal sign as a function, graph and find each intersection. You are asked to solve for x, so the x-values are the solutions.

    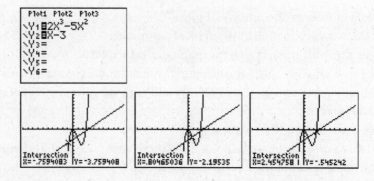

2. $x = -0.582, 0.668$

    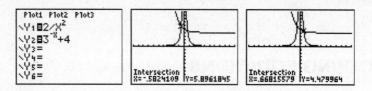

3. $x = -1.67, 1.41$

    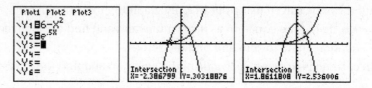

Paper 1 Questions

1. Stephanie is going to study abroad. She has two options for an apartment to rent.

   Apartment 1: monthly rent of \$325 with a \$75 deposit
   Apartment 2: monthly rent of \$250 with a \$600 deposit

   (a) Write an equation in the form $T = am + c$, where T is the total cost for m months rented for
       (i) Apartment 1,
       (ii) Apartment 2.

   (b) Calculate how many months must be rented for the total costs to be the same.

   Stephanie will need to rent the apartment for a year.

   (c) Determine which apartment will be cheapest overall.

2. The graph below shows the function $f(x) = c + qx + 2x^2$, where $c, q \in \mathbb{Z}$. The graph passes through the points $A(-2, -2)$ and $B(0, 6)$.

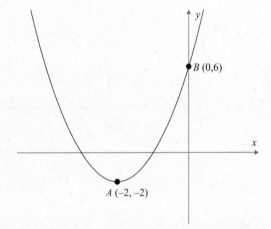

   (a) Write down the value of c.
   (b) Determine the value of q.
   (c) Find the x-intercepts of the graph.
   (d) Write down the range of $f(x)$.

3. CalCo is a company that manufactures calculators for educators. The company has determined the monthly revenue, R, based on n calculators sold and follows the quadratic function $R(n) = 150n - 1.5n^2$.

   (a) Calculate the number of calculators that must be sold to maximize revenue.
   (b) Determine the maximum amount of revenue.

   The monthly cost, C, for CalCo to manufacture n calculators is represented by $C(n) = 792 + 9n$.

(c) Find the maximum number of calculators CalCo must produce in order to make a profit.

4. Marie takes one 500 mg dose of penicillin. The amount of penicillin remaining in her system after t hours follows the exponential model $A = 500(2)^{-2t}$.

The table displays some values of A rounded to the nearest hundredth.

t	0	2	4	5	6
A	500	c	1.95	d	0.12

(a) Rounded to the nearest hundredth, find the value of
   (i) c,
   (ii) d.

(b) Calculate the number of hours needed for there to be less than 0.002 mg of penicillin remaining in Marie's system.

5. The graph displays the function $g(x) = 3 + b(r)^x$.

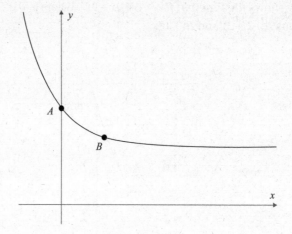

The graph passes through point $A(0, 5)$.

(a) Write down the value of b.

The graph also passes through point $B(2, 3.5)$.

(b) Calculate the value of r.

(c) Write down the equation of the horizontal asymptote.

6. Two functions $f(x)$ and $g(x)$ are given by

$$f(x) = \frac{1}{3}e^{1.1x} + 2 \qquad g(x) = \frac{x^2}{2} + \frac{1}{x^3}, \; x \neq 0$$

(a) Sketch the graphs of $f(x)$ and $g(x)$ on the same axes with $-5 \leq x \leq 3$ and $-1 \leq y \leq 14$.

(b) Write down the number of solutions to the equation $\frac{1}{3}e^{1.1x} + 2 = \frac{x^2}{2} + \frac{1}{x^3}$.

(c) Write down one solution to the given equation in part b.

SOLUTIONS

1. (a) (i) $T = 325m + 75$

    (ii) $T = 250m + 600$

    The deposit is a one-time payment and is the y-intercept. The rent is the slope, as Stephanie will pay that amount each month.

    (b) **7 months**

    You can either graph both lines and find the intersection, or find where the function values are the same on the table.

    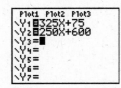 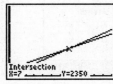 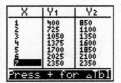

    (c) **Apartment 2**

    Apartment 1 will cost $T = 325(12) + 75 = \$3\,975$.

    Apartment 2 will cost $T = 250(12) + 600 = \$3\,600$.

    Since $3\,600 < 3\,975$, apartment 2 will be cheaper for one year.

2. (a) $c = 6$

    Since the y-intercept of the graph is $(0, 6)$, you know the constant is 6. You could also simply plug $x = 0$ and $y = 6$ into the function.

    $$6 = c + q(0) + 2(0)^2$$
    $$6 = c$$

    (b) $q = 8$

    The axis of symmetry passes through the vertex, so using $x = -\dfrac{b}{2a}$, you have $-2 = -\dfrac{q}{2(2)}$. Thus, $q = 8$.

    You could have also plugged the second point $(-2, -2)$ into the function.

    $$-2 = 6 + q(-2) + 2(-2)^2$$
    $$-2 = 14 - 2q$$
    $$-16 = -2q$$
    $$8 = q$$

    (c) $x = -3, -1$ or $(-3, 0)$ or $(-1, 0)$

    Now that all missing parts of the equation have been found, the actual quadratic is $f(x) = 6 + 8x + 2x^2$.

You can find the *x*-intercepts by graphing and finding each zero.

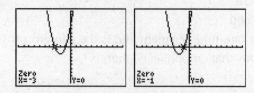

You could also look at the table for *x* when *y* is zero,

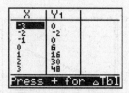

or solve using PLYSMLT.

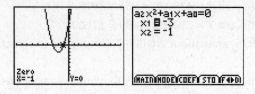

(d) **[–2, ∞) or $y \geq -2$**

The range begins at the vertex, when $y = -2$, and continues to infinity.

3. (a) **50 calculators**

The maximum revenue occurs at the vertex.

$$x = -\frac{b}{2a} \text{ so } n = \frac{-150}{2(-1.5)} = 50$$

(b) **3 750**

Plugging in $n = 50$ gives:

$$R(50) = 150(50) - 1.5(50)^2$$
$$R(50) = 3\,750$$

(c) **CalCo must produce at most 87 calculators.**

If Revenue = Cost, then the company has made zero profit. Find where $R(n) = C(n)$ to determine the breakeven point.

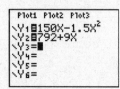 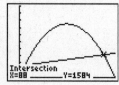

Mathematical Models

The maximum number of calculators cannot be 88, or else the company will make zero profit. Therefore, they should produce 87.

This is also illustrated by the table. When $x = 87$, the revenue function is larger. When $x = 88$, the two functions are equal. When $x = 89$, the cost function is larger.

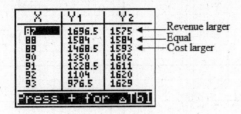

—Revenue larger
—Equal
—Cost larger

4. (a) (i) **31.25 mg**

(ii) **0.49 mg**

To find c and d, plug the given time into the equation.

$$A = 500(2)^{-2(2)} = 31.25$$
$$A = 500(2)^{-2(5)} = 0.49$$

(b) **9 hours**

Using the table, let $Y_1 = 500(2)^{-2x}$ and look for the first value smaller than 0.002.

X	Y₁
4	1.9531
5	.48828
6	.12207
7	.03052
8	.00763
9	.00191
10	4.8E-4

X=9

5. (a) **$b = 2$**

Plug the given point into $g(x)$.

$$5 = 3 + b(r)^0$$
$$5 = 3 + b$$
$$2 = b$$

(b) **$r = 0.5$**

Using the value of b, plug in the point (2, 3.5) and solve for r.

$$3.5 = 3 + 2(r)^2$$
$$0.5 = 2r^2$$
$$0.25 = r^2$$
$$0.5 = r$$

(c) **$y = 3$**

The horizontal asymptote is always $y = $ constant. Since the equation has a constant of 3, the horizontal asymptote is $y = 3$.

6. (a)

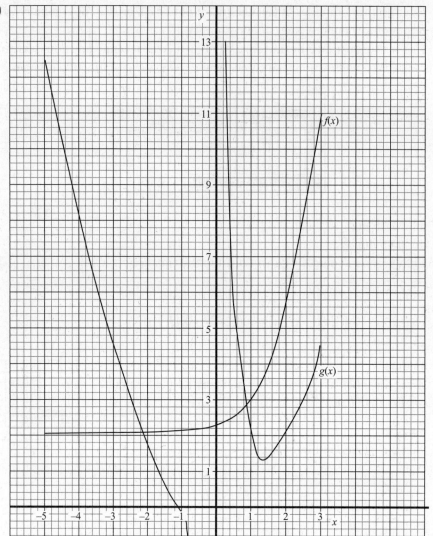

Using the GDC, locate the following key features for $f(x)$:

- x-intercept $(-1.14, 0)$
- minimum $(1.25, 1.29)$
- vertical asymptote $x = 0$
- endpoints $(-5, 12.5)$ and $(3, 4.54)$

Locate the following key features for $g(x)$:

- horizontal asymptote $y = 2$
- y-intercept $(0, 2.33)$
- endpoint $(3, 11)$

(b) **two solutions**

The two graphs intersect each other twice. Therefore, there are two solutions to the equation.

(c) $x = -2.07$ or $x = 0.739$

Using the GDC, find one point of intersection. There are two possible answers.

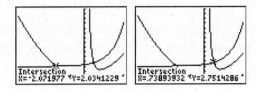

Paper 2 Questions

1. A small jewelry store sells designer watches for \$110 each. The owner believes that for each decrease of \$5 in the price, the store will sell more watches.

   (a) Write an expression for the selling price of the designer watch after x price decreases.

   The store normally sells 20 watches each week. For every \$5 decrease in price, the store expects to sell 5 more watches.

   (b) Write an expression for the number of designer watches sold each week after x price decreases.

   The total revenue generated from selling the designer watches is determined by multiplying the number of watches sold by the selling price.

   (c) Show that the total revenue of the store is $R = 2\,200 + 450x - 25x^2$.

   (d) Determine the maximum number of watches that will be sold each week.

   The jewelry store has a contract with a third-party business to buy the designer watch at a lower fixed price. The jewelry store purchases 15 watches for a cost of \$1\,345. The store must pay a flat fee of \$220 in order to keep the contract.

   (e) Show that the total cost for buying x watches is represented by $C = 75x + 220$.

   (f) Calculate the store's cost for buying 10 watches.

   To stay in business, the jewelry store must make at least \$1\,500 in profit from selling the designer watches.

   (g) Determine the maximum number of designer watches that must be sold in order to stay in business.

2. Shelby sells hot chocolate at a concession stand during a lacrosse game. The temperature of the hot chocolate after being poured into an insulated cup follows the exponential model $T(m) = 18 + 60(1.95)^{-0.3m}$, where $T(m)$ is the temperature in °C of the hot chocolate m minutes after being poured.

   (a) Write down the temperature of the hot chocolate when it is first poured into the cup.

   The table below displays several temperatures of the hot chocolate rounded to the nearest degree Celsius.

M	1	5	10	15	20
$T(m)$	67	a	26	b	19

(b) Calculate, to the nearest degree Celsius, the value of
   (i) *a*,
   (ii) *b*.

(c) Draw the graph of $T(m)$ using the values from the table, along with the initial temperature found in part *a*. Let 2 cm represent 5 minutes on the horizontal axis and 1 cm represent 10 degree Celsius on the vertical axis.

(d) Using the graph or otherwise, find
   (i) the temperature, to the nearest degree Celsius, of the hot chocolate 7 minutes after being poured,
   (ii) how long it will take for the hot chocolate to reach 45°C.

(e) Determine the temperature that the hot chocolate will reach long after being poured.

(f) Write down the equation of the horizontal asymptote.

3. A function $g(x)$ is defined by $g(x) = \dfrac{3}{x} - x^2$, $x \neq 0$.

   The table of values, rounded to the nearest hundreth, of $g(x)$ for $-3 \leq x \leq 4$ is shown below.

x	-3	-1	1	3	4
$g(x)$	-10	a	2	b	-15.25

(a) Calculate the values of *a* and *b*.

(b) Sketch the graph of $g(x)$ for $-3 \leq x \leq 4$ and $-25 \leq y \leq 25$, letting 2 cm represent 1 unit on the horizontal axis and 1 cm represent 5 units on the vertical axis.

(c) Using your graph or otherwise, find the approximate value of x, when $g(x) = 4$.

A second function $h(x)$ is defined by $h(x) = 5^{-0.5x} + 4x - 4$ over the same domain.

(d) Draw the graph of $h(x)$ on the same coordinate grid as $g(x)$.

(e) Write down the number of solutions to the equation $5^{-0.5x} + 4x - 4 = \dfrac{3}{x} - x^2$.

(f) Find **one** approximate solution to $5^{-0.5x} + 4x - 4 = \dfrac{3}{x} - x^2$, rounded to two significant figures.

SOLUTIONS

1. (a) **$110 - 5x$**

   The selling price starts at \$110, but then you must subtract $5x$ for the unknown number of \$5 decreases.

   (b) **$20 + 5x$**

   The store normally sells 20 watches, but with each x pay decrease, it will sell 5 more. Thus, add $5x$ to represent the extra number of watches sold to 20.

   (c) **$R = (110 - 5x)(20 + 5x)$**

   Since revenue is determined by the selling price times the number of watches sold, multiply the expression for selling price by the expression for watches sold.

(d) **9 watches**

The maximum revenue occurs at the vertex. To find the number of watches, use the axis of symmetry equation.

$$x = -\frac{450}{2(-50)} = 9$$

$$m = \frac{1345 - 220}{15 - 0} = 75$$

(e) $c = 220$

$C = 75x + 220$

Since the cost is fixed, the equation is a linear function. The information gives two points, (15, 1345) and (0, 220), which is also the y-intercept. The only missing piece is the gradient, which can be determined using the formula.

(f) $C = 75(10) + 220$

$C = 970$

To determine the cost for 10 watches, let $x = 10$ and simplify.

(g) $P = (2\,200 + 450x - 25x^2) - (75x + 220)$

$1\,500 = 2\,000 + 375 - 25x^2$

$0 = 500 + 375x - 25x^2$

$x = 16.2$

The maximum number of watches that can be sold is 16.

Profit is what is left after the cost is subtracted from the revenue.

First, you must find the profit equation by subtracting cost from revenue. Next, set that equation equal to the profit required.

$$P = (2\,200 + 450x - 25x^2) - (75x + 220)$$
$$1\,500 = 2\,000 + 375 - 25x^2$$

To solve a quadratic, the equation must equal zero.

$$0 = 500 + 375x - 25x^2$$

Now, solve the equation using either PLYSMLT or graph and find the x-intercept.

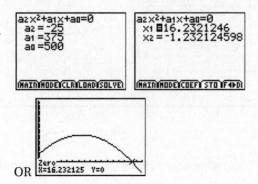

If the company sells more than 16.2 watches, the profit will be less than $1\,500. The profit must be at least $1\,500, so round down to 16 watches.

Mathematical Models

This can also be confirmed by looking at the table of values of the profit equation. When x is no more than 16, the profit is positive. After 16, the profit is negative.

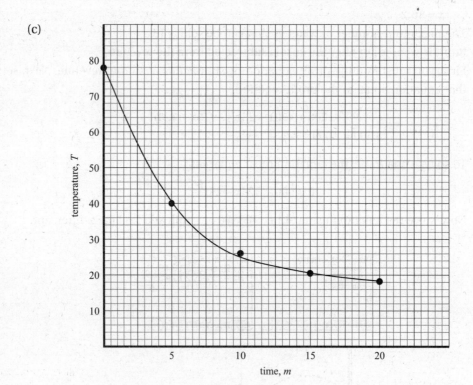

2. (a) **78°C**

When the cup is first poured, $m = 0$.

$$T(0) = 18 + 60(1.95)^{-0.3(0)}$$
$$T(0) = 18 + 60 = 78$$

(b) (i) **40**

(ii) **21**

Plug in the value of m into the equation to determine the temperature. Be sure to round to the nearest degree Celsius.

$$T(0) = 18 + 60(1.95)^{-0.3(5)} \qquad\qquad T(0) = 18 + 60(1.95)^{-0.3(15)}$$
$$T(0) = 40.034 \qquad\qquad\qquad\quad T(0) = 20.972$$

(c)

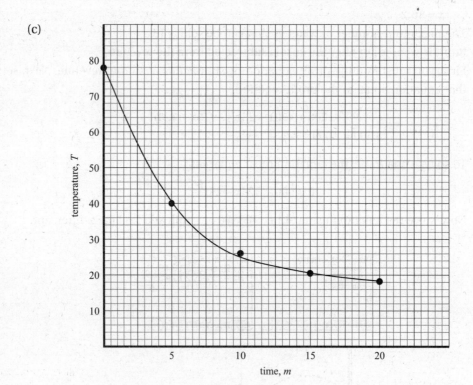

Be sure to clearly show each point from the table along with the y-intercept of $(0, 78)$.

(d) (i) **33°C**

Either plug in $m = 7$ and simplify or use the graph.

$$T(0) = 18 + 60(1.95)^{-0.3(7)}$$
$$T(0) = 32.76$$

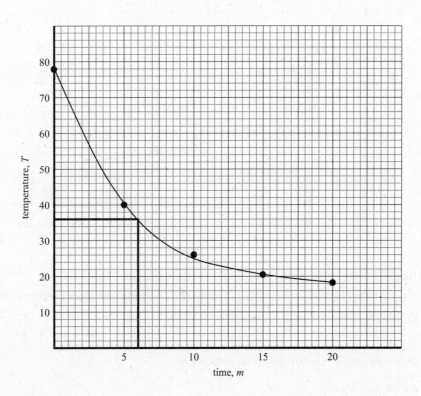

(ii) **4 minutes**

Either graph the function as Y_1 and 45 as Y_2 and find the intersection or use the graph.

 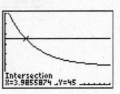

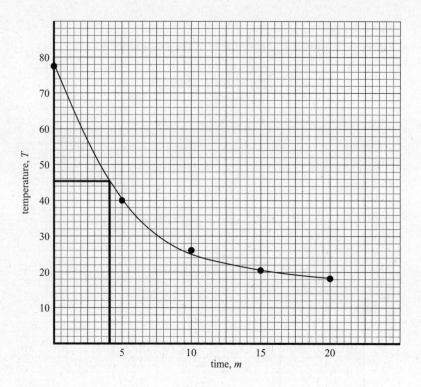

(e) **18°C**

The constant is the horizontal asymptote and the value the graph reaches after a long period of time.

(f) **T = 18**

The constant is the horizontal asymptote.

3. (a) **a = −4 and b = −8**

Enter the given function into Y_1 and use the table of values to locate a and b.

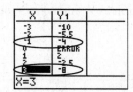

(b)

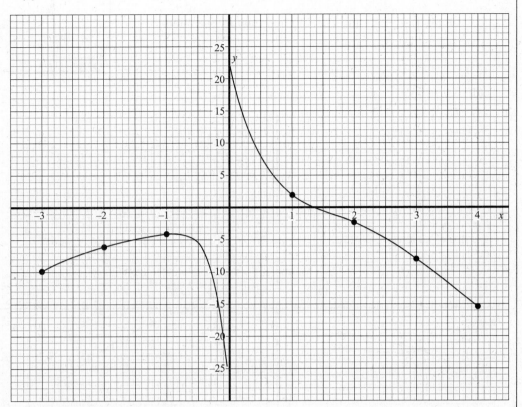

Ensure the axes have been labeled and scaled correctly.

All points from the provided table should be clearly marked.

The x-intercept is (1.44, 0)

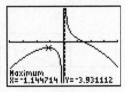

The maximum is (−1.14, 3.93), which is really close to the point (−1, −4).

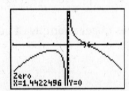

The vertical asymptote of $x = 0$ should be visible.

(c) **x = 0.674**

Graph $Y_1 = \dfrac{3}{x} - x^2$ and $Y_2 = 4$ to find their intersection.

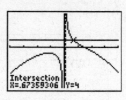

(d)

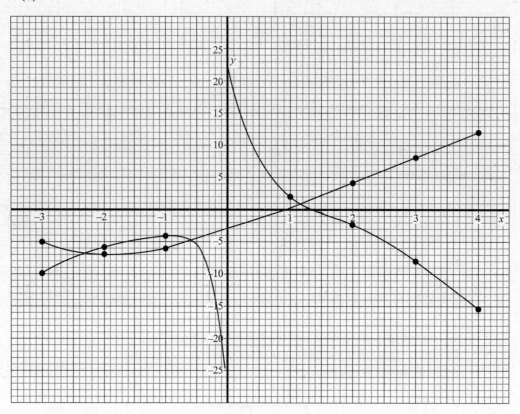

Again, use the table of values to get a sample of points. The y-intercept is $(0, -1)$, and the x-intercept is $(0.877, 0)$.

(e) **3**

The graphs intersect each other at three points, so there will be 3 solutions.

(f) **x = −2.3, x = −0.67, OR x = 1.2**

Use the GDC to find **one** intersection.

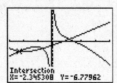 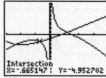 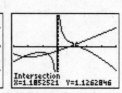

CHAPTER OBJECTIVES

Before you move on to the next chapter, you should be able to:

- ☐ Identify the domain and range of a function
- ☐ Write the equation of a line that models a given scenario
- ☐ Graph a linear function
- ☐ Use a linear model to determine a certain value
- ☐ Graph a quadratic function
- ☐ Identify the axis of symmetry, vertex, and axial intercept of a quadratic function
- ☐ Model a given scenario using a quadratic function
- ☐ Calculate the maximum or minimum value of a quadratic function
- ☐ Find the y-intercept of an exponential function
- ☐ Write down the horizontal asymptote of an exponential function
- ☐ Given an exponential model and a point it passes through, determine the exponential rate
- ☐ Calculate the value of an exponential model at various points
- ☐ Graph an exponential function
- ☐ Use the graph of an exponential function to determine when a certain function value occurs
- ☐ Graph a polynomial function
- ☐ Determine the value of a polynomial function at various points
- ☐ Solve a polynomial equation by graph or using the PLYSMLT app
- ☐ Graph a rational function
- ☐ Identify the vertical and horizontal asymptotes of a rational function
- ☐ Using the GDC, determine the x-intercept(s) of a rational function
- ☐ Using the GDC, identify key features of a function to accurately draw a graph (axial intercepts, maximum or minimum values, asymptotes, several points off GDC, general shape)
- ☐ Solve an unfamiliar function by graphing on the GDC

Mathematical Models

Introduction to Differential Calculus

- **FIRST DERIVATIVE:** instantaneous rate of change; $f(x) = ax^n \Rightarrow f'(x) = n(ax^{n-1})$; gradient of a curve at any point on $f(x)$

- **TANGENT LINE:** line that touches a curve only at one point; $y = mx + c$, where m is the value of the first derivative at x

- **INCREASING AND DECREASING:** $f'(x) > 0$, when $f(x)$ is increasing; $f'(x) < 0$, when $f(x)$ is decreasing

- **STATIONARY POINTS:** occur when $f'(x) = 0$; either a local maximum or minimum

- **OPTIMIZATION:** using the first derivative to locate a maximum or minimum in a real-world context

7.1 CONCEPT OF DERIVATIVES

The gradient of the straight line between any two points can be calculated using the formula $m = \dfrac{y_2 - y_1}{x_2 - x_1}$. The gradient of a straight line is an average rate of change, which is consistent through every point.

For example, the line passing through the points $(-3, -2)$ and $(0, -1)$ has a gradient of $m = \dfrac{-1 - (-2)}{0 - (-3)} = \dfrac{1}{3}$. The line consistently rises one unit while moving right three units.

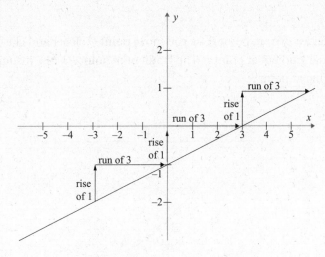

Differential Calculus

Since gradient refers to linear models, can curves also have a gradient? Yes, however the gradient of a curve is considered the instantaneous rate of change, meaning the gradient is the change of the function at one exact point.

In a function modeling distance travelled over time, the instantaneous rate of change would represent the velocity at a specific time.

To calculate gradient, two points are needed. The instantaneous rate of change, or the first derivative, occurs when the two points are so close together that they become the same point.

Consider the graph of $f(x)$ below with point P marked.

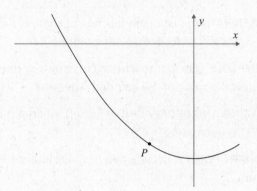

To determine the gradient at point P, consider another point, Q.

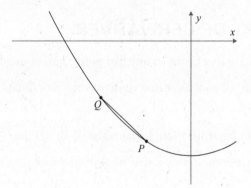

Point Q is rather far away from point P. As you move point Q closer and closer to point P, you approach the actual gradient at point P. The gradient at point P is the instantaneous rate the function is changing at point P.

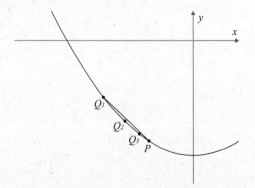

When point Q and point P get so close that they become the same, the gradient at point P is found. This creates a *tangent line*, which is a line that touches the graph only at point P. The gradient of the tangent line is the instantaneous rate of change at point P.

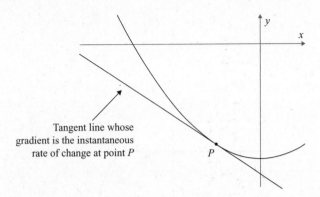

The gradient at point P is negative because the tangent line has a negative gradient. Consider the graph below.

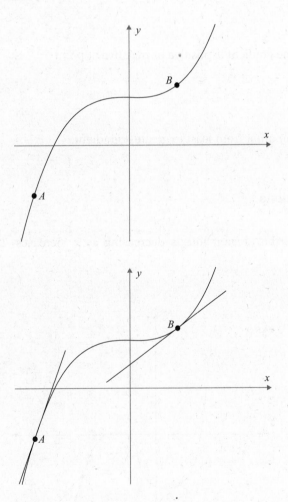

Both tangent lines increase as x increases; therefore, the gradient at each point is positive. The tangent line to point A has a steeper gradient. Thus, the gradient at point A is larger than the gradient at point B.

➡ EXAMPLE 7.10 _____

Consider the graph of $f(x)$ below.

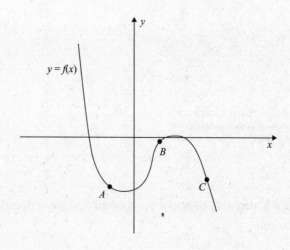

1. Write down if the gradient is positive or negative at point
   (a) A
   (b) B
   (c) C

2. List the points in order from least to greatest gradient.

Answer Explanations

1. (a) **Negative**

   At point A, the tangent line is decreasing as x increases. Thus, the gradient is negative.

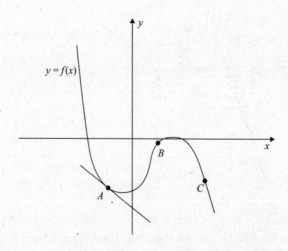

(b) Positive

At point B, the tangent line is increasing as x increases. Thus, the gradient is positive.

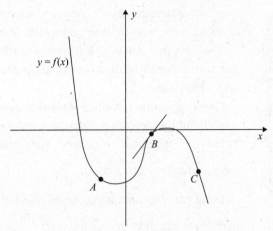

(c) Negative

At point C, the tangent line is decreasing as x increases. Thus, the gradient is negative.

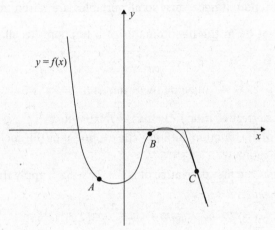

2. **C, A, B**

Judging by the steepness of the tangent lines, point C has the largest gradient, but since the gradient is negative, it will be the smallest of the three.

Point A is the next smallest, since the gradient at point A is also negative. Point B is the largest gradient, as point B has the only positive gradient.

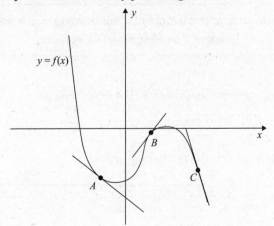

7.2 DERIVATIVES

The instantaneous rate of change, which is the gradient at a single point on a curve, is determined by the first derivative.

The first derivative of a function is also called the gradient function, since the derivative calculates the gradient at any point on a given curve.

Graphically, the steepness and sign of a gradient can be estimated, but the derivative calculates the exact numerical value.

For an algebraic function, the first derivative can be represented by $f'(x)$, $g'(x)$, or $\frac{dy}{dx}$.

For a problem involving area A, related to width, w, the first derivative, might be represented by $\frac{dA}{dw}$.

To Determine the First Derivative

1. Rewrite the function, if necessary, so all variables are raised to an integer exponent. Variables cannot be in the denominator of a fraction. Recall, $\frac{a}{x^n}$ can be rewritten as ax^{-n}.

   Example: $f(x) = 2x^3 + \frac{5}{x^2}$ must be rewritten as $f(x) = 2x^3 + 5x^{-2}$.

2. Apply the first derivative rule: $f(x) = ax^n \Rightarrow f'(x) = n(ax^{n-1})$
   For each term of the function, multiply the exponent by the coefficient. Then, subtract one from the exponent.

   Example: To find the first derivative of $f(x) = 2x^3 + 5x^{-2}$, apply the rule to each term.

   multiply down

   $2x^3 \leftarrow$ *then subtract one* $= 3(2x^{3-1}) = 6x^2$

   multiply down

   $5x^{-2} \leftarrow$ *then subtract one* $= -2(5x^{-2-1}) = -10x^{-3}$

   Use caution with negative exponents. Subtracting one from a negative creates a smaller negative number.

3. Return any negative exponents back to the denominator.

   Example: $f'(x) = 6x^2 - 10x^{-3}$ becomes $f'(x) = 6x^2 - \frac{10}{x^3}$

Remember, the first derivative is the gradient function.

Differential Calculus

The gradient of $f(x) = 2x^3 + \dfrac{5}{x^2}$, when $x = 2$ is calculated by substituting 2 for x in the first derivative.

$$f'(2) = 6(2)^2 - \frac{10}{(2)^3}$$
$$f'(2) = 22.75$$

Therefore, the gradient at $x = 2$ is 22.75.

The chart below displays several common functions with an example.

Function	Example	First Derivative
Constant	$h(x) = -6$	$h'(x) = 0$ • Rewrite $h(x) = -6$ as $h(x) = -6x^0$, since there are no x variables in the constant equation. • Apply the derivative rule: $\quad 0(-6x^{0-1}) = 0x^{-1} = 0$ • The first derivative of a constant is always zero.
Linear	$y = \dfrac{1}{2}x + 2$	$\dfrac{dy}{dx} = \dfrac{1}{2}$ • $\dfrac{1}{2}x^1 \Rightarrow 1\left(\dfrac{1}{2}x^{1-1}\right) = \dfrac{1}{2}x^0 = \dfrac{1}{2}$ Variables raised to the zero power equal one. • The first derivative of a linear term is always the coefficient. • Since 2 is a constant, the first derivative is always zero.
Quadratic	$g(x) = 12 - x - 4x^2$	$g'(x) = -1 - 8x$ • The constant 12 will become zero. • The linear term $-x$ becomes the coefficient of -1. • $-4x^2 \Rightarrow 2(-4x^{2-1}) = -8x$
Cubic	$V = 4x^3 + \dfrac{1}{2}x^2 - 7x + 2$	$\dfrac{dV}{dx} = 12x^2 + x - 7$ • $4x^3 \Rightarrow 3(4x^{3-1}) = 12x^2$ • The quadratic, linear, and constant term follow the same pattern as before.
Rational	$h(x) = \dfrac{2}{x} - \dfrac{10}{x^3}$	$h'(x) = -\dfrac{2}{x^2} + \dfrac{30}{x^4}$ • First, rewrite as $h(x) = 2x^{-1} - 10x^{-3}$. • $2x^{-1} \Rightarrow -1(2x^{-1-1}) = -2x^{-2}$ • $-10x^{-3} \Rightarrow -3(-10x^{-3-1}) = 30x^{-4}$ • Return negative exponents to the denominator.

➡ **EXAMPLE 7.20**

For each of the following, find the first derivative.
1. $y = 2x - 1$
2. $f(x) = 3x^5 - 2x + 14$
3. $g(x) = \dfrac{3}{x^2} - x^4$

Answer Explanations

1. $\dfrac{dy}{dx} = 2$

   $2x \Rightarrow 1(2x^{1-1}) = 2x^0 = 2$ [The first derivative of a linear term is always the coefficient.]

   The first derivative of a constant is always zero.

2. $f'(x) = 15x^4 - 2$

   $$3x^5 \Rightarrow 5(3x^{5-1}) = 15x^4$$

   The first derivative of the linear term $-2x$ is the coefficient of -2.

   The first derivative of the constant is always zero.

3. $g'(x) = -\dfrac{6}{x^3} - 4x^3$

   First rewrite the function as $g(x) = 3x^{-2} - x^4$.

   $$3x^{-2} \Rightarrow -2(3x^{-2-1}) = -6x^{-3}$$

   $$-1x^4 \Rightarrow 4(-1x^{4-1}) = -4x^3$$

   Therefore, $g'(x) = -6x^{-3} - 4x^3$, but you must return all negative exponents to the denominator using the formula $g'(x) = -\dfrac{6}{x^3} - 4x^3$.

7.3 TANGENT LINES

The first derivative gives a quick and convenient way of calculating the gradient of a curve at any point. First differentiate $f(x)$ to get the gradient function (the first derivative). Then, substitute in the given value of x to determine the gradient.

➡ **EXAMPLE 7.30**

For each of the following, calculate the gradient of the curve when $x = -3$.
1. $y = 2x - 1$
2. $f(x) = 3x^5 - 2x + 14$
3. $g(x) = \dfrac{3}{x^2} - x^4$

Answer Explanations

The functions were differentiated in example 7.20.

1. **2**

    Since the first derivative is $\dfrac{dy}{dx} = 2$, the gradient at all points is 2.

2. **$f'(-3) = 1213$**

    To find the gradient when $x = -3$, substitute -3 in for x into the first derivative.

    $$f'(-3) = 15(-3)^4 - 2$$
    $$f'(-3) = 1213$$

3. **$g'(-3) = 108.\bar{2} = \dfrac{974}{9}$ (exact) OR $g'(-3) = 108$ (3 significant figures)**

    To find the gradient when $x = -3$, substitute -3 in for x into the first derivative.

    $$g'(-3) = -\frac{6}{(-3)^3} - 4(-3)^3$$
    $$g'(-3) = 108.\bar{2} = \frac{974}{9}$$

The gradient of a function at any given point can also be determined using the GDC. Type the given function in for Y_1 and graph. Then, using the (CALC) menu, choose option 6: dy / dx. Remember dy / dx is a notation for the first derivative.

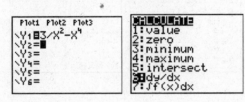

Next, enter the x-value for which the gradient is to be calculated and press (ENTER). The gradient is displayed.

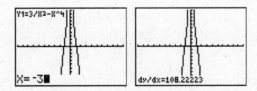

Since the first derivative is the instantaneous rate of change, real-world situations can be solved using the first derivative.

For example, suppose a ball is thrown in the air. The height in meters, h, after t seconds can be modeled by $h(t) = -4.9t^2 + 50t$. Since the function represents distance, the first derivative represents velocity. At any given time, the <u>function</u> calculates the <u>distance</u> traveled, but the <u>instantaneous rate of change</u> is the <u>speed</u> of the ball. Let's find the speed of the ball after 4 seconds.

First, differentiate the function: $h'(t) = -9.8t + 50$. This represents the velocity after t seconds. The speed of the ball after 4 seconds would be $h'(4) = -9.8(4) + 50 = 10.8ms^{-1}$. Therefore, the ball is travelling 10.8 meters per second.

➡️ **EXAMPLE 7.31** _____

James is blowing up a beach ball at a constant rate. The volume, V cm³, after b breaths of air is represented by $V(b) = 25b^2 + 15b$.

1. Determine the rate of increase of the volume after 2 breaths.
2. Find how many breaths James has blown when the rate of increase of the volume is 315 cm³ per breath.
3. Hence, calculate the volume of the beach ball when the rate of increase is 315 cm³ per breath.

Answer Explanations

1. **115 cm³ per breath**

   The rate of increase is the instantaneous rate of change, which is the first derivative.

   $$V'(b) = 50b + 15.$$

   Now, substitute 2 in for b.

   $$V'(2) = 50(2) + 15 = 115$$

   This can be checked using the GDC.

   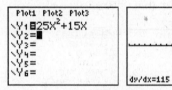

2. **$b = 6$ breaths**

   You are given the rate of increase, which is the value of the first derivative. Set the first derivative equal to the given rate and solve for b.

   $$315 = 50b + 15$$
   $$300 = 50b$$
   $$6 = b$$

   Double check the solution using the GDC.

   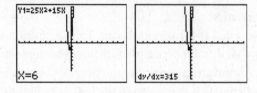

3. **$V = 990$ cm³**

   To calculate the volume, use the original function and let $b = 6$.

   $$V(6) = 25(6)^2 + 15(6) = 990$$

A tangent line to a curve is a line that touches the graph at one, and only one, point. The gradient of the tangent line is equal to the gradient at the point.

To find the equation of the tangent line at a point (x, y)

1. Find the first derivative of the given function.
2. Calculate the gradient of the tangent line by substituting the value of x into the first derivative.
3. If the y-coordinate is not given, plug x into the original function. Remember $f(x) = y$, whereas $f'(x) = m$.
4. Now, substitute y, x, and m into $y = mx + c$ to determine c.

➡ EXAMPLE 7.32

Given $f(x) = 2x^3 - \dfrac{1}{x^2} + 5$:

1. Find the gradient function $f'(x)$.

2. Find the gradient of the curve, where $x = \dfrac{1}{2}$.

3. Write the equation of the tangent to the curve, where $x = \dfrac{1}{2}$.

> **QUICK TIP**
>
> To write the equation of a tangent line, you need a point and a gradient.
>
> **Gradient:** the value of the **first derivative** at a given x.
>
> **Point:** the value of the original function at a given x.
>
> First derivative = Gradient
>
> Original Function = y-value

Answer Explanations

1. $f'(x) = 6x^2 + \dfrac{2}{x^3}$

   First, rewrite the function with integer exponents.
   $$f(x) = 2x^3 - x^{-2} + 5.$$

   Apply the first derivative rule to each term.
   $$2x^3 \Rightarrow 3(2x^{3-1}) = 6x^2$$
   $$-x^{-2} \Rightarrow -2(-x^{-2-1}) = 2x^{-3}.$$

   The constant 5 differentiates to zero.
   Thus, $f'(x) = 6x^2 + 2x^{-3}$, but you must move the negative exponent back to the denominator.

2. $f'\left(\dfrac{1}{2}\right) = 17.5$

   Substitute $\dfrac{1}{2}$ in for x into the first derivative.

   $$f'\left(\frac{1}{2}\right) = 6\left(\frac{1}{2}\right)^2 + \frac{2}{\left(\dfrac{1}{2}\right)^3} = 17.5$$

3. **$y = 17.5x - 7.5$**

    To write the equation of the tangent line, you need y, x, and the gradient m.

    You know $x = \dfrac{1}{2}$ and found $m = 17.5$.

    To determine the value of y, use the original function, since $f(x) = y$.

    $$f\left(\frac{1}{2}\right) = 2\left(\frac{1}{2}\right)^3 - \frac{1}{\left(\frac{1}{2}\right)^2} + 5 = 1.25$$

    $y = 1.25$, $x = \dfrac{1}{2}$, and $m = 17.5$. Substitute these values into $y = mx + c$ and solve for c.

    $$1.25 = 17.5\left(\frac{1}{2}\right) + c$$

    $$1.25 = 8.75 + c$$

    $$-7.5 = c$$

    Therefore, the equation of the tangent line is $y = 17.5x - 7.5$.

Another important linear function related to the tangent line is the normal line. A normal line is perpendicular to the tangent line at a given point.

The graph below displays the function $f(x)$, with the tangent line L_1 at point P.

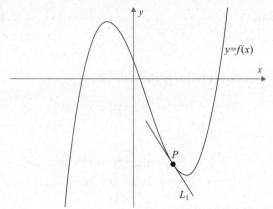

The normal line, which is perpendicular to the tangent line, also passes through point P. The graph below shows the tangent line and normal line with a closer scale.

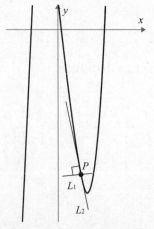

The gradient of the normal line is the opposite reciprocal of the tangent line gradient. Perpendicular lines are covered in section 5.1.

➥ EXAMPLE 7.33 _____

Let $h(x) = \frac{1}{2}x^4 - \frac{1}{3}x^3 + 2x - 7$.

1. Differentiate $h(x)$ with respect to x.

The gradient of the tangent line at point P is 3.

2. Determine the x-coordinate at point P.
3. Hence, find the value of y at point P.
4. Write the equation of the normal line at point P.

QUICK TIP

The **normal line** is perpendicular to the **tangent line**.

Perpendicular lines have opposite reciprocal gradients.

First derivative = Gradient

Original Function = y-value

Answer Explanations

1. $h'(x) = 2x^3 - x^2 + 2$

    Differentiate means to find the first derivative.

    $$\frac{1}{2}x^4 \Rightarrow 4\left(\frac{1}{2}x^{4-1}\right) = 2x^3$$

    $$-\frac{1}{3}x^3 \Rightarrow 3\left(-\frac{1}{3}x^{3-1}\right) = -x^2$$

    The derivative of the linear term is always the coefficient, and the constant differentiates to zero.

2. $x = 1$

    The gradient at point P is 2, so set the first derivative equal to 3 and solve for x.

    $$3 = 2x^3 - x^2 + 2$$

    You can solve by graphing both sides of the equation and finding the intersection.

    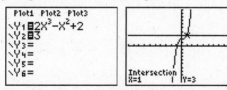

    Another option is to set the equation equal to zero and solve using PLYSMLT.

    $$0 = 2x^3 - x^2 - 1$$

    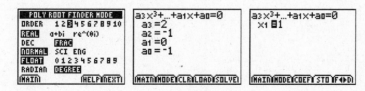

3. $y = -4.83$ **(3 significant figures) OR** $y = -\frac{29}{6}$ **(exact)**

    Substitute 1 in for x into the <u>original</u> function to find the value of y.

    $$h(1) = \frac{1}{2}(1)^4 - \frac{1}{3}(1)^3 + 2(1) - 7 = -\frac{29}{6}$$

4. $y = -\dfrac{1}{3}x - \dfrac{9}{2}$ **(exact) OR** $y = -0.333x - 4.5$ **(3 significant figures)**

Since the normal line is perpendicular to the tangent line, the gradient is the opposite reciprocal of 2. The normal and tangent lines both pass through point P.

$$-\frac{29}{6} = -\frac{1}{3}(1) + c$$

$$-\frac{29}{6} = -\frac{1}{3} + c$$

$$-\frac{9}{2} = c$$

7.4 INCREASING AND DECREASING FUNCTIONS

A function is increasing if the y-values increase as the x-values increase.

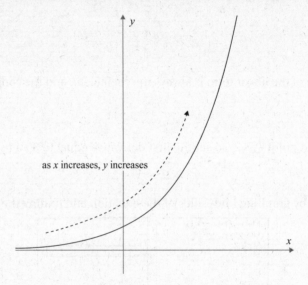

as x increases, y increases

A function is decreasing if the y-values decrease as the x-values increase.

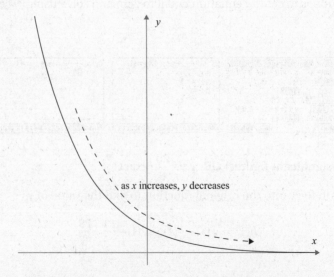

as x increases, y decreases

When a function $f(x)$ is increasing, the values of $f'(x)$ are all positive, meaning $f'(x) > 0$. The first derivative is positive on an increasing function because the tangent lines would all be increasing as well.

When a function $f(x)$ is decreasing, the values of $f'(x)$ are all negative, meaning $f'(x) < 0$. The first derivative is negative on a decreasing function because the tangent lines would all be decreasing as well.

➡ EXAMPLE 7.40

Consider the graph of $g(x)$ shown below.

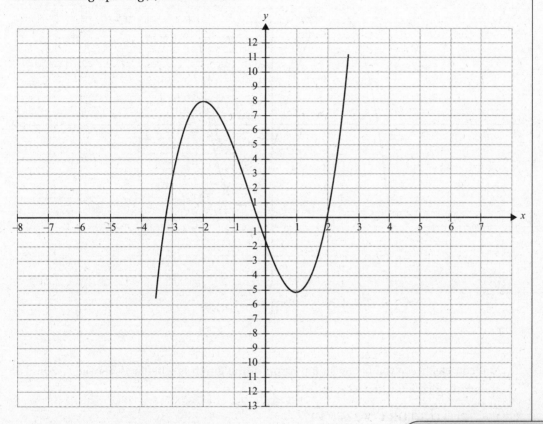

Write down the intervals in which
1. $g'(x) < 0$
2. $g'(x) > 0$

Answer Explanations

1. **(−2, 1) OR −2 < x < 1**

   When $g'(x) < 0$, the function $g(x)$ will be decreasing.

> **QUICK TIP**
>
> When the first derivative is **positive**, the function is **increasing**.
>
> $f(x)$ is increasing when $f'(x) > 0$
>
> When the first derivative is **negative**, the function is **decreasing**.
>
> $f(x)$ is decreasing when $f'(x) < 0$

An easy way to envision increasing or decreasing is to imagine a small ant climbing on the graph from left to right. If the ant is climbing uphill, the function is increasing. If the ant is climbing downhill, the function is decreasing.

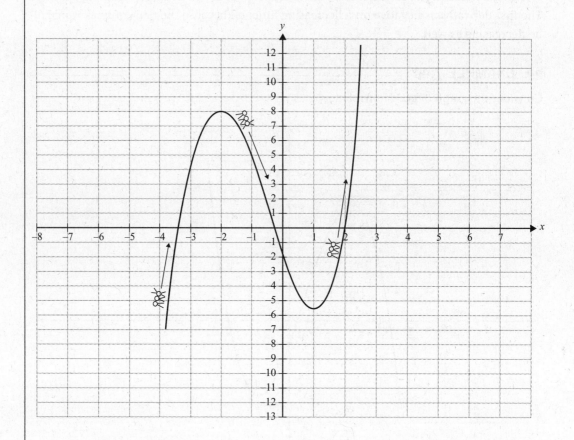

You can easily see the decreasing portion of the graph is when x is between -2 and 1; therefore, $g'(x) < 0$ when $-2 < x < 1$.

2. **$(-\infty, -2) \cup (1, \infty)$ OR $x < -2$ or $x > 1$**

The function is increasing from the very beginning up until the peak at $x = -2$. Then, the function begins increasing again in the valley when $x = 1$ until the end. Therefore, $g'(x) > 0$ when $x < -2$ or $x > 1$.

When a function switches from increasing to decreasing, a local maximum occurs. A local minimum occurs when the functions change from decreasing to increasing.

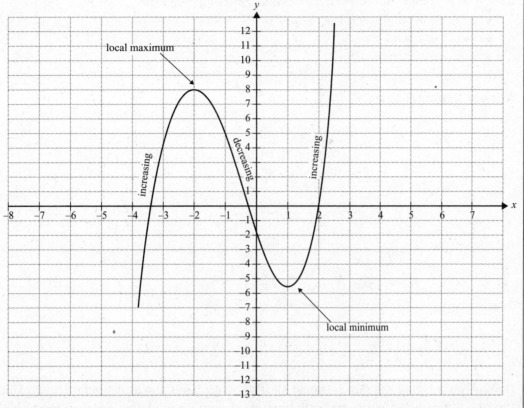

The gradient at a local maximum or minimum is always zero since the tangent line is a horizontal line.

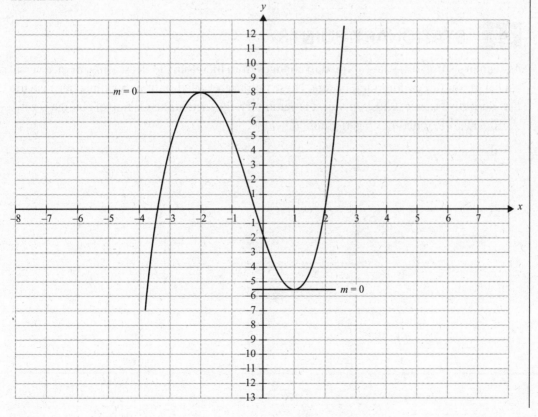

Differential Calculus

➡ EXAMPLE 7.41 _____

Draw a sketch of the function $h(x)$ that has the following properties:

- $h'(x) < 0$, when $-1 < x < 1$
- $h'(x) > 0$, when $1 < x < 5$
- $h'(x) = 0$ at the point $(1, -2)$

Answer Explanations

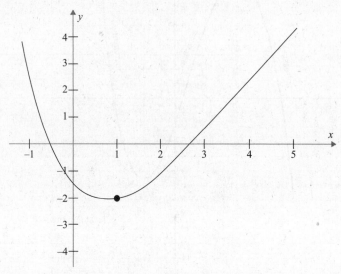

Graphs will vary, but the sketch should clearly show a minimum at the point $(1, -2)$. The function is decreasing (going downhill) from $x = -1$ until the minimum and then increasing (going uphill) from the minimum until $x = 5$.

7.5 STATIONARY POINTS

As seen in section 7.4, when a function changes from increasing to decreasing, or vice versa, a local maximum or minimum occurs. These points are called stationary points because the first derivative is zero. When an object is stationary, the instantaneous rate of change is zero.

Consider the function $y = g(x)$ graphed below.

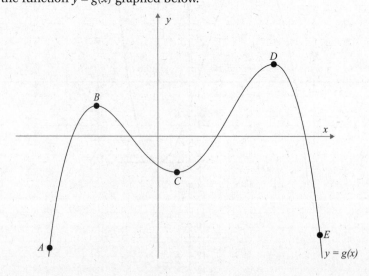

Differential Calculus

A description of the function and its first derivative is summarized in the chart below.

Interval	Function Behavior	First Derivative Behavior
From point A to point B	Increasing	$g'(x) > 0$
At point B	Local maximum	$g'(x) = 0$
From point B to point C	Decreasing	$g'(x) < 0$
At point C	Local minimum	$g'(x) = 0$
From point C to point D	Increasing	$g'(x) > 0$
At point D	Local maximum	$g'(x) = 0$
From point D to point E	Decreasing	$g'(x) < 0$

Graphically, stationary points are easy to identify. A local maximum is when a stationary point occurs at a peak, while a local minimum occurs in a valley.

Algebraically, stationary points can be identified through the first derivative. In the example above, when $\dfrac{dy}{dx} > 0$, and then changed to $\dfrac{dy}{dx} < 0$, a local maximum occurred. However, when $\dfrac{dy}{dx} < 0$, and then changed to $\dfrac{dy}{dx} > 0$, a local minimum occurred.

Sign of first derivative	Positive	Zero	Negative
Tangent line direction	/	—	\

Result:

Maximum

Sign of first derivative	Negative	Zero	Positive
Tangent line direction	\	—	/

Result:

Minimum

➡ **EXAMPLE 7.50**

Find and label the coordinates of the stationary points on the curve represented by $y = 4x^3 + 9x^2 - 12x + 6$.

Answer Explanations

Maximum at $(-2, 34)$

Minimum at $\left(\dfrac{1}{2}, \dfrac{11}{4}\right)$ **OR** $(0.5, 2.75)$

To locate and identify the stationary points, find the first derivative.

$$\frac{dy}{dx} = 12x^2 + 18x - 12$$

QUICK TIP

To determine stationary points:

1. Find the first derivative, $f'(x)$

2. Solve $f'(x) = 0$ (use PLYSMLT)

3. Determine the sign of $f'(x)$ on either side of each x-value.

   Negative, zero, positive = **minimum**

   Positive, zero, negative = **maximum**

4. Find the coordinates by finding the value of the function at each x.

Differential Calculus

Stationary points occur when the gradient is zero, so set the first derivative equal to zero.

$$0 = 12x^2 + 18x - 12$$

> This equation MUST be written down to earn full marks on Paper 2 of the exam.

Now, solve for x. PLYSMLT is the most convenient method.

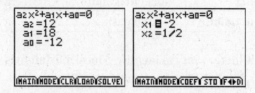

You know that when $x = -2$ or $x = \dfrac{1}{2}$, the first derivative will be zero. However, you need to determine the sign of the first derivative for everywhere else.

Let $Y_1 = 12x^2 + 18x - 12$ and check the table of values. The table must be set up so that x changes by $\dfrac{1}{2}$.

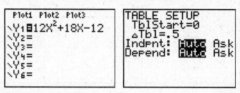

When $x < -2$, the first derivative is positive.

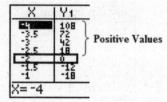

When x is between -2 and $\dfrac{1}{2}$, the first derivative is negative.

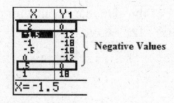

After $\dfrac{1}{2}$, the first derivative is positive again.

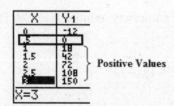

To summarize, place the information on a number line.

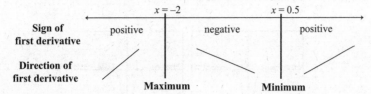

The number line clearly shows the type of stationary points.

The remaining piece to the stationary points is the value of y for each x.

The original function gives the value of y at a specific x, so substitute each x value in to determine the partnering y value.

$y = 4(-2)^3 + 9(-2)^2 - 12(2) + 6 = -14$. The maximum occurs at $(-2, 14)$.

$y = 4\left(\dfrac{1}{2}\right)^3 + 9\left(\dfrac{1}{2}\right)^2 - 12\left(\dfrac{1}{2}\right) + 6 = \dfrac{11}{4}$. The minimum occurs at $\left(\dfrac{1}{2}, \dfrac{11}{4}\right)$.

➡ EXAMPLE 7.51

Let $g(x) = x + \dfrac{9}{x}$, $x \neq 0$.

   (a) Determine $g'(x)$.

   (b) Find the values of x for which $g'(x) = 0$.

   (c) Find the coordinates of the stationary points and determine whether the point is a maximum or minimum.

Answer Explanations

(a) $g'(x) = 1 - \dfrac{9}{x^2}$

    First, rewrite the function as $g(x) = x + 9x^{-1}$.

    Apply the first derivative rule.

    The first derivative of x is 1.

$$9x^{-1} \Rightarrow -1(9x^{-1-1}) = -9x^{-2}$$

    Thus, $g'(x) = 1 - 9x^{-2}$. Finish by moving the negative exponent back to the denominator.

(b) $x = -3, 3$

    Set $g'(x) = 0$.

$$0 = 1 - \frac{9}{x^2}$$

    Solve algebraically by solving for x.

$$-1 = -\frac{9}{x^2}$$
$$x^2 = 9$$
$$x = \pm 3$$

Or, solve graphically by finding the zeros of $Y_1 = 1 - \dfrac{9}{x^2}$.

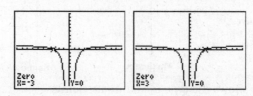

(c) **(–3, –6): maximum**

(3, 6): minimum

In part b, you found the x-coordinates of the stationary points. To determine if they are a maximum or minimum, you need to know the sign of the first derivative everywhere else.

As in example 7.50, place the x-coordinates on a number line and use the table to find the sign of the first derivative.

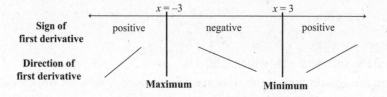

FEATURED QUESTION

A function is defined by $f(x) = \dfrac{5}{x^2} + 3x + c$, $x \neq 0$, $c \in \mathbb{Z}$.

(a) Write down an expression for $f'(x)$.

Consider the graph of f. The graph of f passes through the point $P(1, 4)$.

(b) Find the value of c.

(c) There is a local minimum at the point Q.
    (i) Find the coordinates of Q.
    (ii) Find the set of values of x for which the function is decreasing.

Let T be the tangent to the graph of f at P.

(d) (i) Show that the gradient of T is -7.
    (ii) Find the equation of T.

(e) T intersects the graph again at R. Use your graphic display calculator to find the coordinates of R.

(See page 450 for Solutions)

7.6 OPTIMIZATION

The use of the first derivative to locate a maximum or minimum in a real-world context is called optimization. These types of problems are generally found on Paper 2 of the IB exam.

Optimization problems often involve geometric shapes, which are covered in section 5.5.

EXAMPLE 7.60

Karly has a sheet of cardboard measuring 100 cm by 150 cm from which she is going to cut a square of side x cm from each corner. She will fold the sides up to create an open-top box.

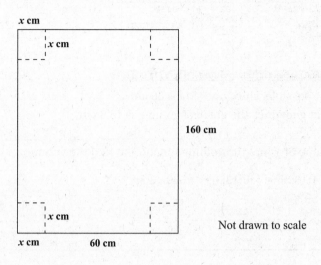

x cm

160 cm

60 cm

Not drawn to scale

QUICK TIP

Key words such as **greatest, least, maximum,** or **minimum** indicate stationary points.

Thus, find the first derivative.

Set it equal to zero.

Solve for x.

(a) Show that the volume of the box is $V = 4x^3 - 440x^2 + 9\,600x$.

(b) Find $\dfrac{dV}{dx}$.

(c) Find the value of x that maximizes the volume.

(d) Determine the maximum volume.

Answer Explanations

(a) $V = x(60 - 2x)(160 - 2x)$

Remember that $V = l \times w \times h$.

To create the box, a square of side x will be cut from each corner. This means each dimension will be reduced by $2x$.

The length is $60 - 2x$ and the width is $160 - 2x$. When the sides are folded up, the height will be x.

(b) $\dfrac{dV}{dx} = 12x^2 - 880x + 9\,600$

Apply the first derivative rule to each term.

$$4x^3 \Rightarrow 3(4x^{3-1}) = 12x^2$$
$$-440x^2 \Rightarrow 2(-440x^{2-1}) = -880x$$

The linear term $9\,600x$ differentiates to $9\,600$.

Differential Calculus

(c) **$x = 13.3$ cm**

To maximize the volume, find the stationary point of the function.

$$0 = 12x^2 - 880x + 9\,600$$

The quickest method to solve for x is to use PLYSMLT.

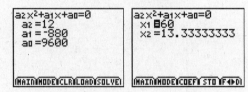

The solution of $x = 60$ is not possible, since one side is 60 cm.

Therefore, the value of x that generates the greatest volume is 13.3 cm.

(d) **$V = 59\,300$ cm³**

Substitute the calculated value of x into the volume function to find the volume.

$$V = 4(13.3)^3 - 440(13.3)^2 + 9\,600(13.3)$$
$$V = 59\,300$$

➡ EXAMPLE 7.61

A manufacturing company is designing a cylindrical tank for a local farm as shown below. The tank must hold 100 cubic meters of water.

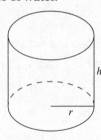

(a) Show that $h = \dfrac{100}{\pi r^2}$.

The company wants to minimize the amount of material used to manufacture the water tank.

(b) Show that the surface area, S, is represented by $S = 2\pi r^2 + \dfrac{200}{r}$.

(c) Find $\dfrac{dS}{dr}$.

(d) Find the value of r that minimizes the surface area.

(e) Calculate the minimum surface area of the water tank.

Answer Explanations

(a) **$100 = \pi r^2 h$**

$$\frac{100}{\pi r^2} = h$$

The volume of a cylinder is found using the formula $V = \pi r^2 h$. You are given the volume of 100 m³ and can solve for h.

(b) $S = 2\pi r^2 + 2\pi rh$

$$S = 2\pi r^2 + 2\pi r\left(\frac{100}{\pi r^2}\right)$$

$$S = 2\pi r^2 + \frac{200}{r}$$

Surface area of a cylinder is made up of the area of two circular bases, each πr^2, and also the area of the curved surface, $2\pi r^2 h$.

Since $h = \dfrac{100}{\pi r^2}$, $S = 2\pi r^2 + 2\pi r\left(\dfrac{100}{\pi r^2}\right)$

Now, simplify.

(c) $\dfrac{dS}{dr} = 4\pi r - \dfrac{200}{r^2}$

First, rewrite the surface area so r is not in the denominator of a fraction.

$$S = 2\pi r^2 + 200r^{-1}$$

Now, apply the first derivative rule to each term.

$$2\pi r^2 \Rightarrow 2(2\pi r^{2-1}) = 4\pi r$$
$$200r^{-1} \Rightarrow -1(200r^{-1-1}) = -200r^{-2}$$

Then, place the term with the negative exponent back in the denominator.

$$\frac{dS}{dr} = 4\pi r - \frac{200}{r^2}$$

(d) $r = 2.52$ m

The minimum occurs when the first derivative is zero.

$$0 = 4\pi r - \frac{200}{r^2}$$

To solve, graph the first derivative and find the zero of the curve.

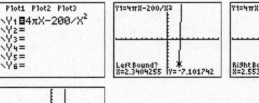

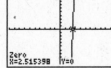

(e) 120 m² (3 significant figures)

To minimize, surface area r will equal 2.52. Substitute 2.52 in for r into the surface area equation and simplify.

$$S = 2\pi(2.52)^2 + \frac{200}{2.52}$$

$$S = 119.90$$

$$S = 12\overline{0}\ m^2$$

TOPIC 7 PRACTICE

Paper 1 Questions

1. Consider the function $g(x) = x^3 - kx^2 + 2x - 1$, where $k \neq 0$.

   (a) Find $g'(x)$.

   The gradient at $x = 2$ is 11.

   (b) Determine the value of k.

   (c) Find $g'(-2)$ and explain what the value represents.

2. The diagram below shows part of the graph of $f(x)$.

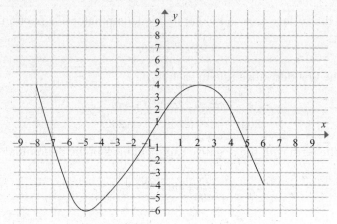

   State the intervals or x-value(s) in which

   (a) $f'(x) < 0$

   (b) $f'(x) = 0$

   (c) $f'(x) > 0$

3. A function is defined by $h(x) = \dfrac{1}{x} - \dfrac{2}{x^2}$, when $x > 0$.

   (a) Find $h'(x)$.

   The graph of $y = h(x)$ has a tangent line L at the point where $x = 2$.

   (b) Determine the gradient of line L.

   (c) Write the equation of L in the form $y = mx + c$.

4. Given $g(x) = \dfrac{1}{2}x^4 - x^3 + 2x - 1$,

   (a) Write down $g'(x)$.

   (b) Determine the value of $g'(x)$, when x is
       (i) -2
       (ii) 1

   (c) Determine if $g(x)$ is increasing or decreasing, when x is
       (i) -2
       (ii) 1

Differential Calculus

5. Consider the function $f(x) = \dfrac{4}{x^2} + x - 1$, $x \neq 0$.

(a) Write the function $f(x)$ in the form $f(x) = 4x^a + x - 1$, where $a \in \mathbb{Z}$.

(b) Write down $f'(x)$.

The graph of $y = f(x)$ has a horizontal tangent at point P.

(c) Determine the coordinates of point P.

(d) Determine if point P is a local maximum or minimum.

SOLUTIONS

1. (a) $g'(x) = 3x^2 - 2kx + 2$

Apply the first derivative rule to each term.

$$x^3 \Rightarrow 3(x^{3-1}) = 3x^2$$
$$-kx^2 \Rightarrow 2(-kx^{2-1}) = -2kx$$
$$2x \Rightarrow 2$$

(b) $k = \dfrac{3}{4}$ or $k = 0.75$

Since the gradient is 11 when $x = 2$,

$$11 = 3(2)^2 - 2k(2) + 2$$
$$11 = 12 - 4k + 2$$
$$11 = 14 - 4k$$
$$-3 = -4k$$
$$\dfrac{3}{4} = k$$

(c) $g'(-2) = 17$. **The gradient of the tangent line where $x = -2$ is 17.**

$$g'(-2) = 3(-2)^2 - 2\left(\dfrac{3}{4}\right)(-2) + 2$$
$$g'(-2) = 17$$

The first derivative is the gradient function, so the value of $g'(-2)$ is the gradient of the tangent line where $x = -2$.

2. (a) $-8 < x < -5$ or $2 < x < 6$

OR $(-8, -5) \cup (2, 6)$

The first derivative is less than zero when the function is decreasing. The function $f(x)$ is decreasing from -8 to -5 and then again from 2 to 6.

(b) $x = -5, x = 2$

The first derivative is zero at a local maximum or minimum; the graph has both. The local minimum occurs when $x = -5$ and the local maximum occurs when $x = 2$.

(c) $-5 < x < 2$

OR $(-5, 2)$

The first derivative is greater than zero when the function is increasing. The function $f(x)$ is increasing from -5 to 2.

3. (a) $h'(x) = -\dfrac{1}{x^2} + \dfrac{4}{x^3}$

First, rewrite $h(x)$ as $h(x) = x^{-1} - 2x^{-2}$.

Apply the first derivative rule to each term.

$$x^{-1} \Rightarrow -1(x^{-1-1}) = -x^{-2}$$
$$-2x^{-2} \Rightarrow -2(-2x^{-2-1}) = 4x^{-3}$$
$$h'(x) = -x^{-2} + 4x^{-3}$$

Finally, rewrite the first derivative with all positive exponents.

(b) $\dfrac{1}{4}$

The gradient of the tangent line is the value of the first derivative, when $x = 2$.

$$h'(2) = -\dfrac{1}{2^2} + \dfrac{4}{2^3} = \dfrac{1}{4}$$

(c) $y = \dfrac{1}{4}x - \dfrac{1}{2}$

In order to write the equation of the tangent line, you must have x, y, and the gradient. You know $x = 2$ and the gradient is $\dfrac{1}{4}$. To find y, substitute $x = 2$ in the original function.

$$h(2) = \dfrac{1}{2} - \dfrac{2}{2^2}$$
$$h(2) = 0$$

Now, substitute those values into $y = mx + c$.

$$0 = \dfrac{1}{4}(2) + c$$
$$0 = \dfrac{1}{2} + c$$
$$-\dfrac{1}{2} = c$$

Therefore, the equation of the tangent line is $y = \dfrac{1}{4}x - \dfrac{1}{2}$.

4. (a) $g'(x) = 2x^3 - 3x^2 + 2$

Apply the first derivative rule to each term.

$$\frac{1}{2}x^4 \Rightarrow 4\left(\frac{1}{2}x^{4-1}\right) = 2x^3$$
$$-x^3 \Rightarrow 3(-x^{3-1}) = -3x^2$$
$$2x \Rightarrow 2$$

The constant differentiates to zero.

(b) (i) **−26**

(ii) **1**

Substitute the given value of x into the first derivative.

$$g'(-2) = 2(-2)^3 - 3(-2)^2 + 2 \qquad g'(1) = 2(1)^3 - 3(1)^2 + 2$$
$$g'(-2) = -26 \qquad\qquad\qquad g'(1) = 1$$

(c) (i) **decreasing**

(ii) **increasing**

When the first derivative is negative, the function is decreasing.

When the first derivative is positive, the function is increasing.

5. (a) $f(x) = 4x^{-2} + x - 1$

When terms in the denominator of a fraction are moved to the numerator, the exponent becomes negative.

(b) $f'(x) = -\dfrac{8}{x^3} + 1$

Apply the first derivative rule to each term.

$$4x^{-2} \Rightarrow -2(4x^{-2-1}) = -8x^{-3}$$

The linear term differentiates to the coefficient 1.

The constant term differentiates to zero.

(c) **(2, 2)**

The gradient is zero at the horizontal tangent.

$$0 = -\frac{8}{x^3} + 1 .$$

This can be solved algebraically or graphically.

$$0 = -\frac{8}{x^3} + 1$$
$$-1 = -\frac{8}{x^3}$$
$$-x^3 = -8 \qquad \text{OR}$$
$$x^3 = 8$$
$$x = 2$$

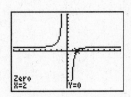

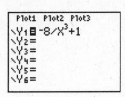

To find the value of y when $x = 2$, use the original function.

$$f(2) = \frac{4}{(2)^2} + 2 - 1$$
$$f(2) = 2$$

(d) **Local minimum**

To determine if the point (2, 2) is a minimum or maximum, you must find the sign of the first derivative on either side.

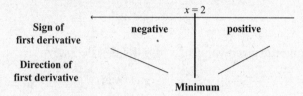

Paper 2 Questions

1. Consider the function $h(x) = x^3 + 6x^2 - 15x + 4$.

   (a) Determine $h(1)$.
   (b) Write down $h'(x)$.
   (c) Find the gradient of the graph of $h(x)$ where $x = -2$.

   The graph of $h(x)$ has a local maximum at point P and a local minimum at point Q.

   (d) Using $h'(x)$, determine the coordinates of
      (i)   point P,
      (ii) point Q.

   When $x \geq 0$, the normal line at the point M is represented by $y = \frac{1}{15}x + 4$.

   (e) Write down the gradient of the tangent line at point M.
   (f) Determine the x-coordinate of point M.
   (g) Write the equation of the tangent line at point M.

2. Quantasia is creating an open-top box from a piece of cardboard by cutting out a square from each corner and folding the resulting sides.

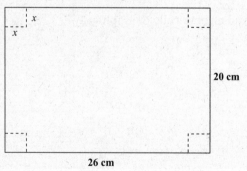

(a) Write down an expression, in terms of x, for the
  (i) length of the box,
  (ii) width of the box.

(b) Show that the volume of the resulting box is $V = 4x^3 - 92x^2 + 520x$.

(c) Find $\dfrac{dV}{dx}$.

Quantasia wants to cut out the square that will maximize the volume of the box.

(d) Find the value of x that maximizes the volume of the box.

(e) Determine the maximum volume of the box.

After the box is assembled, Quantasia will fill the box with sand. The amount of sand, S, in the box measured by cubic centimeters after t seconds can be determined by $S = 1.3t^2 + 17.5t$.

(f) Determine the amount of sand in the box after 6 seconds.

(g) The rate the sand is being poured, $\dfrac{dS}{dt}$, at k seconds is 43.5 cm^3/sec. Find k.

(h) Determine how long it will take to fill the box with sand.

SOLUTIONS

1. (a) **$h(1) = -4$**

   Substitute 1 for x and simplify.

   $$h(1) = 1^3 + 6(1)^2 - 15(1) + 4$$
   $$h(1) = -4$$

   (b) **$h'(x) = 3x^2 + 12x - 15$**

   Apply the first derivative rule to each term.

   $$x^3 \Rightarrow 3(x^{3-1}) = 3x^2$$
   $$6x^2 \Rightarrow 2(6x^{2-1}) = 12x$$

   The linear term differentiates to the coefficient -15.
   The constant term differentiates to zero.

   (c) **$h'(-2) = -27$**

   Substitute -2 for x into the first derivative.

   $$h'(-2) = 3(-2)^2 + 12(-2) - 15$$
   $$h'(-2) = -27$$

   (d) (i) **$(-5, 104)$**
       (ii) **$(1, -4)$**

   To determine where the graph has a maximum or minimum, first find the x-coordinates when the gradient is zero.

   $$0 = 3x^2 + 12x - 15$$

Using PLYSMLT:

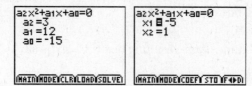

Next, determine the sign of the first derivative on either side of each x value.

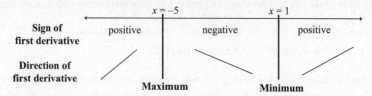

Finally, determine the value of y by substituting each x back in to the original function.

$$h(-5) = (-5)^3 + 6(-5)^2 - 15(-5) + 4 = 104$$
$$h(1) = (1)^3 + 6(1)^2 - 15(1) + 4 = -4$$

The coordinates $(1, -4)$ were found in part a.

(e) **$m = -15$**

The normal line is perpendicular to the tangent line, so $\dfrac{1}{15} \times m = -1$. Thus, $m = -15$.

(f) **$x = 0$**

The gradient at the point is -15, so let $h'(x) = -15$ and solve using PLYSMLT.

$$-15 = 3x^2 + 12x - 15$$
$$0 = 3x^2 + 12x$$

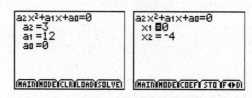

The problem states $x \geq 0$, so the value of -4 cannot be included.

(g) **$y = -15x + 4$**

To write the equation of the tangent, you need the gradient, x, and y. You know the gradient and the x-coordinate. To find the y-coordinate, substitute 0 for x into the original function.

$$h(0) = (0)^3 + 6(0)^2 - 15(0) + 4$$
$$h(0) = 4$$

Now, using $y = mx + c$:

$$4 = -15(0) + c$$
$$4 = c$$
$$y = -15x + 4$$

2. (a) (i) $l = 26 - 2x$

   (ii) $w = 20 - 2x$

   Since each side will have a square cut from two corners, both the length and the width will decrease by $2x$.

(b) $V = x(26 - 2x)(20 - 2x)$

   $V = 4x^3 - 92x^2 + 520x$

   $$V = l \times w \times h$$

   The length is $26 - 2x$, the width is $20 - 2x$, and the height is x.

(c) $\dfrac{dV}{dx} = 12x^2 - 184x + 520$

   Apply the first derivative rule to each term.

   $$4x^3 \Rightarrow 3(4x^{3-1}) = 12x^2$$
   $$-92x^2 \Rightarrow 2(-92x^{2-1}) = -184x$$

   The linear term differentiates to the coefficient.

(d) $x = 3.74$ cm

   The maximum occurs where the first derivative is zero.

   $$0 = 12x^2 - 184x + 520$$

   Using PLYSMLT:

   The value of 11.6 is not possible given the original dimensions.

(e) $V = 867$ cm^3

   To determine the maximum volume, substitute 3.74 for x into the volume equation.

   $$V = 4(3.74)^3 - 92(3.74)^2 + 520(3.74)$$
   $$V = 867.195$$

   Therefore, the volume is 867 when rounded to 3 significant figures.

(f) $S = 151.8$ cm^3 (or 152 cm^3)

   To determine the amount of sand, substitute 6 in for t and simplify.

   $$S = 1.3(6)^2 + 17.5(6)$$
   $$S = 151.8$$

Differential Calculus

(g) **k = 10**

First, determine $\dfrac{dS}{dt}$ by applying the first derivative rule to each term.

$$\frac{dS}{dt} = 2.6t + 17.5$$

You know $\dfrac{dS}{dt} = 43.5$, so solve for t.

$$43.5 = 2.6t + 17.5$$
$$26 = 2.6t$$
$$10 = t$$

(h) **20.0 seconds**

The volume of the box is 867 cm³, which means $S = 867$.

$$867 = 1.3t^2 + 17.5t$$

To solve using PLYSMLT, set the equation equal to zero.

$$0 = 1.3t^2 + 17.5t - 867$$

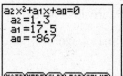 

CHAPTER OBJECTIVES

Before you move on to the Practice Tests, you should be able to:

☐ Apply the first derivative rule to find the first derivative (or gradient function) of a given function

☐ Rewrite rational functions using negative exponents

☐ Determine the gradient of a tangent line to a function at a certain point

☐ Determine where the tangent line of a function has a certain gradient

☐ Write the equation of a tangent line to a curve at a certain point

☐ Write the equation of the normal line to a curve at a certain point

☐ Determine coordinates on a curve whose tangent lines are parallel or perpendicular

☐ Determine the intervals in which a function is increasing or decreasing

☐ Find and label all stationary points of a function

☐ Find maximum or minimum values of real-world situations using the first derivative

Differential Calculus

PRACTICE TEST 1

PRACTICE TEST 1

Paper 1

Each question is worth 6 marks. You do not have to show work to earn full marks.

1. Given $f(x) = \dfrac{5}{x^3} - \dfrac{2}{x^2} + \dfrac{3}{x} - 6x + 1,$

   (a) Calculate $f'(x)$.

   (b) Find $f'(-1)$.

   (c) Explain what the value of $f'(-1)$ represents.

2. Let p stand for the proposition "I will wear a hat". Let q stand for the proposition "it is cloudy".

   (a) Write the following statements in symbolic logic form:

   (i) "I will wear a hat if and only if it is not cloudy".

   (ii) "Either I will not wear a hat or it will be cloudy, but not both".

   (b) Write down, in words, the contrapositive of the statement, "If it is cloudy, then I will wear a hat".

3. The table below shows the number of men and women in a small town who voted in a local election.

	Voted	Did not vote	Total
Male	57	10	67
Female	45	18	63
Total	102	28	130

   (a) If a person was selected at random from the town, find the probability that the person

   (i) voted and is female,

   (ii) did not vote,

   (iii) did not vote, given that the person selected is male.

   (b) If two randomly selected people were selected from the town, find the probability that both voted.

4. A bag contains 6 red and 4 green candies.

Paulina randomly selects one sweet out of the bag and eats it. Then she randomly selects a second sweet. Below is a tree diagram showing Paulina's possible choices. Two of the probability values are missing.

(a) Fill in the missing probability values on the tree diagram.

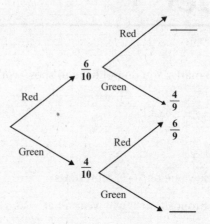

(b) If Paulina eats two candies, what is the probability that she will eat at least one green?

(c) What is the probability Paulina will select a red candy given the first was green?

5. Paige is training for an endurance cycling challenge. She rides 1.5 kilometers in her first week of training, 2.25 kilometers during the second week, 3 kilometers during the third, and so on.

(a) Calculate the number of kilometers Paige rides during her tenth week of training.

(b) Find the total number of kilometers ridden by the end of her tenth week of training.

The cycling challenge involves riding 125 kilometers in one day.

(c) Determine the number of training weeks Paige must ride before she has ridden a **total** of 125 kilometers.

6. The data below display the number of cars parked at a shopping center on 21 randomly selected days last month.

5	8	13	16	17	18	19	20	21	22	22
22	28	31	33	35	36	37	40	41	41	

(a) Determine the

    (i) lower quartile,

    (ii) median,

    (iii) upper quartile.

(b) Complete the frequency table below.

Number of Cars	Frequency
1–10	
11–20	
21–30	
31–40	
41–50	

(c) State whether the data is discrete or continuous.

7. A professor surveyed 200 recent college graduates to determine if the degree obtained was independent of employment status. The majors were engineering, education, marketing, accounting, and computer science. The graduates were either employed or unemployed. A χ^2 test was conducted at the 5% significance level.

(a) Write down the null hypothesis.
(b) Find the number of degrees of freedom for this test.
(c) If the calculated p-value was 0.032, determine if the degree obtained is independent of employment status. Give a clear reason for your answer.

8. Let X be normally distributed with a mean of 75 and a standard deviation of 10.

(a) On the diagram below, shade the area representing $P(X < 65)$.

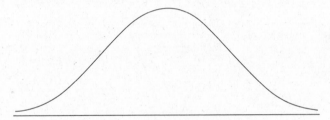

(b) Calculate the area of the shaded region above.
(c) Find the $P(65 < X < 95)$.

9. A boat is 450 meters from the base of a cliff. The angle of elevation from the boat to the top of the cliff is 25°.

(a) Draw a diagram representing the situation. Clearly label the distance and the angle given.
(b) Find the height of the cliff in **kilometers**.

The boat moves closer to the base of the cliff such that the angle of elevation increases to 30°.

(c) Determine the distance traveled by the boat.

10. Ariana was in charge of a game for children at a school festival. Children tossed a bean bag from a set distance and tried to land the bean bag in a small basket. They could toss the bean bag as many times as they needed in order to win a piece of candy.

During a one-hour period, she recorded the number of tosses it took each child before winning.

The results are shown in the frequency table below.

Number of Tosses	Frequency
1–3	3
4–6	6
7–9	9
10–12	5
13–15	2

(a) Calculate an approximate

   (i) mean,

   (ii) standard deviation.

(b) What is the probability a child would need more than nine tosses to win?

(c) Suppose the number of tosses follows a normal distribution. What is the probability a randomly selected child would need fewer than five tosses?

11. An ice-cream cone with one scoop of ice cream, as shown below, has a vertical angle of 20° and a slant height of 15 cm.

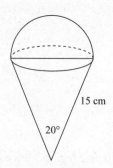

15 cm

20°

(a) Calculate the length of the radius of the ice-cream cone.

(b) Calculate the volume of the cone with the scoop of ice cream. Round the answer to the nearest cubic centimeter.

12. The diagram below shows the graphs of two functions, $f(x) = 4x^2 - 1$ and $g(x) = 2^x$.

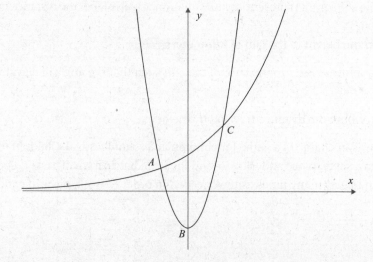

(a) Write down the coordinates of B, the y-intercept of $f(x) = 4x^2 - 1$.

(b) Determine the x-coordinates for the two points, A and C, where the graphs intersect.

(c) Determine the values of x for which $4x^2 - 1 < 2^x$.

13. Genghis invests 3 100 Chinese Yuan in an account that earns 3.25% interest per annum compounded quarterly.

(a) Calculate, to the nearest Yuan, how much Genghis will have in his account after 5 years.

When Genghis has saved 3 800 Yuan, he will vacation in Italy.

(b) Determine the number of years before Genghis will have saved enough money for his vacation.

(c) Before he leaves, Genghis converts 2 000 Yuan to euro for spending money. If the exchange rate is 1 Yuan = 0.12 Euro, how many euros will Genghis receive?

14. Consider the function $g(x) = \dfrac{1}{2}x^2 - 2 + \dfrac{4}{x}$.

(a) On the grid below, sketch the graph of $g(x)$ for the domain $-4 \leq x \leq 5$.

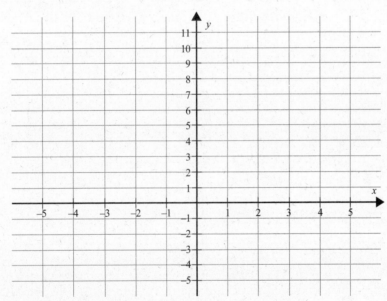

(b) Write down the value of x-intercept to the nearest tenth.

(c) On the same grid, sketch the graph of $h(x) = 2^{-x} + 5$ for the domain $-4 \leq x \leq 5$.

(d) Hence or otherwise, solve $g(x) = h(x)$.

15. The truth table shows part of the truth values for the compound statement $[\neg q \wedge (p \Rightarrow q)] \Rightarrow \neg p$

p	q	$\neg q$	$p \Rightarrow q$	$\neg q \wedge (p \Rightarrow q)$	$\neg p$	$[\neg q \wedge (p \Rightarrow q)] \Rightarrow \neg p$
T	T	F	T	F	F	T
T	F	T			F	
F	T		T		T	
F	F	T	T	T	T	T

(a) Fill in the missing truth values on the table.

(b) Determine if the argument $[\neg q \wedge (p \Rightarrow q)] \Rightarrow \neg p$ is logically valid.

Paper 2

Each question is worth a different amount of marks. You must show **all** work to earn full marks.

1. Below are the scores for 15 students on their final examination for twelfth grade math and English.

Math (x)	68	88	92	55	72	79	81	64	32	95	71	82	86	61	78
English (y)	72	86	92	65	68	85	85	70	43	99	79	80	82	70	78

    (a) Draw a scatter diagram to show the data above. Let 1 cm represent 10 on both the horizontal and vertical axes.

    (4)

    (b) Write down the Pearson's product-moment correlation coefficient, r, for the data.

    (1)

    (c) Describe what the correlation coefficient suggests about the relationship between the two variables.

    (2)

    (d) Calculate the mean point.

    (2)

    (e) Write the equation of the regression line for y on x in the form $y = ax + b$.

    (2)

    (f) Draw the regression line on the scatter diagram.

    (2)

    (g) Use your equation for the regression line to predict the math score for a student who scored a 25 on the English examination.

    (2)

    (h) Determine if the prediction from part g is inappropriate. Justify your answer.

    (2)

    (Total 17 marks)

2. A sheet of square cardboard with side length 12 cm will be made into an open box by cutting equal-sized squares from each corner and folding up the four edges, as shown below.

    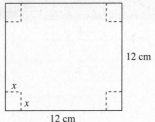

    12 cm

    x

    x

    12 cm

    Not drawn to scale

(a) Write an expression in terms of x for the side length of the box.

(1)

(b) Show that the volume, $V(x)$, of the box will be $V(x) = 4x^3 - 48x^2 + 144x$.

(2)

(c) Find $V'(x)$.

(2)

(d) Find the height of the box that yields the greatest volume.

(3)

(e) Determine the maximum volume possible.

(2)

A sheet of cardstock is to be rolled such that a cylinder is formed. The cylinder has the same volume as the open-top box and the height of the cylinder is 8 cm.

(f) Determine the radius of the cylinder. Round your answer to the nearest centimeter.

(4)

(Total 14 marks)

3. Sixty high school students are asked of which of the following academic clubs they are a member: Science Club (S), Math Club (M), or History Club (H).

   28 students are members of the Science Club
   28 students are members of the Math Club
   26 students are members of the History Club
   14 students are members of the Science and the Math Club
   11 students are members of the Science and the History Club
   10 students are members of the Math and the History Club
   1 student is a member of all three

(a) Represent the above information on a Venn diagram. Clearly label each region with the number of students.

(4)

(b) Find the number of students that are not members of any of the academic clubs.

(2)

(c) A student is chosen at random. Find the probability that the student is a member of:

   (i) the History Club or the Math Club, but not both;
   (ii) the Science Club, given the student is a member of the Math Club.

(4)

The students were also asked to identify their postsecondary education plans. The results are displayed below and categorized based on gender.

	Post Secondary Plans			
	Attend 2 year college	Attend 4 year college	Begin Working	Total
Male	10	14	8	32
Female	14	9	5	28
Total	24	23	13	60

A chi-squared test of independence is to be performed to determine if there is any association in postsecondary plans and gender.

(d) State the null hypothesis.

(1)

(e) Show that the number of degrees of freedom is 2.

(1)

(f) Write down the:
   (i) chi-squared calculated value,
   (ii) p-value.

(2)

At the 5% significance level, $\chi^2 = 7.81$.

(g) Determine the conclusion of the chi-squared test. Justify your answer.

(2)

(Total 16 marks)

4. *VABCD* is a square-based pyramid, and *V* is directly above the center of the square.

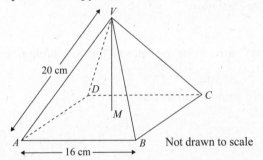

16 cm

20 cm

Not drawn to scale

(a) Find:

   (i) the length *AM*,
   (ii) the height *VM*.

(4)

(b) Calculate the size of the angle *VAM*.

(2)

(c) Calculate the height of one of the pyramid's triangular faces.

(2)

(d) Calculate the total surface area of the pyramid.

(3)

A square cuboid is to be made with the same surface area as the pyramid.

(e) Determine the length of the side, *s*, of the cuboid.

(3)

(Total 14 marks)

5. Thomas has a custom furniture business, and he is planning to launch a website to market his products worldwide. In 2010, he invested 10 000 USD in a savings account that earns 4.5% interest per annum compounded monthly.

(a) How much **interest** will Thomas have earned after 5 years? Give your answer to the nearest dollar.

(3)

Thomas needs 15 000 USD to launch his website.

(b) Determine the number of years before Thomas will have enough money to launch his website.

(3)

In 2010 Thomas purchased a table saw for 1 200 USD, but the table saw depreciates at a rate of 9.8% per annum.

(c) Find the value of the table saw after 5 years. Round answer to the nearest hundredth.

(2)

The sofa table is the most popular piece of furniture ordered by customers. Thomas sells the sofa table for 350 USD plus an 8% shipping and handling fee.

(d) Determine the cost a customer pays for one sofa table.

(2)

(e) A customer in Italy purchases a sofa table. Find her cost in euro if 1 euro = 1.30 USD. Round answer to the nearest euro.

(2)

The monthly profit Thomas earns from the sales of the sofa table follows the function $P(x) = -2x^2 + 120x - 75$.

(f) Calculate the maximum profit Thomas can earn.

(2)

(Total 14 marks)

6. The quadrilateral *ABCD* graphed below has vertices at *A*(–3, 3), *B*(0, 6), *C*(2, 2), and *D*(–2, 1).

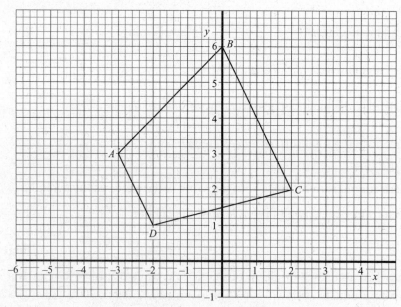

(a) Calculate the gradient of line *AD*.

(2)

(b) Show that line *AD* is parallel to line *BC*.

(2)

(c) Write the equation of line *AB* in the form $ax + by + d = 0$, where $a, b, c, \in \mathbb{Z}$.

(2)

The line *AB* intersects line *CD* at point *P*.

(d) Determine the coordinates of point *P*.

(2)

Consider the triangle *BCP*.

(e) Calculate the length of side:

  (i) *BP*,

  (ii) *CP*.

(4)

The length of side *BC* is $\sqrt{20}$.

(f) Calculate the size of angle *BPC*.

(3)

(Total 15 marks)

SOLUTIONS

PAPER 1

1. (a) $f'(x) = -\dfrac{15}{x^4} + \dfrac{4}{x^3} - \dfrac{3}{x^2} - 6$

First, rewrite function using integer exponents. Next, apply the first derivative rule to each term. Finally, put all negative exponents back in fraction form.

$f(x) = 5x^{-3} - 2x^{-2} + 3x^{-1} - 6x + 1$

$5x^{-3} \Rightarrow -3(5x^{-3-1}) = -15x^{-4}$

$-2x^{-2} \Rightarrow -2(-2x^{-2-1}) = 4x^{-3}$

$3x^{-1} \Rightarrow -1(3x^{-1-1}) = -3x^{-2}$

$-6x$ differentiates to the coefficient of -6.
1 differentiates to zero.

C3 for all four terms of the derivative correct

OR

- **M1** for rewriting $f(x)$ using integer exponents
- **A1** for correctly writing $f'(x)$ using negative exponents
- **A1** for three or more terms correct

(b) $f'(-1) = -28$

Substitute -1 in for x and simplify.

$f'(-1) = -\dfrac{15}{(-1)^4} + \dfrac{4}{(-1)^3} - \dfrac{3}{(-1)^2} - 6$

$f'(-1) = -28$

C2 for correct answer

OR

- **M1** for substituting -1 into $f'(x)$
- **A1** for correct answer

(c) **−28 represents the gradient of the tangent line when $x = -1$.**

The value of the first derivative always represents the gradient of the tangent line at a specific x.

C1 for correct answer that includes "gradient" and "at $x = -1$"

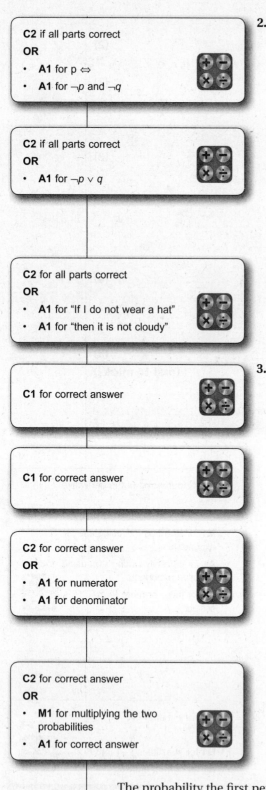

C2 if all parts correct
OR
- **A1** for p ⇔
- **A1** for ¬p and ¬q

C2 if all parts correct
OR
- **A1** for ¬p ⊻ q

C2 for all parts correct
OR
- **A1** for "If I do not wear a hat"
- **A1** for "then it is not cloudy"

C1 for correct answer

C1 for correct answer

C2 for correct answer
OR
- **A1** for numerator
- **A1** for denominator

C2 for correct answer
OR
- **M1** for multiplying the two probabilities
- **A1** for correct answer

---

2. (a) (i) $p \Leftrightarrow \neg q$
   - "if and only if" is the symbol ⇔.
   - "wearing a hat" is proposition p.
   - "not cloudy" is the negation of proposition q, ($\neg q$).

   (ii) $\neg p \veebar q$
   - "Either . . . or . . . but not both" is exclusive disjunction, ⊻.
   - "not wear a hat" is the negation of proposition p, ($\neg p$).
   - "be cloudy" is proposition q.

   (b) **If I do not wear a hat, then it is not cloudy.**
   The contrapositive is the negation of the implication while switching the order.

   Implication: $p \Rightarrow q$ Contrapositive: $\neg q \Rightarrow \neg p$

3. (a) (i) $\dfrac{45}{130}$ or $\dfrac{9}{26}$ or **0.346**

   45 people voted AND were female, and there were a total of 130 people.

   (ii) $\dfrac{28}{130}$ or $\dfrac{14}{65}$ or **0.215**

   The number of people who did not vote was 28 out of a total of 130 people.

   (iii) $\dfrac{10}{67}$ or **0.149**

   The randomly selected person is male, so the total number of outcomes is now just 67 (total number of males). From there you want only non-voters, so 10 males did not vote, hence $\dfrac{10}{67}$.

   (b) $\dfrac{1717}{2795}$ or **0.614**

   Two people are being selected, so you must multiply the probabilities together. However, you are selecting **without** replacement, since the same person cannot be selected twice.

   The probability the first person voted is $\dfrac{102}{130}$.

   The probability the second person voted is $\dfrac{101}{129}$, since the first person also voted.

   The total probability is $\dfrac{102}{130} \times \dfrac{101}{129} = 0.614$.

4. (a)

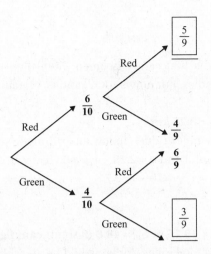

Since Paulina is eating the candies, the second time she reaches into the bag there is one less candy, so the total decreases from 10 to 9.

- If she selected a red candy first, then there would only be 5 remaining.
- If she selected a green candy first, there would only be 3 remaining.

(b) $\frac{2}{3}$ **or 0.667**

There are two ways to calculate the probability.

Method 1: Pauline wants at least one green. You can find the complement to the event, which is no green.

Then, 1—P(no green) is the answer.

In order for her to get no green candies, she must select 2 red candies.

$$1 - \frac{6}{10}\left(\frac{5}{9}\right)$$

$$= \frac{2}{3} \text{ or } 0.667$$

Method 2: There are three ways to get at least one green candy:

- Red then green $\frac{6}{10} \times \frac{4}{9}$
- Green then red $\frac{4}{10} \times \frac{6}{9}$
- 2 green $\frac{4}{10} \times \frac{3}{9}$

The solution is the sum of each probability.

$$\frac{6}{10}\left(\frac{4}{9}\right) + \frac{4}{10}\left(\frac{6}{9}\right) + \frac{4}{10}\left(\frac{3}{9}\right)$$

$$= \frac{2}{3} \text{ or } 0.667$$

C1 for correct answer

C2 for correct answer

OR

- **M1** for using arithmetic sequence
- **A1** for correct answer

C2 for correct answer

OR

- **M1** for using arithmetic sum
- **A1** for correct answer

C2 for correct answer

OR

- **M1** for using arithmetic sum
- **A1** for correct answer

C3 for all three correct

OR

A1 for each correct answer

(c) $\dfrac{6}{9}$

If the first candy is green, then there are still 6 red candies, but only 9 total candies remain.

5. (a) $u_{10} = 8.25$ **km**

The miles ridden follow an arithmetic sequence.

$u_1 = 1.5$ and $d = 2.25 - 1.5 = 0.75$

$u_{10} = 1.5 + (10 - 1)(0.75)$

$u_{10} = 8.25$

(b) $S_{10} = 48.75$ or $S_{10} = 48.8$ **(3 significant figures)**

The total miles ridden would be an arithmetic sum.

$$S_{10} = \frac{10}{2}(1.5 + 8.25) \quad \textbf{OR} \quad S_{10} = \frac{10}{2}(2(1.5) + (10 - 1)(0.75))$$

$$S_{10} = 48.75 \qquad\qquad\qquad S_{10} = 48.75$$

(c) $n = 17$ **weeks**

Paige must ride a total of 125 kilometers, so you know the sum is 125. You must find n.

$$125 = \frac{n}{2}(2(1.5) + (n - 1)(0.75))$$

You can solve by letting $Y_1 = \dfrac{x}{2}(2(1.5) + (x - 1)(0.75))$ and looking at the table for the first value larger than 125.

You could also solve by simplifying the equation, setting equal to zero, and using PLYSMLT.

$$125 = \frac{n}{2}(3 + 0.75n - 0.75)$$

$$250 = n(0.75n + 2.25)$$

$$0 = 0.75n^2 + 2.25n - 250$$

In an arithmetic sequence, n must be a natural number, so $n = 17$.

6. (a) (i) $Q_1 = 17.5$
   (ii) **Median** $= 22$
   (iii) $Q_3 = 35.5$

First, enter the data into L_1 and calculate 1-VAR STATS.

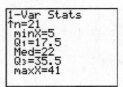

(b)

Number of Cars	Frequency
1–10	2
11–20	6
21–30	5
31–40	6
41–50	2

C2 for all 6 correct frequencies

OR

A1 for 4–5 correct frequencies

Count up the number of data within each interval.

(c) **discrete**

The data represent the count of cars, which is discrete.

C1 for correct answer

7. (a) **H_0: Degree obtained and employment status are independent.**

The null hypothesis of a chi-squared test is always that the two factors are independent. Since the professor was comparing the degree obtained with employment status, those were the two factors.

C1 for correct hypothesis
(key word = independent)

(b) **$df = 4$**

Degrees of freedom are found by:

$df = (\text{rows} - 1)(\text{columns} - 1)$

No chart was given, but there were 5 majors and 2 employment statuses, $df = (5 - 1)(2 - 1)$, which is 4.

C2 for correct answer

OR

- **M1** for using df formula
- **A1** for the correct answer

(c) **$0.032 < 0.05$ so reject H_0. Degree obtained and employment status are dependent.**

The p-value was smaller than the significance, so reject the null hypothesis. Since the null hypothesis is rejected, you conclude the two factors are dependent.

C3 for correct answer of "dependent" with justification

OR

- **A1** for $0.032 < 0.05$
- **A1** for reject H_0
- **A1** for conclusion of dependent

8. (a) $P(X < 65)$ is the area from 65 to the left.

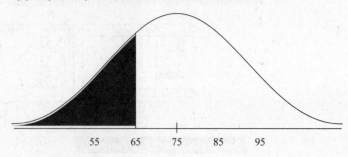

C1 for correct shading

(b) **0.159 or 15.9% (0.16 or 16% are also accepted)**

Based on the shaded diagram in part *a*, you find NORMALCDF($-$E99, 65, 75, 10).

An alternate solution is to use the 68 – 95 – 99% rule. Since 65 is one standard deviation below the mean, the area below the line is $\frac{100-68}{2} = 16\%$.

C2 for correct answer

OR

- **M1** for using NORMALCDF
- **A1** for the correct answer

(c) **0.819 or 81.9% (0.815 or 81.5% are also accepted)**

You want the area between 65 and 95, so find NORMALCDF (65, 95, 75, 10).

An alternate solution is to use the 68 – 95 – 99% rule.

C3 for correct answer

OR

- **M1** for using NORMALCDF
- **A1** for correct values used
- **A1** for the correct answer

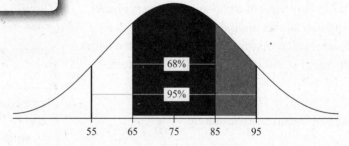

The bulk of the shading is between 65 and 85, which you know is 68%. You have the grey portion remaining, so subtract 68% from 95% and cut that in half. Putting these two areas together gives 81.5%

9. (a)

C1 for correct diagram with the side, angle, and right angle clearly identified

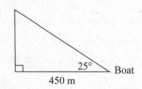

The boat is 450 m from the base of the cliff, so the base of the triangle is 450 m. The angle of elevation from the boat to the cliff refers to the angle at the boat. This is a right triangle.

(b) **0.210 km**

This is a right triangle, and you are looking for the side opposite 25°. You know the side adjacent, so the tangent ratio is used.

$$\tan 25 = \frac{x}{450}$$

$$450 \tan 25 = x$$

$$x = 210 \text{ m}$$

The answer must be in kilometers, so move the decimal three times to the left.

(c) **86.3 m (or 0.0863 km)**

You determined the height of the cliff to be 210 m and the angle of elevation is now 30°.

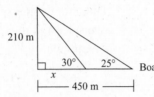

$$\tan 30° = \frac{210}{x}$$

$$x = 210 / \tan 30°$$

$$x = 363.7$$

The distance traveled is the difference between 450 m and 363.7 m.

10. (a) (i) 7.64

   (ii) 3.32

To find an approximate mean and standard deviation, calculate the mid-interval values. These are the averages of the endpoints.

The mid-interval values are 2, 5, 8, 11, and 14. Enter the mid-interval values into L_1 and the frequencies into L_2.

(1-VAR STATS) L_1, L_2 gives the mean and standard deviation.

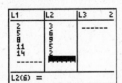 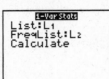 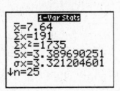

(b) $\frac{7}{25}$ **or 0.28**

The probability a child would need more than 9 tosses would be the intervals 10–12 and 13–15. Adding the frequencies gives 7 chances and there are a total of 25 tosses recorded.

C2 for correct answer

OR

- **M1** for using [NORMALCDF]
- **A1** for correct answer

(c) **0.213**

A sketch of the scenario would be

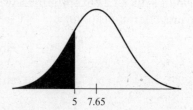

5   7.65

Calculate the approximate mean and standard deviation in part *a*.

Calculate [NORMALCDF]($-$E99, 5, 7.64, 3.32) $= 0.213$.

11. (a) **2.60 cm**

The slant heights of the cone are the same, so you have two sides and the included angle. Thus, solve using the cosine rule.

$$d^2 = 15^2 + 15^2 - 2(15)(15)\cos 20$$
$$d^2 = 27.138$$
$$d = 5.21$$

The diameter is 5.21 cm, so the radius is half of 5.21 cm, 2.60 cm.

C3 for correct answer

OR

- **M1** for using cosine rule
- **A1** for diameter length
- **A1** for correct radius

(b) **142 cm³**

In order to find the volume of the cone, you need the height of the cone.

Using the Pythagorean theorem, $h^2 + (2.60)^2 = 15^2$ gives a height of 14.8 cm.

C3 for correct answer

OR

- **A1** for correct cone volume
- **A1** for correct ice cream volume
- **A1** for correct total volume

The volume of a cone is $V = \frac{1}{3}\pi r^2 h$

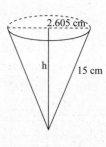

2.605 cm

h   15 cm

$$V = \frac{1}{3}\pi(2.60)^2(14.8) = 105.$$

The volume of a sphere is $V = \frac{4}{3}\pi r^3$

$$V = \frac{4}{3}\pi(2.60)^3 = 73.6.$$

You need half of the sphere.

The volume of the ice cream is $\frac{73.6}{2} = 36.8$.

Thus, the total volume is $105 + 36.8 = 141.8$, which rounds to 142 cm³.

C1 for correct answer

12. (a) **(0, −1)**

The *y*-intercept is where $x = 0$. Substituting 0 in for x gives $y = 4(0)^2 - 1 = -1$.

(b) **$x = -0.641, 0.834$**

Using you GDC, find where the graphs intersect.

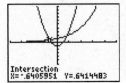

(c) **$(-0.641, 0.834)$ or $-0.641 < x < 0.834$**

The graph $f(x) = 4x^2 - 1$ is less than $g(x) = 2^x$ when the curve falls below. This area is highlighted on the graph. Therefore, the interval when $4x^2 - 1 < 2^x$ is true is between the values of x of the intersection.

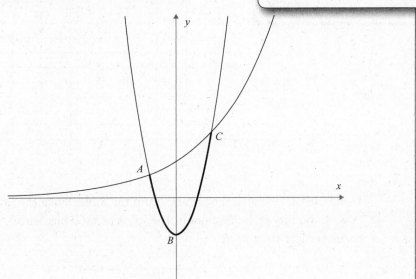

13. (a) **3 645 Yuan**

The interest is compounded quarterly, so $k = 4$.

$$FV = 3100\left(1 + \frac{3.25}{100 \times 4}\right)^{4 \times 5}.$$

(b) **7 years**

The final value is 3 800, thus $FV = 3800$.

$$3800 = 3100\left(1 + \frac{3.25}{100 \times 4}\right)^{4n}$$

Let $Y_1 = 3100\left(1 + \frac{3.25}{400}\right)^{4x}$ and look at the table. When the y-value is larger than 3 800, the coordinate x-value is the answer.

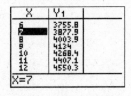

C1 for correct answer

(c) **240 Euro**

$$2\,000\,\text{Yuan} \times \frac{0.12\,\text{Euro}}{1\,\text{Yuan}} = 240\,\text{Euro}$$

14. (a)

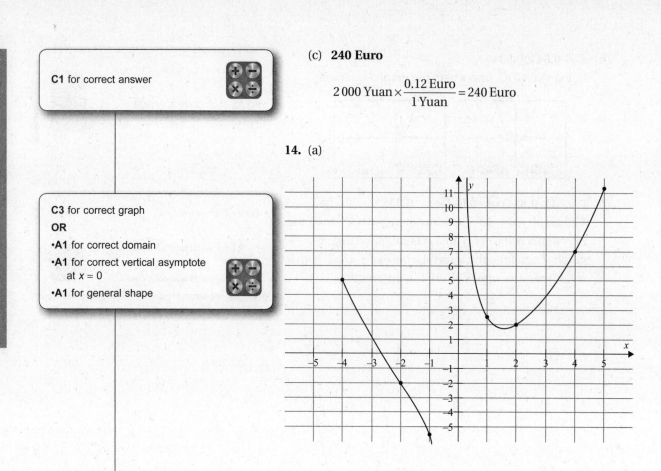

C3 for correct graph

OR

•**A1** for correct domain

•**A1** for correct vertical asymptote at $x = 0$

•**A1** for general shape

Using the GDC, plot several points from the table. Since the domain begins when $x = -4$, use this as the first point. The value of $x = 0$ has "error", which corresponds to the vertical asymptote.

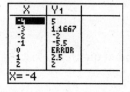

C1 for correct answer

(b) **−2.6**

Using the GDC, find the zero of the function

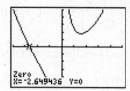

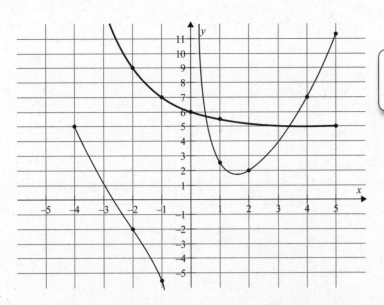

C1 for correct answer

(c) Using the GDC, you can plot several points off the table. Since the domain begins when $x = -4$, use this as the first point but the y-value is too large. The most reasonable point to graph first is $(-2, 9)$. Choose a few more points to get the shape correct.

(d) $x = 0.530, 3.44$

Using your GDC find where the graphs intersect.

C1 for both correct answers

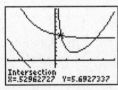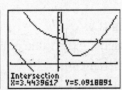

15. (a)

p	q	$\neg q$	$p \Rightarrow q$	$\neg q \wedge (p \Rightarrow q)$	$\neg p$	$[\neg q \wedge (p \Rightarrow q)] \Rightarrow \neg p$
T	T	F	T	F	F	T
T	F	T	**F**	**F**	F	**T**
F	T	**F**	T	**F**	T	**T**
F	F	T	T	T	T	T

C4 for all 6 correct values

OR

- **A3** for 5 correct
- **A2** for 4 correct
- **A1** for 3 correct

- $\neg q$ are the opposite truth values of q.
- $p \Rightarrow q$ is only **false** when p is true but q is false.
- $\neg q \wedge (p \Rightarrow q)$ can only be true when $\neg q$ and $p \Rightarrow q$ are both true.
- $\neg p$ should be the opposite truth values of p.
- $[\neg q \wedge (p \Rightarrow q)] \Rightarrow \neg p$ is only **false** when $\neg q \wedge (p \Rightarrow q)$ is true but $\neg p$ is false.

(b) **It is logically valid because it is a tautology.**

Since all truth values of the answer column were true, the argument is a tautology and logically valid.

C2 for correct answer

OR

- **A1** for "logically valid"
- **A1** for tautology

PAPER 2

1. (a)

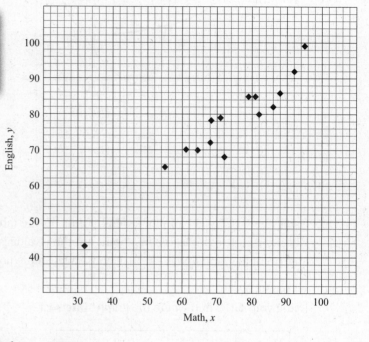

(b) $r = 0.961$

To calculate the value of r, enter the data into L_1 and L_2 in the STAT menu.

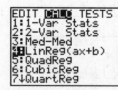

Next, find the linear regression.

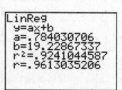

Remember, diagnostics must be "On" to see the correlation coefficient in the output.

(c) **There is a very strong, positive linear relationship between the variables.**

The closer r is to 1 or −1, the stronger the linear relationship is between the data. Thus, $r = 0.961$ indicates a very strong relationship. Since r is a positive value, the data have a positive relationship.

(d) **(73.6, 76.9)**

Using the GDC, calculate 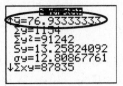.

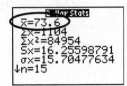

(e) $y = 0.784x + 19.2$

The equation was already generated in the output from part *c*.

(f)

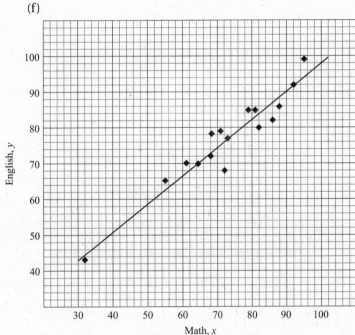

The regression line must pass through the mean point. Since the scale did not begin at 0 on the *y*-axis, you cannot use the *y*-intercept at the second point. Draw the line so roughly half the points lie above and half below.

(g) $25 = 0.784x + 19.2$

$7.40 = x$

Since 25 is an English score, then $y = 25$. Substitute 25 in for *y* and solve.

(h) Predicting the math score when a student scored a 25 in English is inappropriate, since 25 is too far outside the data range.

The English scores range from 43 to 99, so estimating with a score of 25 is inappropriate.

A1 for correct expression

- **M1** for multiplying $l \times w \times h$
- **A1** for correct values of 3 dimensions

- **A2** for all three terms correct
- **OR A1** for two terms correct

- **M1** for setting $V'(x) = 0$
- **A2** for correct value of 2
- (**Only A1** if answer was $x = 6, 2$)

- **M1** for substituting 2 for x into $V(x)$
- **A1** for correct answer

- **M1** for using volume of cylinder formula
- **A1** for substituting 128 in for V and 8 in for h
- **M1** for solving the resulting equation
- **A1** for correct answer

2. (a) $l = 12 - 2x$

Since a square is being removed from each corner, the length and the width will be reduced by two x's. Thus, the new dimension is subtracted by $2x$.

(b) $V(x) = (12 - 2x)(12 - 2x)x$
$V(x) = (144 - 48x + 4x^2)x$
$V(x) = 4x^3 - 48x^2 + 144x$

$V = l \times w \times h$ and the side lengths were determined in part a. The height is x, since that is the portion which folds upward to make the actual box. Substitute in each piece and multiply.

(c) $V'(x) = 12x^2 - 96x + 144$

Apply the first derivative rule to each term.

(d) $0 = 12x^2 - 96x + 144$
$x = 2$

First set $V'(x) = 0$ and solve using [PLYSMLT].

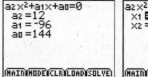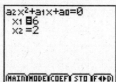

Disregard the answer of 6, since the side length of 12 cannot have two squares of size 6 removed. It is impossible.

(e) $V(2) = 4(2)^3 - 48(2)^2 + 144(2)$
$V = 128 \text{ cm}^3$

The maximum volume occurs when $x = 2$ as was determined in part d.

(f) $128 = \pi r^2(8)$
$16 = \pi r^2$
$2.26 \text{ cm} = r$

The volume of a cylinder is $V = \pi r^2 h$, since the area of the base is the area of a circle.
The cylinder must have the same volume as the box, thus $V = 128$. The height of 8 is given. Substitute in those values and solve for r.

3. (a)

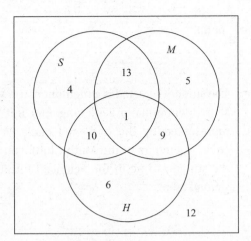

Label the innermost region first. This is represented by the 1 student who is a member of all 3 clubs. Then subtract 1 from:

- 14 students are a member of the Science and the Math Club
- 11 students are a member of the Science and the History Club
- 10 students are a member of the Math and the History Club

This gives the other 3 numbers in the overlapping sections. Finally, subtract the three labeled numbers in each circle from the provided totals to determine the last 3 places in the circles.

(b) $60 - (28 + 14 + 6) = x$
$12 = x$

Since there were 60 total students, you can find the "leftover" amount by adding the numbers inside the Venn diagram and subtracting from 60.

(c) (i) $\dfrac{34}{60}$ or $\dfrac{7}{15}$ or **0.567**

You need all the students who are members of History or Math Club, but not both (the overlapping parts of the two circles). The shaded regions identify those students.

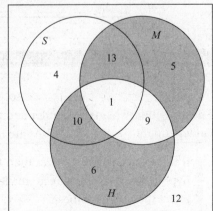

- **A1** for correct numerator
- **A1** for correct denominator

(ii) $\dfrac{14}{28}$ or $\dfrac{1}{2}$ or **0.5**

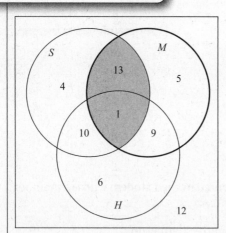

The student selected is a member of the Math Club, so the denominator is no longer 60, it is 28. This is represented by the darkened circle. Within that circle (members of the Math Club), you also need the student to be in the Science Club. This is the shaded area.

A1 for correct hypothesis
(key word "independent")

(d) **H₀: Gender and postsecondary plans are independent.**

The null hypothesis is always that the two factors are independent.

M1 for $(2 - 1)(3 - 1)$

(e) $df = (2 - 1)(3 - 1) = 1 \times 2 = 2$

Degrees of freedom are found by $df = (r - 1)(c - 1)$, where r is the number of rows and c is the number of columns.

- **A1** for correct chi-squared value
- **A1** for correct p-value

(f) (i) $\chi^2_{calc} = \mathbf{2.19}$

(ii) $p = \mathbf{0.335}$

Enter the observed data into matrix A and perform the χ^2 test. The output gives the chi-squared calculated value and the p-value.

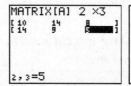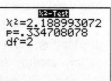

(g) Two possible answers:

Option 1	Option 2
$2.19 < 7.81$	$0.335 > 0.05$
Accept H₀	Accept H₀
Gender and postsecondary plans are independent	Gender and postsecondary plans are independent

- **A1** for accept H₀ (or independence)
- **R1** for correct inequality justification

If the χ^2 calculated is larger than the χ^2 critical, reject the null hypothesis. If the p-value is smaller than the significance level, reject the null hypothesis.

4. (a) (i) $AM^2 = 8^2 + 8^2$

$AM = 11.3$ or $\sqrt{128}$

AM is the hypotenuse of the right triangle shown below.

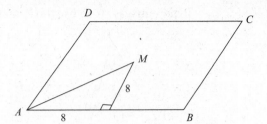

Since *M* is in the center of the base, the length *AB* is divided in half when the perpendicular is dropped.

(ii) $20^2 = 11.3^2 + VM^2$

$VM = 16.5$ or $\sqrt{272}$

VM is the height of the right triangle shown below.

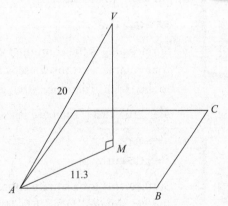

We can use the Pythagorean theorem to find *VM*.

(b) $\cos\theta = \dfrac{11.3}{20}$

$\theta = \cos^{-1}\left(\dfrac{11.3}{20}\right)$

$\theta = 55.6$

Using the triangle *VAM*, you were given the hypotenuse of 20 and you have already calculated *AM* to be 11.3. This means you know the adjacent side and the hypotenuse. Therefore, the appropriate trig function to use is cosine.

- **M1** for using Pythagorean theorem
- **A1** for correct answer

(c) $20^2 = 8^2 + x^2$

$x = 18.3 \text{ or } \sqrt{336}$

The height of each triangular face is the same, and the height is also the perpendicular bisector to the base. The hypotenuse is 20, so use the Pythagorean theorem to solve.

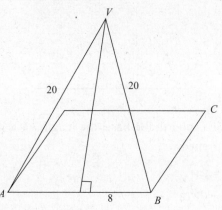

- **A1** for $\left(\dfrac{1}{2}\right)(16)(18.3)$
- **A1** for 16^2
- **A1** for correct answer

(d) $SA = 4\left(\dfrac{1}{2}\right)(16)(18.3) + (16)^2$

$SA = 842 \text{ cm}^2$

Surface area is the combined area of each face. There are 4 triangular faces whose base is 16, and you found the height to be 18.3. Therefore, the area of all the triangles is $4\left(\dfrac{1}{2} \times 16 \times 18.3\right)$. The base is a square, so its area is 16×16.

- **M1** for using surface area of cuboid
- **A1** for $6s^2 = 842$
- **A1** for 11.9

(e) $6s^2 = 842$

$s = 11.8 \text{ cm}$

Surface area for a square cuboid is made up by the area of the six squares' faces, $6s^2$. This value is the same as the pyramid, thus $6s^2 = 842$. Now solve for s.

5. (a) $10\,000\left(1 + \dfrac{4.5}{100 \times 12}\right)^{5 \times 12} = 12\,518$

Interest: 12 518 – 10 000 = 2 518 USD

- **M1** for using compound interest formula
- **A1** for 12 518
- **A1** for correct answer

Using the compound interest formula, $PV = 10\,000$, $r = 4.5$, $k = 12$ (monthly), $n = 5$.

- **M1** for using compound interest formula
- **A1** for substituting correct values
- **A1** for correct answer

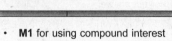

(b) $15\,000 = 10\,000\left(1 + \dfrac{4.5}{100 \times 12}\right)^{12n}$

$n = 10$

Using the compound interest formula, $FV = 15\,000$, $PV = 10\,000$, $r = 4.5$, $k = 12$ (monthly).

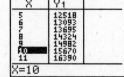

Using the GDC, let $Y_1 = 10\,000(1.00375)^{12x}$ and look at the table for when Y_1 is larger than 15 000. This occurs when $x = 10$ on the GDC, but the variable in the question is n, thus $n = 10$.

(c) $1\,200\left(1-\dfrac{9.8}{100}\right)^{5}=716.50$

Using the compound interest formula, $PV = 10\,000$, $r = 4.5$, $k = 1$, $n = 5$, but since the table saw is depreciating in value, you must subtract the rate from 1.

- **M1** for using compound interest formula
- **A1** for correct answer

(d) $350 + 350(0.05) = 378$ **USD**

Customers will pay \$350 plus an additional 8%, so multiply 350(0.08) to determine the shipping and handling fee. This is part of the total cost, so add the fee of \$28 to the price: \$350 + \$28 = \$378.

- **A1** for calculating 8% of 350
- **A1** for correct answer

(e) $\dfrac{378\,\text{USD}}{1} \times \dfrac{1\,\text{EUR}}{1.30\,\text{USD}} = 291\,\text{EUR}$

OR $\dfrac{378\,\text{USD}}{x\,\text{EUR}} = \dfrac{1.30\,\text{USD}}{1\,\text{EUR}}$

- **M1** for dividing conversion factor into 378
- **A1** for correct answer

To perform a currency conversion, either multiply by the conversion factor or set up a proportion.

- In the first method, you are starting with USD, so multiply by a factor where USD is in the denominator. This will divide out the USD currency and leave Euro.
- In the second method, set up a proportion, cross multiply, and solve.

(f) $-2(30)^{2} + 120(30) - 75 = 1\,725$

The maximum value is the y-value of the vertex.
There are two methods to finding the vertex:

- **M1** for substituting 30 in for x
- **A1** for correct answer
- **OR A2** for 1725 using GDC

Algebraically: Find the axis of symmetry first:

$x = -\dfrac{120}{2(-2)} = 30$. Now, substitute 30 in for x into the quadratic. This gives a maximum profit of \$1\,725.

Graphically: Let $Y_1 = -2x^2 + 120x - 75$, select an appropriate window, and calculate the maximum.

 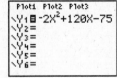 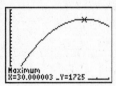

6. (a) $m = \dfrac{1-3}{-2-(-3)}$

$m = -2$

The gradient of any line is determined by $m = \dfrac{y_2 - y_1}{x_2 - x_1}$.

Substitute the values of y and x from points A and D and simplify.

- **M1** for using gradient formula
- **A1** for correct answer

- **A1** for correct gradient of line BC
- **R1** for correct reason (gradients are equal)

(b) **Gradient of BC:** $m = \dfrac{6-2}{0-2}$

$$m = -2$$

Since the lines have the same gradients, the lines are parallel.

Parallel lines have the same gradient. First determine the gradient of line BC. Since $-2 = -2$, the lines are parallel.

- **A2** for all 3 correct terms
- **OR A1** for 2 correct terms

(c) $0 = x - y + 6 \text{ OR } -x + y - 6 = 0$

Looking at the provided graph, line AB has a y-intercept of 6 since the graph intersects the y-axis at the point (0, 6). The line is rising one vertical unit for every one horizontal unit. Thus, the gradient is one.

The line must be written in the form $ax + by + d = 0$, where $a, b, c, \in \mathbb{Z}$, so subtract y over.

- **A1** for correct x-coordinate
- **A1** for correct y-coordinate

(d) **(–6, 0)**

The best way to find point P is to extend the lines on the graph and find where they intersect. The gradient of line AB is one, so continue that pattern. The gradient of line CD is $\dfrac{1}{4}$. The two lines intersect at point (–6, 0).

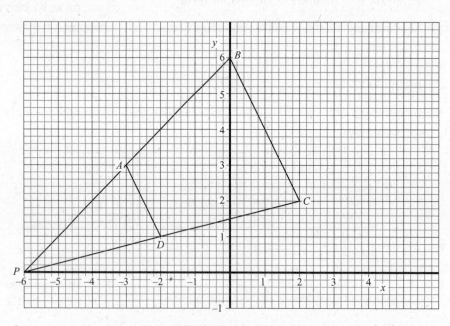

(e) (i) $BP = \sqrt{(0-(-6))^2 + (6-0)^2} = \sqrt{72}$ **or 8.49**

(ii) $CP = \sqrt{(2-(-6))^2 + (2-0)^2} = \sqrt{68}$ **or 8.25**

For both i and ii:
- **M1** for distance formula
- **A1** for correct answer

Since each side of the triangle is defined by two points, the length will be determined by the distance formula: $d = \sqrt{(x_1 - x_2)^2 + (y_1 - y_2)^2}$. Substitute the values from the endpoints of the sides and simplify.

$$\cos\theta = \frac{\left(\sqrt{72}\right)^2 + \left(\sqrt{68}\right)^2 - \left(\sqrt{20}\right)^2}{2\left(\sqrt{72}\right)\left(\sqrt{68}\right)}$$

(f) $\theta = \cos^{-1}(0.85749)$

$\theta = 31.0$

- **M1** for use of cosine rule
- **A1** for correct substituted values
- **A1** for correct answer

You have three sides to the triangle, so to determine the angle use the cosine rule, $\cos A = \dfrac{b^2 + c^2 - a^2}{2bc}$.

Remember, side a is opposite angle A.

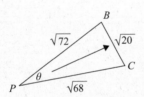

PRACTICE TEST 2

Paper 1

Each question is worth 6 marks. You do not have to show work to earn full marks.

1. The diagram shows a part of the curve $y = f(x)$.

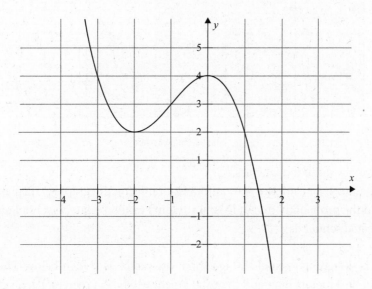

   (a) Write down the values of x, when $f'(x) = 0$.
   (b) Write down the interval of x for which $f'(x) > 0$.
   (c) Determine if the gradient of the tangent line at $x = -3$ will be positive or negative. Justify your answer.

2. The third term of a geometric sequence is 56, and the sixth term is 7.

   (a) Determine the value of the common ratio.
   (b) Calculate the value of the first term.
   (c) Find the sum of the geometric sequence for the first 10 terms.

3. Rachel recorded the number of students, t, late to school each day for an entire month. She displayed the results in the histogram below. She used the intervals $0 \leq t < 2$, $2 \leq t < 4$, etc.

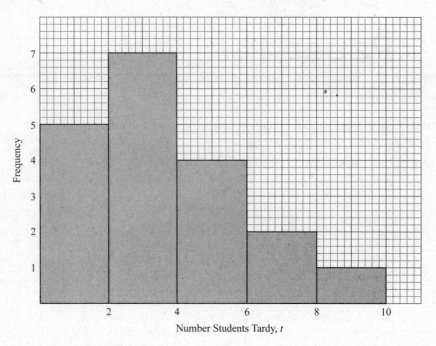

(a) Write down the modal class.

(b) Write down an estimate standard deviation using the graphics display calculator.

(c) Find the probability that at least 6 students would be late to school on a randomly selected school day.

4. Garrett is designing a small storage shed; the sketch is shown below. The base of the shed will be 3 meters long and the total height will be 2.4 meters. Each side of the roof has a small overhang of 11 cm. The roof makes a 26.5° angle with the shed.

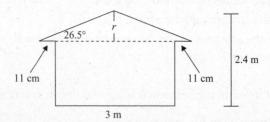

(a) Calculate the height of the roof, r, in meters.

(b) Determine the area of the triangular roof.

Garrett is going to paint the front rectangular face of the shed light grey before installing a door.

(c) Determine the number of square meters he will paint light grey.

5. Consider each of the following statements:

p: Sam passes math class
q: Sam passes history
r: Sam studies each night

(a) Write the following compound statement in words.

$$\neg(p \vee q) \Rightarrow \neg r$$

(b) Complete the truth table for the compound statement in part (a).

p	q	r	$\neg r$	$p \vee q$	$\neg(p \vee q)$	$\neg(p \vee q) \Rightarrow \neg r$
T	T	T	F			
T	T	F	T			
T	F	T	F			
T	F	F	T			
F	T	T	F			
F	T	F	T			
F	F	T	F			
F	F	F	T			

(c) Write down one example of when $\neg(p \vee q) \Rightarrow \neg r$ is false.

6. The figure below is a triangular prism, where $QT = 10$ cm, $QW = 7.5$ cm, and $ST = 18$ cm.

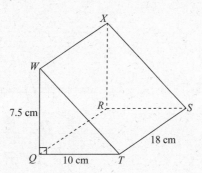

(a) Calculate the length of TW.
(b) Determine the size of the angle between TW and the base.
(c) Find the surface area for the prism.

7. Consider the function $f(x) = x^3 - x^2$.

(a) Write down $f'(x)$.

When $x = a$, the gradient of the tangent line is 8.

(b) Determine if $f(x)$ is increasing or decreasing at $x = a$. Justify your answer.
(c) Calculate the value of a if $a \in \mathbb{Z}$.

8. Devin invested 1 500 euro into an account with a 2.5% nominal annual interest rate compounded quarterly.

    (a) Calculate the value of her investment after 10 years. Round answer to the nearest euro.

    After those 10 years, Devin's job relocates her to Japan. Her new bank will convert her investment to Japanese yen (JPY) but charges a 3% transaction fee.

    (b) Determine the amount of euro Devin will lose when switching to the Japanese bank. Round answer to the nearest hundredth.

    (c) Calculate to the nearest yen the amount Devin will have in her new account if the conversion rate is 1 euro = 130 JPY.

9. *A* and *B* are subsets of a universal set *U*.

    $$U = \{x: x \in \mathbb{N}, 1 \le x < 13\}, A = \{\text{multiples of 3}\}, \text{ and } B = \{\text{factors of 24}\}$$

    (a) List the elements of sets:
    (i) *A*
    (ii) $A \cup B$
    (iii) $A' \cap B$

    (b) Represent sets *U*, *A*, and *B* on a Venn diagram.

10. The function $f(x) = \frac{1}{2}x^2 + 3x - 8$ intersects the *x*-axis at points *A* and *B* and has a vertex at point *V*.

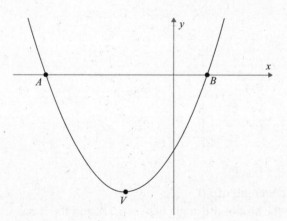

    (a) Determine the coordinates of point *V*.
    (b) Write down the axis of symmetry for $f(x)$.
    (c) Find the coordinates of point *A*.
    (d) Write down the range of $f(x)$.

11. On average, the moon is 384 400 km from Earth.

    (a) Express the distance from Earth to the moon in the form $a \times 10^k$, where $1 \le a < 10$, $k \in \mathbb{Z}$.

    A probe was launched to the moon. The journey took 8 hours and 45 minutes.

    (b) Express the time the probe took to reach the moon in minutes.
    (c) Calculate the average speed of the probe in **meters per minute**. Write the answer in the form $a \times 10^k$, where $1 \le a < 10$, $k \in \mathbb{Z}$.

12. The probability Lindsay will win her first tennis match is 0.65. If she wins her first match, the probability she will win the second is 0.78. If she does not win her first match, the probability she will win the second is 0.42.

    (a) Complete the tree diagram shown below.

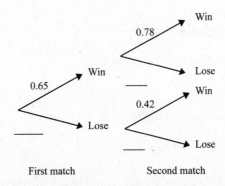

    (b) Calculate the probability Lindsay wins her second match.
    (c) Given that Lindsay won her second match, find the probability she lost the first match.

13. The exam scores of Mr. Anderson's IB Physics class were normally distributed with a mean of 67 and standard deviation of 3.75.

    (a) Sketch a normal curve showing the distribution of exam scores.

    A student is randomly selected from Mr. Anderson's class. Find the probability the student scored:

    (b) below 58,
    (c) between 60 and 73,
    (d) higher than 77.

14. The function $f(x) = 1 + a^x$, where $a \in \mathbb{Z}$ is shown in the graph below. The graph crosses the y-axis at the point B.

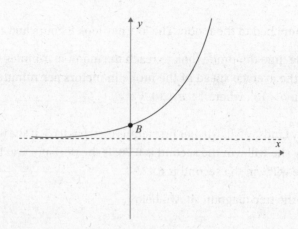

(a) Write down the coordinates of point B.

The graph of $f(x)$ also contains the point $(3, 9)$.

(b) Calculate the value of a.

The graph has a horizontal asymptote $y = q$.

(c) Write down the value of q.

15. The parent organization at a local high school asked the members to select one activity for which they would be willing to volunteer. The results were then grouped based on the type of job the member holds.

		Selected Activity		
		Book Fair	Student Dance	Bake Sale
Type of Job	Full Time	18	12	8
	Part Time	13	15	16
	Stay at Home	11	7	21

The parent organization will carry out a χ^2 test for independence to determine if the selected activity was dependent on the type of job.

(a) Write down the null hypothesis for the χ^2 test.
(b) Show that the number of degrees of freedom is 4.
(c) Write down the χ^2 calculated value.

The χ^2 critical value at the 5% significance level is 9.488.

(d) State the conclusion to the test. Justify your answer.

Paper 2

Each question is worth a different amount of marks. You must show **all** work to earn full marks.

1. A math consulting company is researching what time of day students learn math best. Data from five local high schools was collected focusing on what time of day each student took math along with the student's final grade. The results are displayed in the table below.

<table>
<tr><td></td><td></td><td colspan="4" align="center">Final Math Grade</td></tr>
<tr><td></td><td></td><td>A/B</td><td>C/D</td><td>F</td><td>Total</td></tr>
<tr><td rowspan="4">Time of
Math Class</td><td>Morning</td><td>75</td><td>85</td><td>40</td><td>200</td></tr>
<tr><td>Lunchtime</td><td>50</td><td>50</td><td>25</td><td>125</td></tr>
<tr><td>Afternoon</td><td>50</td><td>55</td><td>70</td><td>175</td></tr>
<tr><td>Total</td><td>175</td><td>190</td><td>135</td><td>500</td></tr>
</table>

The math consulting company carries out a chi-squared test for independence at the 5% significance level with the data above.

(a) State the null hypothesis for the test.

(1)

(b) Show that the number of degrees of freedom is 4.

(1)

(c) Calculate the expected number of students who took math in the afternoon and made an A or B.

(2)

(d) Using your graphics display calculator, find:
   (i) the χ^2 calculated value,
   (ii) the p-value.

(2)

The χ^2 critical value at the 5% significance level is 9.488.

(e) State the conclusion of the chi-squared test. Justify your answer.

(2)

After further research, the company discovered the exam scores for ninth graders were normally distributed with a mean score of 77.6 and standard deviation of 5.8.

43% of the ninth graders scored higher than x.

(f) Determine the value of x.

(3)

Ninth graders whose exam score was above an 85 were placed into honors math the following year. There were a total of 600 ninth graders.

(g) Find how many ninth graders are expected to be placed into honors math.

(3)

(Total 14 marks)

2. The incoming freshmen group of Mount Hill College were asked to select the factor(s) that influenced their decision to attend Mount Hill: small class size (C), scholarship opportunities (S), or the study abroad program (P).

53 selected small class size **only**
41 selected scholarship opportunities **only**
42 selected the study abroad program **only**
29 selected small class size and the study abroad program **only**
38 selected small class size and scholarship opportunities **only**
41 selected scholarship opportunities and the study abroad program **only**
15 selected none of the three factors

(a) Draw a Venn diagram to represent the given information.

(3)

There were 275 incoming freshmen surveyed.

(b) Calculate the number of freshmen who selected all three factors.

(2)

(c) Shade the area on the Venn diagram that corresponds to the freshmen who selected small class size or scholarship opportunities but not the study abroad program.

(2)

One freshman student is randomly selected from the group.

(d) Given that the student selected the study abroad program, find the probability the student also selected scholarship opportunities.

(2)

Two freshmen students are randomly selected from the group.

(e) Find the probability that:
(i) neither student selected the study abroad program,
(ii) at least one of the students selected scholarship opportunities.

(5)

(Total 14 marks)

3. Consider the function $f(x) = \dfrac{3x-1}{x-2}$.

(a) Write down the equation of the vertical asymptote.

(1)

(b) Write down the domain of the function $y = f(x)$.

(2)

(c) Sketch the graph of the function $y = f(x)$ for $-16 \le x \le 16$. Let 1 cm represent 2 units on the horizontal axis. Clearly show the axial intercepts and asymptotes.

(4)

(d) On the same graph, sketch the function $g(x) = 8 - 0.05x^2$.

(2)

(e) Hence or otherwise, write down the number of solutions to $\dfrac{3x-1}{x-2} = 8 - 0.05x^2$.

(1)

(f) Using your graphics display calculator, write down **one** solution to $\dfrac{3x-1}{x-2} = 8 - 0.05x^2$.
Express your answer correct to four significant figures.

(2)

(g) Differentiate $g(x)$ in terms of x.

(2)

(h) Hence, determine the interval in which $g(x)$ is decreasing.

(2)

(Total 16 marks)

4. Gourmet Chocolate Factory is creating a large chocolate bar to sell for special occasions. The chocolate is in the shape of a triangular prism whose cross section is an isosceles triangle. The diagram below illustrates the chocolate bar. The length of AB is 12 cm and the length of BC is 21 cm. The size of angle AEB is 128°.

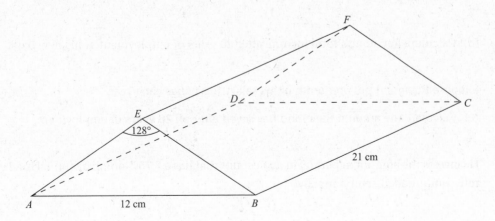

(a) Find the size of angle *EAB*.

(2)

(b) Calculate the length of *EB*.

(3)

(c) Show that the length of *BD* is 24.19 cm, correct to 4 significant figures.

(2)

(d) Calculate the volume of the chocolate bar.

(2)

Gourmet Chocolate Factory sells its chocolates at a rate of $0.05 per cubic centimeter.

(e) Calculate the selling price for the large chocolate bar.

(1)

During a quality check, the bar is melted and poured inside a cylindrical container whose radius is 5 cm.

(f) Determine the height the chocolate reaches in the cylinder.

(3)

(Total 13 marks)

5. Raymond is hired at a bank with a starting salary of 35 000 Great British Pounds. He is promised a raise of 2 800 GBP after each year he is employed.

(a) Calculate Raymond's salary after six years of employment.

(2)

(b) Calculate Raymond's **total** income after six years of employment.

(2)

After those six years, Raymond is employed at a new bank where his starting salary resets to 35 000 GBP. The company gives a 6.5% raise after each year of employment.

(c) Determine the number of years before Raymond earns more at the new bank than he did in part (*a*).

(3)

(d) Calculate Raymond's **total** income after 20 years of employment at his new bank.

(2)

Suppose Raymond put one-tenth of his salary in savings each year.

(e) Calculate the amount Raymond has saved after **all 26 years** of employment.

(3)

He invests the amount in part (*e*) in an account that has a 2.75% nominal annual interest rate compounded semi-annually.

(f) Determine the amount in his account after 4 years.

(2)

(g) Determine how many years it will take before Raymond has 500 000 GBP.

(2)

(Total 16 marks)

6. The coach of Central High School records the weights of two teams of basketball players. The results are displayed in the frequency table below.

Weight, w (kg)	Frequency
$75 \leq w < 80$	10
$80 \leq w < 85$	12
$85 \leq w < 90$	15
$90 \leq w < 95$	9
$95 \leq w < 100$	4
Total	50

(a) Write down the mid-interval value for $95 \leq w < 100$.

(1)

(b) Using your graphics display calculator, write down an estimate
   (i) mean weight,
   (ii) standard deviation.

(2)

(c) Construct a cumulative frequency chart for the weights of the basketball players.

(2)

(d) Draw a cumulative frequency curve. Let 2 cm represent 5 kg on the horizontal axis and 1 cm represent 5 players on the vertical axis.

(4)

(e) Use your graph to find:
   (i) median weight,
   (ii) interquartile range,
   (iii) the weight of the heaviest 10%.

(5)

(f) Draw a box and whisker plot for the weights of the basketball players.

(3)

(Total 17 marks)

PAPER 1

C2 for all values are correct

OR

• **A1** for one correct

1. (a) $x = -2, 0$

The first derivative is zero when the function has a maximum or minimum. The graph has a minimum when $x = -2$ and a maximum when $x = 0$.

C2 for correct intervals or inequalities

OR

• **A1** for correct values

• **A1** for correct symbols

(b) **(−2, 0) OR**

$-2 < x < 0$

The first derivative is positive when the gradients of the tangent lines are positive. Thus, you find the interval where the function is increasing.

C2 for correct answer plus:

reason $f(x)$ is decreasing or $f'(x) < 0$.

OR

• **A1** for "negative"

• **R1** for the reason

(c) **The gradient would be negative since the function is decreasing at $x = -3$.**

When $x = -3$, the function is decreasing, which means $f'(x) < 0$. Thus, the gradient would be negative.

C2 for correct value

OR

• **M1** for setting up a solvable equation

• **A1** for correct value

2. (a) $r = \dfrac{1}{2}$

You know $u_3 = 56$ and $u_6 = 7$. Set up two equations using the given information. The sixth term is r^3 units away from the third term.

Therefore, $\begin{aligned} 56r^3 &= 7 \\ 56r^3 - 7 &= 0 \end{aligned}$

This can be solved using $\boxed{\text{PLYSMLT}}$.

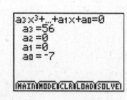 

C1 for correct value

(b) **224**

Since you know $r = \dfrac{1}{2}$, you can determine u_1 using either of the given terms.

$$56 = u_1\left(\dfrac{1}{2}\right)^2$$

$$u_1 = 224$$

(c) **448**

The sum formula for a geometric sequence is

$S_n = \dfrac{u_1(1 - r^n)}{1 - r}$, since our common ratio is less than 1.

$$S_{10} = \dfrac{224\left(1 - \left(\dfrac{1}{2}\right)^{10}\right)}{1 - \dfrac{1}{2}} = \dfrac{223.78125}{0.5} = 447.5625$$

C3 for correct value

OR

- **M1** for the sum formula used correctly
- **A1** for substituting $n = 10$, $u_1 = 224$, and $r = \dfrac{1}{2}$
- **A1** for correct answer

3. (a) **$2 \le t < 4$**

The modal class is the most frequent, which is the highest bar on a histogram.

C2 for correct interval

OR

- **A1** for correct values
- **A1** for correct inequality symbols

(b) **2.25**

You need the mid-interval value for each interval to use as an estimate along with its frequency. Then, enter the data into L_1 and the frequencies into L_2 to calculate the standard deviation.

C2 for correct value

Interval	Mid-Interval Value	Frequency
$0 \le t < 2$	1	5
$2 \le t < 4$	3	7
$4 \le t < 6$	5	4
$6 \le t < 8$	7	2
$8 \le t < 10$	9	1

 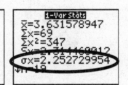

(c) **$\dfrac{3}{19}$ or 0.158**

There are a total of 19 days ($n = 19$ on the $\boxed{\text{1-VAR STATS}}$ output). For at least 6 students to be tardy, add the heights of the last two bars, which yields 3.

C2 for correct value

OR

- **A1** for correct numerator
- **A1** for correct denominator

4. (a) **0.803 m**

The base of the shed is 3 meters, which means from r to the start of the overhang is 1.5 m. The overhang is measured in centimeters, and you need meters. 11 cm = 0.11 m.

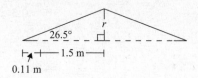

Thus, given the angle of 26.5° and the base measurement of 1.61 m, set up a tangent ratio to solve for r.

$$\tan 26.5° = \frac{r}{1.61}$$
$$1.61 \tan 26.5° = r$$
$$0.803 = r$$

(b) **1.29 m²**

You know the base of the triangle (3 meters plus two 0.11 meter overhangs), and you just found the height, so using the formula $A = \frac{1}{2}bh$ gives:

$$A = \frac{1}{2}(3.22)(.803) = 1.29$$

(c) **4.79 m²**

The total height of the shed is 2.4 meters, while the height of the roof is 0.803 m. To find the height of just the rectangular face, subtract the two measurements. The area he will paint is $3 \times 1.597 = 4.791$.

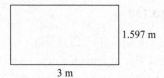

5. (a) **If it is not the case that Sam passes math or history, then she did not study each night. (Equivalent answer: If Sam did not pass math and did not pass history, then she did not study each night.)**

The negation of "p or q" can be simply written as "it is not the case that…" The implication arrow means "if, then."

Equivalently, the negation could be distributed to both the p and the q, but the $\vee$ (or) sign must flip to $\wedge$ (and).

$$\neg(p \vee q) = \neg p \wedge \neg q$$

(b)

p	q	r	$\neg r$	$p \vee q$	$\neg(p \vee q)$	$\neg(p \vee q) \Rightarrow \neg r$
T	T	T	F	**T**	**F**	**T**
T	T	F	T	**T**	**F**	**T**
T	F	T	F	**T**	**F**	**T**
T	F	F	T	**T**	**F**	**T**
F	T	T	F	**T**	**F**	**T**
F	T	F	T	**T**	**F**	**T**
F	F	T	F	**F**	**T**	**F**
F	F	F	T	**F**	**T**	**T**

- For $p \vee q$ to be true, either p or q or both must be true.
- $\neg(p \vee q)$ is the negated truth value of $p \vee q$.
- $\neg(p \vee q) \Rightarrow \neg r$ is only <u>false</u> when $\neg(p \vee q)$ is true but $\neg r$ is false.

(c) **$\neg(p \vee q) \Rightarrow \neg r$ is false when Sam does not pass math and does not pass history but she does study each night.**

$\neg(p \vee q) \Rightarrow \neg r$ is only false when p is false, q is false, and r is true.

6. (a) **12.5 cm**

TW is the hypotenuse of right triangle QTW.

$$(10)^2 + (7.5)^2 = TW^2$$
$$156.25 = TW^2$$
$$12.5 = TW$$

(b) **36.9°**

The angle TW with the base forms $Q\hat{T}W$. Triangle QTW is a right triangle, so any of the three trigonometric ratios can be used.

Option 1	Option 2	Option 3
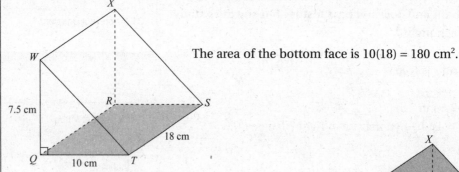		
$\sin\theta = \dfrac{7.5}{12.5}$	$\cos\theta = \dfrac{10}{12.5}$	$\tan\theta = \dfrac{7.5}{10}$
$\theta = \sin^{-1}\!\left(\dfrac{7.5}{12.5}\right)$	$\theta = \cos^{-1}\!\left(\dfrac{10}{12.5}\right)$	$\theta = \tan^{-1}\!\left(\dfrac{7.5}{10}\right)$
$\theta = 36.9°$	$\theta = 36.9°$	$\theta = 36.9°$

(c) $SA = 615$ **cm²**

The prism is constructed from two triangular bases and three rectangular lateral faces.

The area of one triangle is $A = \dfrac{1}{2}(10)(7.5) = 37.5$, but you need two, so the area of the triangles is 75 cm².

The three rectangular lateral faces have different dimensions.

The area of the bottom face is $10(18) = 180$ cm².

The area of the right face is $18(12.5) = 225$ cm².

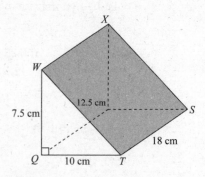

The area of the remaining face is 7.5(18) = 135 cm².

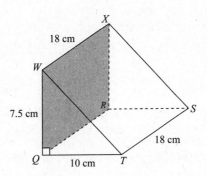

Therefore, $SA = 75 + 180 + 225 + 135$
$SA = 615 \text{ cm}^2$

7. (a) $f'(x) = 3x^2 - 2x.$

Apply the first derivative rule to each term.

$$x^3 \Rightarrow 3(x^{3-1}) = 3x^2$$
$$-x^2 \Rightarrow 2(-x^{2-1}) = -2x$$

(b) **Increasing because the gradient is positive**

The gradient of the tangent line is positive, which means the function is increasing.

(c) $a = 2$

The first derivative is the gradient function, so plug the given point into the first derivative along with the given gradient.

$$8 = 3a^2 - 2a$$

This creates a quadratic equation, which you can solve using the PLYSMLT app.

First set the equation equal to zero: $0 = 3a^2 - 2a - 8$.

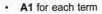

Since the problem states a must be an integer, the only value a can be is 2.

C2 for correct value

OR

- **M1** for using compounded interest formula
- **A1** for correct answer

8. (a) **1 925 euro**

Since Devin's investment is compounded quarterly, $k = 4$.

$$FV = PV \times \left(1 + \frac{r}{100k}\right)^{kn}$$

$$FV = 1\,500\left(1 + \frac{2.5}{100 \times 4}\right)^{10 \times 4}$$

$$FV = 1\,500(1.00625)^{40}$$

$$FV = 1924.54$$

Rounding to the nearest euro gives an answer of 1 925.

C2 for correct value

(b) **57.75 euro**

The bank charges 3%, so you must calculate 3% of 1 925 euro.

$$1\,925(0.03) = 57.75$$

C2 for correct value

OR

- **M1** for using conversion factor
- **A1** for correct answer

(c) **242 745.50 JPY**

Before you can convert to yen, you must subtract the 3% transaction fee.

$$1\,925 - 57.75 = 1\,867.25 \text{ euro}$$

There are two methods to convert the currency.

Method 1	Method 2
Set up a proportion with given currencies and the exchange rate. Then, cross multiply.	Multiply the currency by the conversion factor.
$\dfrac{1\,867.25\,\text{euro}}{x\,\text{JPY}} = \dfrac{1\,\text{euro}}{130\,\text{JPY}}$ $242\,742.50 = x$	$1\,867.25\,\text{euro} \times \dfrac{130\,\text{JPY}}{1\,\text{euro}} = 242\,742.50$

C1 for correct values

9. (a) (i) **{3, 6, 9, 12}**

Set A contains multiples of 3 that are within the universal set. The universal set contains only the natural numbers 1 through 12. The multiples of 3 within that set are 3, 6, 9, and 12.

C1 for correct values

(ii) **{1, 2, 3, 4, 6, 8, 9, 12}**

$A \cup B$ should contain all the elements in set A combined with all the elements in set B.

The factors of 24 within the universal set are 1, 2, 3, 4, 6, 8, and 12.

(iii) **{1, 2, 4, 8}**

$A' \cap B$ should contain the elements NOT in set A that are in set B.

Set B contains 1, 2, 3, 4, 6, 8, and 12, but you cannot include numbers that are also in set A. Therefore, do not include 3, 6, or 12.

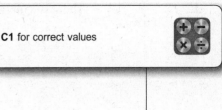

(b)

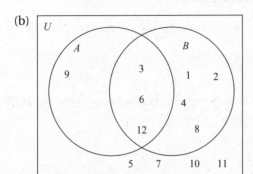

The intersection of sets A and B always goes in the middle (the "football"). The elements that are in both set A and set B are 3, 6, and 12. Do not forget to include the remaining elements in the universal set outside the two circles.

10. (a) **(−3, −12.5)**

The x-coordinate of the vertex is always the value of the axis of symmetry.

$$x = -\frac{b}{2a}$$

$$x = -\frac{3}{2\left(\frac{1}{2}\right)} = -3$$

To find y, plug in x and simplify.

$$f(-3) = \frac{1}{2}(-3)^2 + 3(-3) - 8$$

$$f(-3) = -12.5$$

The vertex can also be found using the graphics display calculator.
    Enter the function into Y_1 and graph, ensuring the vertex is visible.

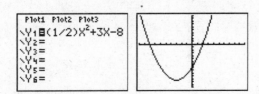

Then, calculate the minimum value.

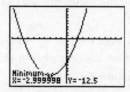

The vertex is (–3, –12.5). Calculator limitations give the x-value as –2.99999 but this rounds to –3.

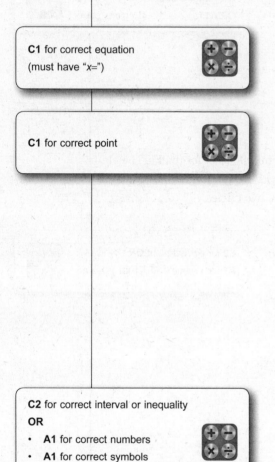
(b) $x = -3$
The axis of symmetry passes through the x-coordinate of the vertex.

(c) **(–8, 0)**
The graph crosses the x-axis at two points, but point A is when the x-value is negative. To find the coordinates of point A, either graph and calculate the zero, look at the table for when $y = 0$, or solve using PLYSMLT.

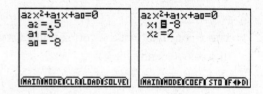

Point A has a negative x-coordinate, so point A is (–8, 0).

(d) **[–12.5, ∞) or $y \geq -12.5$**
Range is all the y-values of the function. The graph begins at –12.5 on the y-axis and continues to positive infinity.

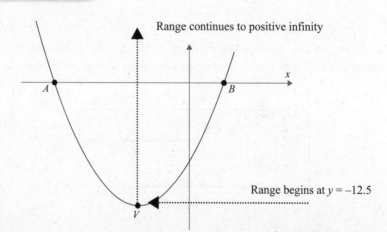

Range continues to positive infinity

Range begins at $y = -12.5$

11. (a) $\mathbf{3.844 \times 10^5}$ **km or** $\mathbf{3.84 \times 10^5}$

To change the distance to standard form (or scientific notation), move the natural resting decimal until there is only one number in front of the decimal.

$384\,400 = 3.844 \times 10^5$, since the decimal was moved

five times to the left.

(b) **525 minutes**

There are 60 minutes in 1 hour, so 8 hours would be $8 \times 60 = 480$ minutes. Now, add on the additional 45 minutes, so the total time in minutes would be $480 + 45 = 525$ minutes.

(c) $\mathbf{7.32 \times 10^5}$ **m/min**

The speed of the probe is to be in meters per minute, so the distance from Earth to the moon must be converted. You are staring at kilometers and moving to the "base unit", which is meter.

10^3	10^2	10^1	10^0	10^{-1}	10^{-2}	10^{-3}
kilo-k	hecto-h	deca-da	Base Unit (Meter, gram, second, etc.)	deci-d	centi-c	milli-m

Changing the original number of 384 400 km means adding three zeros.

$$384\,400 \text{ km} = 384\,400\,000 \text{ m}$$

Or you could change the standard form of the distance, which means multiplying by an additional 10^3 (multiply like bases, add exponents).

$$3.844 \times 10^5 \text{ km} = 3.844 \times 10^8 \text{ m}$$

The rate is meters per minute. The probe traveled 3.844×10^8 meters in 525 minutes, so the rate is $\dfrac{3.844 \times 10^8 \text{ m}}{525 \text{ min}} = 732\,190.47 \text{ m/min}$. Now, convert to standard form.

12. (a)

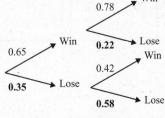

First match   Second match

Each branch of a tree diagram must add to one. Subtract the given probability from one to find the remaining portion.

C1 for correct answer

(b) **0.654**

There are two ways for Christina to win.

- Win the first match, and then win the second:
  $0.65 \times 0.78 = 0.507$
- Lose the first match, and then win the second:
  $0.35 \times 0.58 = 0.147$

The total probability is the sum of the two ways to win: $0.507 + 0.147 = 0.654$

C2 for correct answer

OR

- **M1** for use of conditional probability formula
- **A1** for correct answer

(c) **0.225**

Since you know Christina won her second match, this is a conditional probability problem.

The formula is $P(A|B) = \dfrac{P(A \cap B)}{P(B)}$.

Converting the formula to context and simplifying gives

$$P(\text{lost first} \mid \text{won 2nd}) = \frac{P(\text{lost first AND won 2nd})}{P(\text{won 2nd})} = \frac{0.35 \times 0.42}{0.654} = 0.225$$

C3 for correct curve

OR

- **A1** for mean in center
- **A1** for labeling 3 standard deviations down and up
- **A1** for normal curve

13. (a)

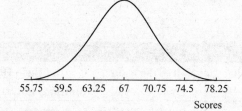

The graph is centered on the mean of 67. Next, add one, two, and three standard deviations to the mean. Then, subtract one, two, and three standard deviations from the mean.

C1 for correct value

(b) **0.00820**

First, sketch the scenario. Then, calculate using NORMALCDF.

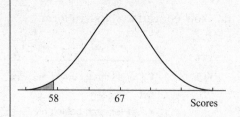

(c) **0.914**

First, sketch the scenario. Then, calculate using NORMALCDF.

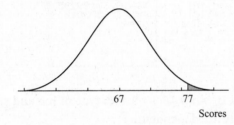

(d) **0.00383**

First, sketch the scenario. Then, calculate using NORMALCDF.

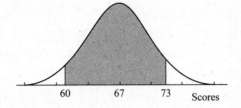

14. (a) **(0, 2)**

The y-intercept has an x-value of zero. Plug in zero for x and simplify. Recall that any number raised to the zero power is one.

$$f(0) = 1 + a^0$$
$$f(0) = 1 + 1$$
$$f(0) = 2$$

(b) **$a = 2$**

Since the function passes through the point $(3, 9)$, substitute 3 in for x and 9 in for y to solve for a.

$$9 = 1 + a^3$$
$$8 = a^3$$
$$\sqrt[3]{8} = \sqrt[3]{a^3}$$
$$2 = a$$

(c) **1**

The horizontal asymptote is always the constant of an exponential function. Since the constant of the equation is 1, the horizontal asymptote would be $y = 1$.

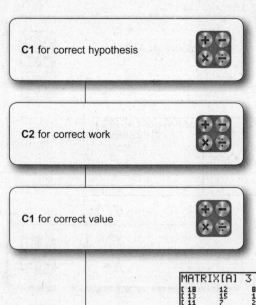

C1 for correct hypothesis

C2 for correct work

C1 for correct value

C2 for correct conclusion with valid reason
OR
- **A1** for reject H_o
- **R1** for valid reason

15. (a) **H_o: Type of job and selected activity are independent.**
The null hypothesis is always that the two categories are independent.

(b) **$(3 - 1)(3 - 1) = 2(2) = 4$**
Degrees of freedom are found by $df = (r - 1)(c - 1)$, where r is the number of rows and c is the number of columns.

(c) **10.3**
Input the observed data into matrix Ⓐ and then perform a χ^2–test.

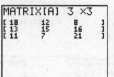 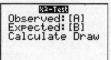 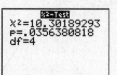

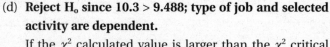

(d) **Reject H_o since 10.3 > 9.488; type of job and selected activity are dependent.**
If the χ^2 calculated value is larger than the χ^2 critical value, reject the null hypothesis.

PAPER 2

1. (a) **H_0: Time of math class and the final grade are independent**
   The null hypothesis is always that the two factors are independent.

   (b) **$(3 - 1)(3 - 1) = 2(2) = 4$**
   Degrees of freedom are found by $df = (r - 1)(c - 1)$, where r is the number of rows and c is the number of columns.

   M1 for correctly showing work

   (c) $\dfrac{175(175)}{500} = \textbf{61.25 or 61.3}$

   The expected number of students who took math in the afternoon and made an A or B is found by multiplying the row total for "afternoon" by the column total of "A/B," and then dividing by the total number of students.

   - **M1** for correctly showing work
   - **A1** for correct answer

   (d) (i) **$\chi^2 = 23.4$**
   (ii) **$p = 1.08 \times 10^{-4}$ or 0.000108**
   Enter the observed data into matrix Ⓐ and perform the χ^2 test.

   - **A1** for correct chi-squared value
   - **A1** for correct p-value

   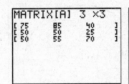 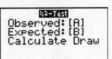 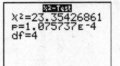

   (e) Two possible answers:

Option 1	Option 2
$23.4 > 9.488$	$1.08 \times 10^{-4} < 0.05$
Reject H_0	Reject H_0
Time of math class and the final grade are dependent	Time of math class and the final grade are dependent

   If the χ^2 calculated is larger than the χ^2 critical, reject the null hypothesis. If the p-value is smaller than the significance level, reject the null hypothesis.

   - **A1** for correct conclusion
   - **R1** for correct reason

- **M1** for correct sketch of normal curve showing desired area
- **A1** for using an area of 0.57
- **A1** for correct answer

(f) **$x = 78.6$**

First draw a sketch showing the given area of 0.43. This area is to the right of x since students must score <u>higher</u> than x. In order to use INVNORM, subtract the given area from one.

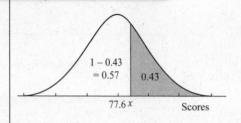

$1 - 0.43$
$= 0.57$ 0.43

77.6 x Scores

- **A1** for finding the correct probability
- **M1** for multiplying by total number of ninth graders
- **A1** for correct answer

(g) **About 61 students**

In order to determine the expected number, find the probability a student scored above an 85. Then, multiply that probability by the 600 ninth graders to determine the expected number.

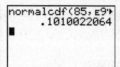
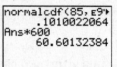

- **A2** for all 7 numbers correctly placed in circles
- **A1** for the rectangle

(**OR A1** for 5–6 numbers placed correctly)

2. (a)

C S

53   38   41

29   41

42

P 15

All the information given contains the word "only," so when drawing the Venn diagrams, the numbers are simply placed in the correct location.

- **M1** for showing the work
- **A1** for correct value

(b) **$53 + 38 + 41 + 29 + 41 + 42 + 15 + x = 275$**
$259 + x = 275$
$x = 16$

Since there were a total of 275 freshmen, the entire Venn diagram must add up to 275. The spot in the middle is the only missing piece, so sum up all the other numbers to determine what is missing.

(c)

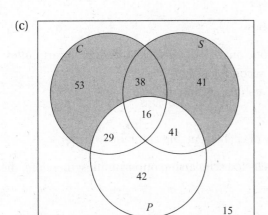

Freshmen who selected class size or scholarship opportunities would be found in either circle *C* or circle *S* (union). However, you must exclude the part of those circles that is also in *P*.

(d) $\frac{57}{128}$ or **0.445**

You know the student selected the study abroad program, so the total possible number of outcomes is 128, as that is the total number of students who selected study abroad. Out of those, 57 (16 + 41) also selected scholarship opportunities.

(e) (i) $\frac{147}{275}\left(\frac{146}{274}\right) = \mathbf{0.285}$

There are a total of 128 students who selected the study abroad program, so there are 147 students who did <u>not</u> selecte the program (275 − 128 = 147). Since you are selecting two students from the same group, you must use probability without replacement. This means the second probability will have one less in the numerator and denominator.

(ii) $\frac{136}{275}\left(\frac{139}{274}\right) + \frac{139}{275}\left(\frac{136}{274}\right) + \frac{136}{275}\left(\frac{135}{274}\right) = \mathbf{0.745}$

OR $1 - \frac{139}{275}\left(\frac{138}{274}\right) = \mathbf{0.745}$

You want at least one of the students to select scholarship opportunities. There are two methods to calculate the answer.

Method 1: Straight Computation

First, list all the ways at least one out of two select scholarship opportunities:
SS' (first selects scholarships, second does not)
S'S (first does not select scholarships, second does)
SS (both select scholarships)

Now, calculate the probability of each event and combine together.

There are 136 students who selected scholarship opportunities; therefore, there are 139 students who did not.

$$SS' = \frac{136}{275}\left(\frac{139}{274}\right)$$
$$S'S = \frac{139}{275}\left(\frac{136}{274}\right)$$
$$SS = \frac{136}{275}\left(\frac{135}{274}\right)$$

The three probabilities sum up to 0.745.

Method 2: Complement Events

When two students are selected there are four possibilities:

Both students select scholarship opportunities, only the first student selects it, only the second student selects it, or neither student selects it.

You need to find the probability that at least one student selects scholarship opportunities, which means the complement event is <u>neither</u> student selected that factor.

You know an event and its complement add to 1.

Thus, the quickest way to calculate the desired event is to find the complement event and subtract from one.

$$1 - P(\text{neither student selects scholarship opportunities}) =$$
$$1 - \left(\frac{139}{275}\right)\left(\frac{138}{275}\right) = 0.745$$

3. (a) $x = 2$

The denominator of the rational function can never equal zero; therefore, the vertical asymptote is found by setting the denominator equal to zero and solving.

$$x - 2 = 0$$
$$x = 2$$

This can also be seen when graphing using the GDC.

(b) $(-\infty, 2) \cup (2, \infty)$
OR $x < 2$ or $x > 2$

The only value x cannot be is 2, so the domain would be all real numbers except 2.

(c)

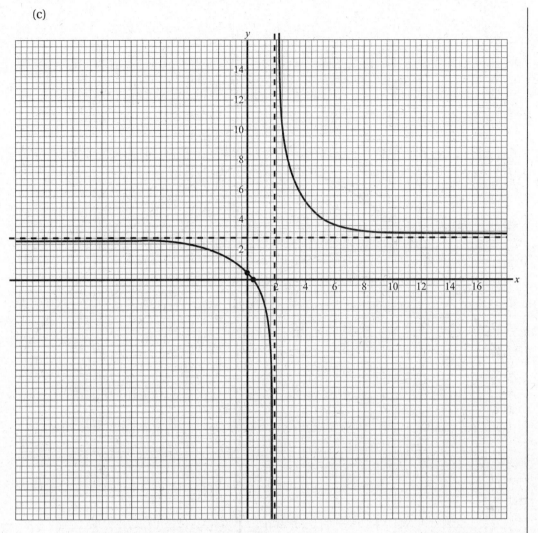

The *x*-axis should start at −16 and end at 16 with the scale of 2. Be sure to label the *x*- and *y*-axes. The vertical asymptote should be clearly drawn at $x = 2$. The intercepts can be found using the GDC as can the shape of the graph.

- **A1** for correct scale and labels
- **A1** for *y*-intercept at (0, 0.5)
- **A1** for *x*-intercept at (0.33, 0)
- **A1** for two curves separated by vertical asymptote

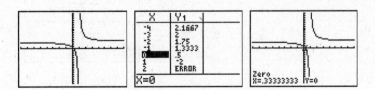

(d)

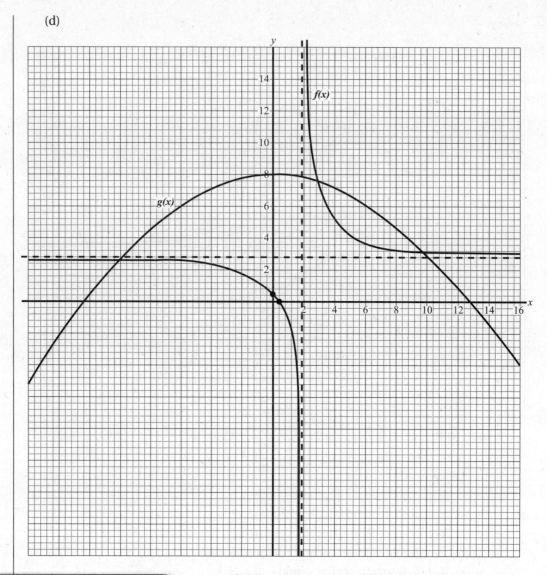

- **A1** for vertex at (0, 8)
- **A1** for correct quadratic curve

The function $g(x)$ is a quadratic, so the key feature to graph is the vertex. The vertex can be found using the GDC or the formula $x = -\dfrac{b}{2a}$.

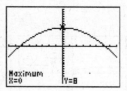

(e) **3**

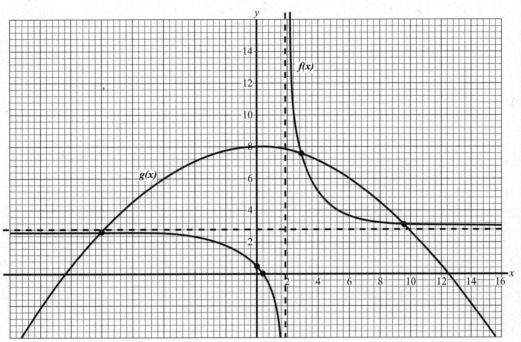

The number of solutions to $\dfrac{3x-1}{x-2} = 8 - 0.05x^2$ would be the number of intersections of the two graphs. They intersect three times.

(f) $\boldsymbol{x = 9.289, 3.107, \text{or} -10.40}$

Only one intersection is needed, so use the
(INTERSECT) option on the GDC.

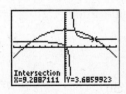 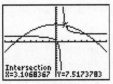 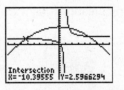

(g) $\boldsymbol{g'(x) = -0.10x}$

When you differentiate $g(x) = 8 - 0.05x^2$, the constant 8 goes to zero. Multiply the exponent of x to its coefficient and subtract one from the exponent.

(h) $\boldsymbol{g(x) \text{ is decreasing on the interval } (0, \infty) \text{ or } x > 0}$

The function is decreasing with $g'(x) < 0$. First, solve $g'(x) = 0$.

$$0 = -0.10x$$
$$x = 0$$

Now, test each region on the number line to determine the sign of $g'(x)$.

Since the first derivative is negative when $x > 0$, that is the interval in which the function is decreasing.

- **M1 for** subtracting 128° from 180°
- **A1** for correct value

4. (a) $180° - 128° = 52°$

$$E\hat{A}B = \frac{52°}{2} = 26°$$

Triangle ABE is isosceles, which means the two base angles have the same measure. The top angle measures 128°, which leaves 52° for the two base angles.

- **M1** for setting up the sine rule
- **A1** for substituting correct values
- **A1** for correct answer

(b) $\dfrac{12}{\sin 128°} = \dfrac{EB}{\sin 26°}$

$12 \sin 26° = EB \sin 128°$

$\dfrac{12 \sin 26°}{\sin 128°} = EB$

$6.68 \text{ cm} = EB$

- **M1** for setting up the cosine ratio
- **A1** for substituting correct values
- **A1** for correct answer

OR

$\cos 26° = \dfrac{6}{EB}$

$EB \cos 26° = 6$

$EB = \dfrac{6}{\cos 26°} = 6.68 \text{ cm}$

To find EB, either use the sine rule with triangle ABE, or drop the perpendicular height creating a right triangle and use the cosine ratio.

- **M1** for using Pythagorean theorem
- **A1** for substituting correct values

(c) $12^2 + 21^2 = BD^2$

$585 = BD^2$

$BD = \sqrt{585} = 24.19 \text{ cm}$

BD is the hypotenuse of triangle DBC. The legs of the triangle are 12 cm and 21 cm.

- **M1** for using volume of a prism
- **A1** for correct answer

(d) $V = \left(\dfrac{1}{2}ab\sin C\right)h$

$V = \left(\dfrac{1}{2}(12)(6.68)\sin 26°\right) \times 21$

$V = 369 \text{ cm}^3$

The volume of a prism is $V = Ah$, where A is the area of a cross section. The cross section is an isosceles triangle. Since we have two sides with the included angle, the quickest way to find the area is using the formula $A = \frac{1}{2}ab\sin C$.

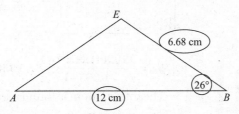

(e) $369 \text{ cm}^3 \times \dfrac{\$0.05}{\text{cm}^3} = \$18.45$

The cost is per cubic centimeter, so multiply the cost by the volume.

(f) $369 = \pi (5)^2 h$

$369 = 25\pi h$

$4.70 \text{ cm} = h$

Since the chocolate bar was melted, the volume will stay the same, but the shape will change. There will still be 369 cm³, but now it will be in the shape of a cylinder. Plug in the volume with the 5 cm radius and solve.

5. (a) $u_6 = 35\,000 + (6 - 1)\, 2\,800$

$u_6 = 49\,000$

Since Raymond earns a constant raise, this creates an arithmetic sequence. The first term, u_1, is his starting salary, n is the years employed, which is six, and the common difference is his raise of 2 800.

(b) $S_6 = \dfrac{6}{2}(35\,000 + 49\,000) = 252\,000$

You need the total amount, which is the sum of the sequence. Since you know the first term and the last term, use the formula $S_n = \dfrac{n}{2}(u_1 + u_n)$.

(c) $49\,000 = 35\,000\,(1.065)^{n-1}$

$1.4 = (1.065)^{n-1}$

$n = 6.34$

or after 6 years

The sequence is now geometric since Raymond's raise is a percentage. Raymond earned 49 000 in part (a), so set the geometric sequence formula equal to this amount and solve.

Before using the GDC, divide both sides by 35 000 so the numbers are smaller.

$$1.4 = (1.065)^{n-1}$$

Now, let $Y_1 = 1.4$ and $Y_2 = (1.065)^{x-1}$ and find their intersection.

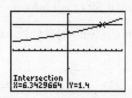

Intersection
X=6.3429664 Y=1.4

(d) $S_{20} = \dfrac{35\,000(1.065^{20}-1)}{1.065-1} = 1\,358\,885.80$

The total amount is found using the sum of a geometric sequence formula.

    Be careful when using the GDC to put parentheses around the numerator and denominator.

- **M1** for using multiplying by one-tenth to each total
- **M1** for adding the two values together
- **A1** for correct answer

(e) $\dfrac{1}{10}(252\,000) + \dfrac{1}{10}(1\,358\,885.80) = 161\,088.58$

Over 26 years Raymond worked at two banks. At one bank he earned a total of 252 000, and at the other he earned 1 358 885.80. If he put one-tenth in savings, multiply the totals by $\dfrac{1}{10}$ and add.

- **M1** for using compound interest formula
- **A1** for correct answer

(f) $FV = 161\,088.58\left(1 + \dfrac{2.75}{100(2)}\right)^{2(4)}$

$FV = 179\,684.95$

Semi-annually means twice a year: $PV = 161\,088.58$, $r = 2.75$, $k = 2$, $n = 4$.

- **M1** for using compound interest formula
- **A1** for correct answer

(g) $500\,000 = 161\,088.58\left(1 + \dfrac{2.75}{100(2)}\right)^{2n}$

$3.10388 = (1.01375)^{2n}$

$n = 41.5$ years

The final value is to be 500 000, so set the equation equal to that amount. To make the numbers more manageable, divide both sides by 161 088.58.

Let $Y_1 = 3.10388$ and $Y_2 = (1.01375)^{2x}$. Graph and find the intersection.

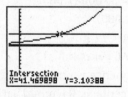

Intersection
X=41.469898 Y=3.10388

6. (a) **97.5**

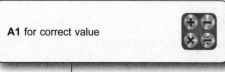

A1 for correct value

The mid-interval value is the midpoint of the interval.

$\dfrac{100+95}{2} = 97.5$

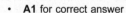

PRACTICE TEST 2

(b) (i) **86**

(ii) **6.02**

Enter the mid-interval values into L_1 and the frequencies into L_2.

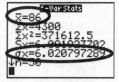

(c)

Weight, w (kg)	Frequency	Cumulative Frequency
$75 \leq w < 80$	10	**10**
$80 \leq w < 85$	12	**22**
$85 \leq w < 90$	15	**37**
$90 \leq w < 95$	9	**46**
$95 \leq w < 100$	4	**50**

Cumulative frequency is the running total, so sum up the frequencies as you go down the chart.

(d)

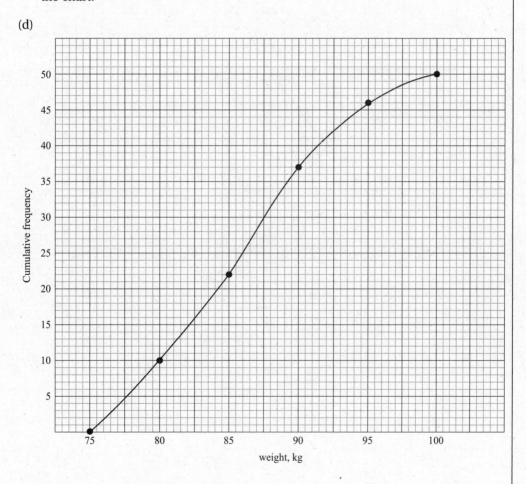

- **A1** for correct scale and labeled axes
- **A2** for 5–6 correct points
  (OR A1 for 3–4 correct points)
- **A1** for smooth curve

A1 for correct value

The x-axis is scaled, so every 2 cm = 5 kg. On the y-axis, every 1 cm = 5 players. The graph always begins on the x-axis, so the first point is (75, 0). From there, graph the end of each interval with the corresponding cumulative frequency, and then draw a smooth curve through the point.

(e) (i) **86 kg (± 0.5 kg)**

The median is the middle weight. Since there are 50 players, the median occurs at player 25.

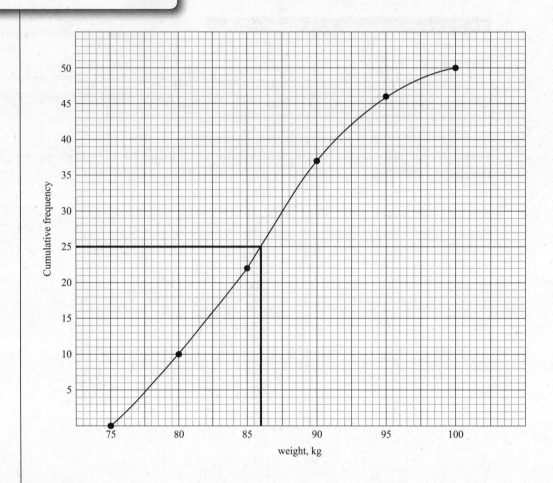

(ii) **90.5 – 81 = 9.5 kg (±1 kg)**

    The interquartile range is $q_3 - q_1$.

- The quartiles are the lower and upper 25%.
- The lower quartile can be found at $50(0.25) = 12.5$.
- The upper quartile can be found at $50(0.75) = 37.5$.

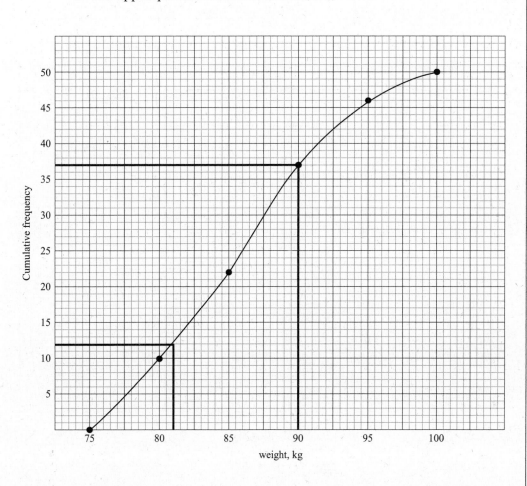

- **M1** for calculating 10% of 50
- **A1** for correct answer

(iii) **50(0.10) = 5**

94.5 kg (± 0.5 kg)

The heaviest 10% represent the top weights. After finding 10% of 50, you know that the top 5 players will be the heaviest. Go to 45 on the graph.

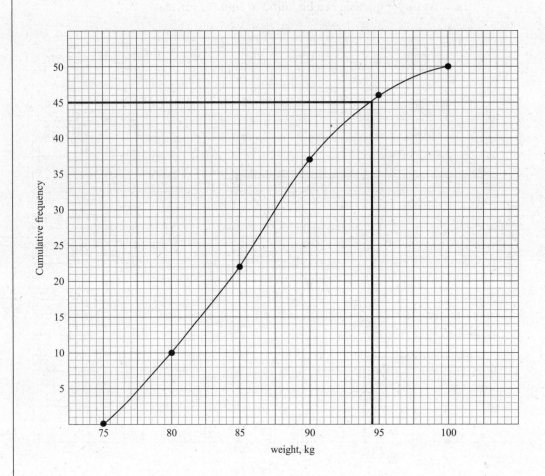

PRACTICE TEST 2

(f)

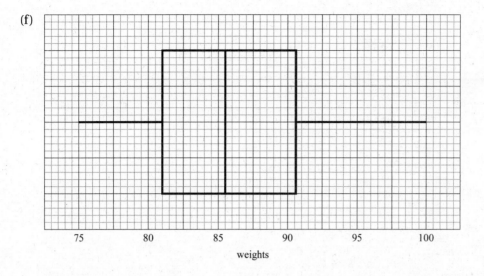

weights

The box and whisker plot shows the five-number summary:

Min, Q1, Med, Q3, and Max.

In previous parts, you identified the median and the quartiles. The min and max come from the frequency table.

PRACTICE
TEST 3

PRACTICE
TEST 3

Paper 1

Each question is worth 6 marks. You do not have to show work to earn full marks.

1. Lauren is completing a homework problem where she must evaluate the expression

   $q = \sqrt{\dfrac{5x^3}{3y}}$, when $x = 2.35$ and $y = 4.75$.

   Lauren first wants to estimate the solution.

   (a) Write down an approximate value for x and y that Lauren can use.
   (b) Calculate her estimate, correct to 2 significant figures.
   (c) Calculate the exact answer, correct to 4 significant figures.
   (d) Find the percent error between her answers from parts a and b written in the form $a \times 10^k$, where $1 \le a < 10$ and k is an integer.

2. Let $U = \{x \mid 0 < x \le 8, x \in \mathbb{N}\}$

   > A and B are subsets of U.
   > $A = \{$factors of 12$\}$
   > $B = \{$odd integers$\}$

   (a) List the elements of:
       (i) A
       (ii) B

   (b) Complete the Venn diagram, placing each element of U correctly.

   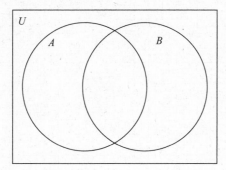

   (c) Write down the elements in $A \cup B'$.

3. Consider the following logic propositions:

p: Coleman will ride his bike.

q: Coleman is late to work.

(a) Write down the following compound propositions in symbolic form:

   (i) Coleman will ride his bike, but he will be late to work.

   (ii) Either Coleman will ride his bike, or he will be on time to work, but not both.

(b) Complete the following truth table.

p	q	$\neg q$	$p \Rightarrow \neg q$
T	T		
T	F		
F	T		
F	F		

(c) Write down if $p \Rightarrow \neg q$ is a tautology, contradiction, or neither.

4. Wesley records the weights of 220 newborn calves on his ranch during one year. The weights of the calves in kilograms are shown in the cumulative frequency graph below.

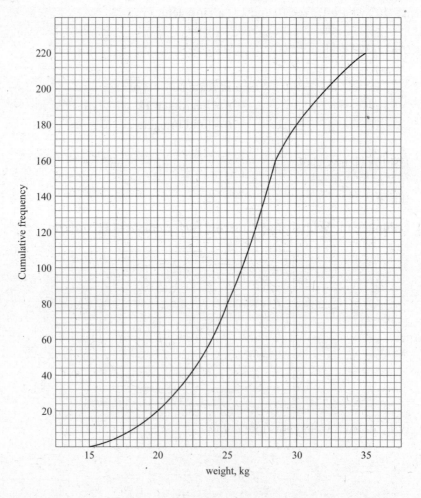

(a) Write down the
   (i) median weight
   (ii) weight of the lower quartile
   (iii) weight of the 90th percentile

(b) Determine the percent of calves that weigh more than 27 kilograms.

5. The probability that Alexandra will go shopping on Saturday is $\frac{3}{4}$. If she goes shopping, the probability that she will buy lunch is $\frac{7}{10}$. If she does not go shopping, the probability that she will buy lunch is only $\frac{1}{5}$.

(a) Complete the tree diagram.

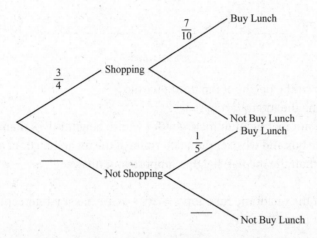

(b) Calculate the probability Alexandra will not buy lunch.
(c) Given Alexandra did not buy lunch, determine the probability she went shopping.

6. The seventh term of an arithmetic sequence is 23. The tenth term is 32.

(a) Find the common difference of the sequence.
(b) Write down the first term of the sequence.

   The sum of the first n terms is 390.

(c) Find the number of terms, n.

7. The heights of 80 adult male giraffes are collected from a random sample of zoos. The box and whisker plot displays the recorded heights in meters.

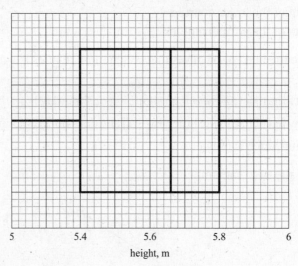

height, m

(a) Write down the height of the third quartile.

(b) Determine the interquartile range.

(c) Find the number of adult male giraffes whose height is less than 5.4 meters.

(d) Using the box and whisker plot, determine if the mean height of adult male giraffes is more than the median height. Support your answer.

8. The graph of the quadratic function $y = ax^2 + 4x + c$ has a y-intercept at point A (0, −3) and vertex at point B (1, −1).

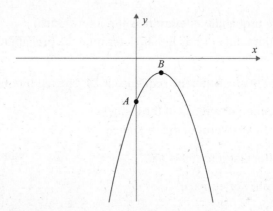

(a) Write down the value of c.

(b) Find the value of a.

(c) Write down the equation of the axis of symmetry.

Point C has the same y-coordinate as point A.

(d) Write down the coordinate of point C.

9. The line R_1 has the equation $3x - 2y = 9$. The point D has coordinates $(1, 1)$.

(a) Determine if point D lies on R_1.

Line R_2 is perpendicular to R_1 and passes through point D.

(b) Find the gradient of R_2.

(c) Write the equation of R_2 in the form $ax + by + d = 0$, where $a, b, d, \in \mathbb{Z}$.

10. Consider the function $y = h(x)$ graphed below.

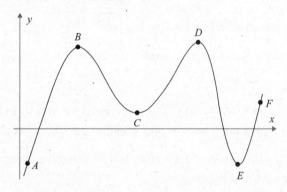

(a) Write down the intervals where
   (i) $h'(x) > 0$
   (ii) $h'(x) < 0$

(b) Write down the points where $h(x) = 0$.

11. The Mumford family is going on vacation to Australia. They have 4000 euro and need to convert them to Australian dollars (AUD).

The exchange rate is 1 euro = 1.10 AUD.

Their bank charges a 3% commission on the transaction.

(a) Calculate the amount, in euro, the Mumfords will convert.

(b) Calculate the amount, in AUD, the Mumfords receive. Round to the nearest dollar.

Once they return home from vacation, they have 50 AUD remaining. They will convert the AUD back to euro at a rate of 1 euro = 1.43 AUD.

(c) Calculate the amount the Mumfords spent on vacation, rounded to the nearest euro.

12. Consider the function $g(x) = \dfrac{a}{x^2} - 2x + 1$, where $a \neq 0$, $x \neq 0$.

(a) Find $g'(x)$.
   The tangent line at $x = -1$ has a gradient of 4.

(b) Determine the value of a.

(c) Write down if $g(x)$ is increasing or decreasing, when $x = -1$. Justify your answer.

13. Ashley is going to invest $1 200 in a savings account. Her bank offers two options. **Round all answers to the nearest integer.**

> **Option A**: 2.75% nominal annual interest compounded monthly
> **Option B**: 2.8% nominal annual interest compounded yearly

(a) Determine the amount Ashley will have after 10 years if she selects:
   (i) Option A
   (ii) Option B

Ashley's goal is to have $2 000 in her savings account, and she selects the option that will help her reach her goal the fastest.

(b) Determine the number of years Ashley must invest her money before reaching her goal.

14. The weights of snack-sized potato chip bags packaged at a local factory are normally distributed with a mean of 9 ounces and standard deviation of 0.25 ounces.

(a) Determine the probability that a randomly selected bag weighs
   (i) between 8.6 and 8.8 ounces
   (ii) less than 8.6 ounces
   (iii) more than 9.35 ounces

The probability a randomly selected bag weighs more than w ounces is 0.12.

(b) Determine the value of w.

15. A ball on a pendulum moves 40 cm on the first swing. On each successive swing, the ball moves 80% of the distance from the previous swing.

(a) Write down the distance the ball travels on the
   (i) first swing
   (ii) third swing

(b) Find the total distance the pendulum traveled after 3 swings.

After s swings, the ball has traveled a total distance of 189 cm.

(c) Determine the value of s.

Paper 2

Each question is worth a different amount of marks. You must show **all** work to earn full marks.

Part A

1. Consider an arithmetic sequence 3, 7, 11, 15, . . .

   (a) Determine which term in the sequence has a value of 79.

   (2)

   (b) Find the sum of the first 10 terms of the sequence.

   (3)

   The sum of the first n terms is 820.

   (c) Determine the value of n.

   (4)

Part B

   A geometric sequence has a third term of 216 and fifth term of 96.

   (a) Find the common ratio of the sequence.

   (2)

   (b) Find the sum of the first 10 terms of the sequence. Round answer to the nearest integer.

   (3)

   A second geometric sequence is represented by $u_n = \dfrac{1}{16}(4)^{n-1}$.

   (c) Determine the value of n for which the two geometric sequences are equal.

   (3)

   (Total 17 marks)

2. Lesley was researching to determine if working part time while attending school has an association with grades earned. Through her research, she recorded the following data from several local high schools in her area.

		Grade Average			
		A/B	C/D	Fail	Total
	Yes	50	70	73	240
Work Part Time	No	76	65	66	160
	Total	126	135	139	400

She plans to carry out a χ^2 test of independence at the 5% significance level.

(a) Write down the null hypothesis, H_o.

(1)

(b) Show that the number of degrees of freedom is 2.

(1)

(c) Determine the expected value of students who worked part time and failed.

(2)

(d) Write down the:
   (i) χ^2 calculated value
   (ii) p-value

(2)

At the 5% significance level, the χ^2 critical value is 5.99.

(e) State the conclusion of the test with a supporting reason.

(2)

Considering the 400 students included in the research, a random student is selected.

(f) Calculate the probability the selected student:
   (i) worked part time and failed
   (ii) does not work part time given that the student has an A/B average

(4)

(g) Two students are selected at random. Find the probability that at least one student failed.

(3)

(Total 15 marks)

3. Andrew heated a bowl of soup to 96°C and is letting the soup cool. The cooling of the soup can be modeled by the function $T = ab^{-m} + 15$, where T is the temperature in °C after m minutes.

(a) Find the value of a.

(2)

After 2 minutes, he measures the temperature to be 51°C.

(b) Determine the value of b.

(3)

(c) Find the temperature of the soup after 4 minutes.

(2)

(d) Write down the temperature the soup will approach if left to cool for a long period of time.

(1)

When the soup reaches 18°C, Andrew will need to reheat the soup.

(e) Determine how many minutes will pass before he will need to reheat the soup. Round answer to the nearest tenth.

(2)

As the soup is being reheated on the stove, Andrew collects the following data.

Time, m (minutes)	0	2	4	6	8	10
Temperature, T (°C)	18	25	30	40	46	54

(f) Write down the equation of the regression line.

(2)

The temperature of the soup while cooling and the temperature of the soup while reheating both reach an equal temperature after n minutes.

(g) Determine the value of n, rounded to the nearest minute.

(3)

(Total 15 marks)

4. Kelly is mapping out a long distance natural hiking trail on a map. He plots the following course, *ABCD*.

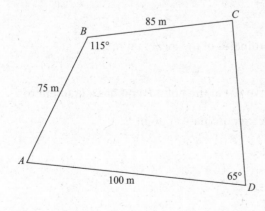

Diagram not drawn to scale

Since hikers will begin the trail at point *A*, he plans to include the path *AC* for those hikers who may want to end the trail early.

(a) Determine the length of path *AC*.

(2)

(b) Show that the size of angle *ACD* is 42.2°.

(2)

(c) Calculate the length of path *CD*.

(3)

Kelly is marking the path with a bright rope. He plans to mark the triangular path *ABC* first.

(d) Write down the total amount of rope Kelly will need if he marks all paths on the trail.

(1)

The area inside the trail is to be used as a campground. Each camper will be given 150 square meters as a campsite.

(e) Determine the total area of the campground.

(4)

(f) Find the maximum number of campers that the campground will hold.

(2)

(Total 14 marks)

5. Consider the function $f(x) = x^2 + \dfrac{10}{x}$, $x \neq 0$.

(a) Sketch the graph of $f(x)$ for $-6 \leq x \leq 9$ and $-40 \leq y \leq 105$. Let 1 cm represent 1 unit on the horizontal axis and 1 cm represent 10 on the vertical axis.

(4)

(b) Find $f'(x)$.

(2)

(c) Solve the equation $f'(x) = 0$.

(2)

(d) Hence, determine the coordinates of the local minimum.

(2)

The line *L* is tangent to the curve $f(x)$ at the point *P* and has a gradient of $\dfrac{3}{2}$.

(e) Determine the value of the *x*-coordinate at point *P*.

(2)

(f) Write the equation of line L.

(3)

(g) Find the equation of the normal line at point P.

(2)

(Total 17 marks)

6. **Express all answers to the nearest currency unit.**

Five years ago George sold his truck for 21 000 USD and invested the money into his son's college fund, which pays 4.5% nominal interest compounded quarterly.

(a) Calculate the current value of the college account.

(2)

George's son is accepted into a university in Switzerland. He converts the college fund into Swiss franc (CHF) at an exchange rate of 1 USD = 0.89 CHF.

(b) Calculate the amount George receives in CHF after the exchange.

(2)

The tuition cost for one year at the university is 2 800 CHF. The cost for room and board is 700 CHF per month.

(c) Calculate the total amount George will pay in CHF if his son attends the university for four full years. Assume all rates stay constant.

(2)

(d) Determine how much **more** George needs in order to pay for all four years.

(1)

To cover the remaining university costs, George invests 18 500 USD in a Swiss bank that pays r% nominal interest annually. The bank has an exchange rate of 1 CHF = 1.12 USD.

(e) Determine the interest rate, r%, if George has four years to save the remaining university fees. Write answer to the nearest hundredth.

(5)

(Total 12 marks)

PAPER 1

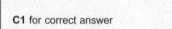

C2 for correct values
OR
- **A1** for x
- **A1** for y

1. (a) **x = 2 and y = 5**

To estimate a solution, Lauren would round x and y to the nearest integer.

C1 for correct answer

(b) **1.6**

To determine the estimate, substitute the values of x and y from part a into the expression and simplify. Be sure to round to 2 significant figures.

$$\sqrt{\frac{5(2)^3}{3(5)}} = 1.6$$

C1 for correct answer

(c) **2.134**

To determine the exact solution, substitute the true values of x and y into the expression and simplify. Be sure to round to 4 significant figures.

$$\sqrt{\frac{5(2.35)^3}{3(4.75)}} = 2.134$$

C2 for correct answer
OR
- **M1** for using percent error formula
- **A1** for correct answer in standard form

(d) **2.50 × 10**

Percent error is found using the formula $\varepsilon = \left| \frac{v_A - v_E}{v_E} \right| \times 100\%$, where the approximate is 1.6 and exact is 2.134.

$$\left| \frac{1.6 - 2.134}{2.134} \right| \times 100\% \approx 25.023$$

The answer must be written in standard form. Move the decimal one place to the left.

C2 for both sets correct
OR
- **A1** for *each* set correct

2. (a) List the elements of:
   (i) $A = \{1, 2, 3, 4, 6\}$
   (ii) $B = \{1, 3, 5, 7\}$

The universal set is only the numbers 1–7, so sets A and B must only contain numbers between 1 and 7, inclusive.

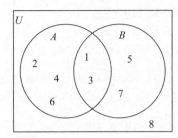

(b) The numbers sets A and B have in common are 1 and 3; these go inside the overlap. The remaining elements are placed in the outside part of the circle. The only number not placed inside set A or B is 8; thus 8 is placed in the outside region.

(c) $A \cup B' = \{1, 2, 3, 4, 6, 8\}$

You need the elements in set A <u>OR</u> **not** in set B. The elements in set A are 1, 2, 3, 4, 6. Union means "all", so combine those elements with the remaining number **not** in B, which is 8.

3. (a) (i) $p \wedge q$

"Coleman will ride his bike" is proposition p; "he will be late to work" is proposition q. The word "but" is like "and", so $\wedge$ joins p and q together.

(ii) $p \veebar \neg q$

"Either . . . or . . . but not both" is the exclusive disjunction, so the symbol $\veebar$ joins the two propositions. "Coleman will ride his bike" is proposition p and "he will be on time to work" is the negation of proposition q.

(b)

p	q	$\neg q$	$p \Rightarrow \neg q$
T	T	**F**	**F**
T	F	**T**	**T**
F	T	**F**	**T**
F	F	**T**	**T**

$\neg q$ is the negated truth values of q.

$p \Rightarrow \neg q$ is only false when p is true but $\neg q$ is false.

(c) **Neither**

Since the answer column is not all true nor all false, the result is neither a tautology nor contradiction.

4. (a) (i) **26.5 kg (±0.5 kg)**

(ii) **23.5 kg (±0.5 kg)**

(iii) **32 kg (±0.5 kg)**

The median occurs in the middle of the data set. Since there are 220 calves, the median is the 110th one.

The lower quartile is 25% of the data, which can be found at $220(0.25) = 55$ calves.

The 90th percentile is found at $220(0.98) = 198$ calves.

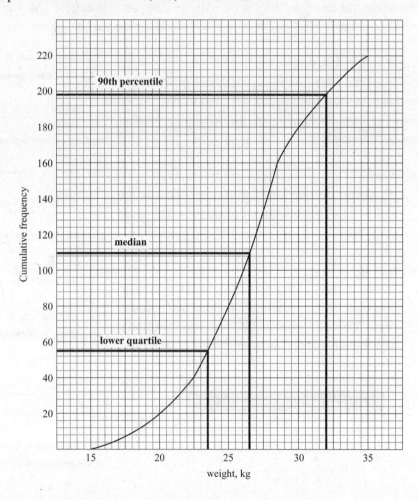

(b) **43.6% (± 2%)**

To find the number of calves that weigh 27 kg or less, draw a vertical line from 27 on the *x*-axis up to the graph, and then a horizontal line to the *y*-axis. This is 124 calves. The question asks for the percent of calves that weigh *more*, so subtract 124 from 220 and then divide by the total.

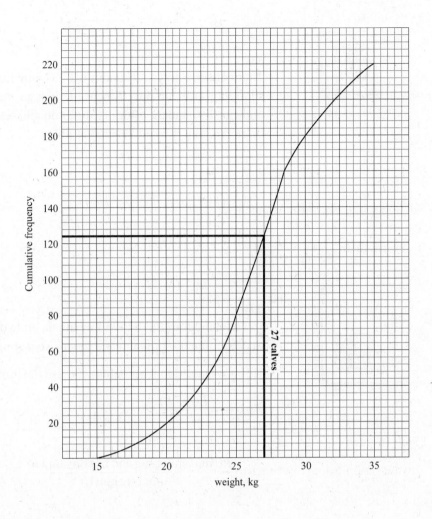

5. (a)

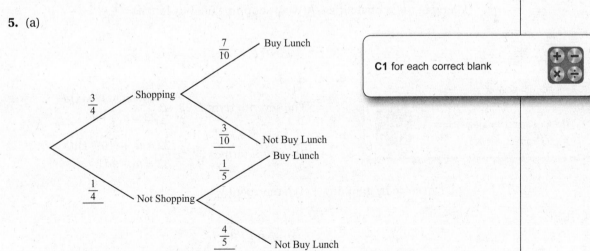

Each branch of a tree diagram must add to one. Subtract each given probability from one to find the missing blank.

(b) $\dfrac{17}{40}$ **or 0.425**

There are two ways Alexandra will not buy lunch: go shopping and not buy lunch OR not go shopping and not buy lunch. Multiply the probabilities of the branches, and then add each possibility.

Shopping and Not Buy Lunch: $\dfrac{3}{4}\left(\dfrac{3}{10}\right)=\dfrac{9}{40}$

Not Shopping and Not Buy Lunch: $\dfrac{1}{4}\left(\dfrac{4}{5}\right)=\dfrac{1}{5}$

Not Buy Lunch: $\dfrac{9}{40}+\dfrac{1}{5}=\dfrac{17}{40}$

(c) $\dfrac{9}{17}$ **or 0.529**

You know Alexandra did not buy lunch, and you need to determine the probability that she went shopping.

This is a conditional probability, so use the formula

$$P(A\,|\,B)=\dfrac{P(A\cap B)}{P(B)}.$$

Using context:

$$P(\text{went shopping}|\text{not buy lunch})=\dfrac{P(\text{went shopping and not buy lunch})}{P(\text{not buy lunch})}$$

Filling in the probabilities: $P(\text{went shopping}|\text{not buy lunch})=\dfrac{\dfrac{3}{4}\times\dfrac{3}{10}}{\dfrac{17}{40}}=\dfrac{9}{17}$

6. (a) $d=3$

The seventh term is 23, so $\begin{array}{l}23=u_1+(7-1)d\\23=u_1+6d\end{array}$

The tenth term is 32, so $\begin{array}{l}32=u_1+(10-1)d\\32=u_1+9d\end{array}$

You can solve by graphing. Let u_1 represent y.

$$u_1=-6d+23$$
$$u_1=-9d+32$$

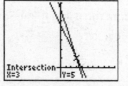

$$d=3$$

(b) $u_1 = 5$

The first term was y in the intersection found in part a.

(c) $n = 15$

The sum is 390, so $S_n = 390$. You found $u_1 = 5$ and $d = 3$. Substitute these values into the arithmetic sum formula, simplify, and set equal to zero.

$$390 = \frac{n}{2}(2(5) + (n-1)(3))$$

$$390 = \frac{n}{2}(10 + 3n - 3)$$

$$780 = 3n^2 + 7n$$

$$0 = 3n^2 + 7n - 780$$

Now solve using PLYSMLT.

The answer of 15 is the only appropriate solution.

7. (a) **5.8 m**

The third quartile is the last edge of the box. This is located at 5.8.

(b) **0.4 m**

The interquartile range is $Q_3 - Q_1$. You determined Q_3 to be 5.8 and Q_1 is 5.4. Thus, the IQR = 5.8 – 5.4 = 0.4 m

(c) **20 giraffes**

25% of the data lie below the first quartile. Since 5.4 is the first quartile, 80(0.25) = 20 giraffes would be below 5.4 meters.

(d) **The mean is not more, since the median is closer to the third quartile than the center.**

If the median was in the center of the box and whisker plot, then the mean and median would be relatively the same. Since the median is closer to the third quartile, there are smaller data values that drag the mean down. Thus, the mean would be smaller than the median.

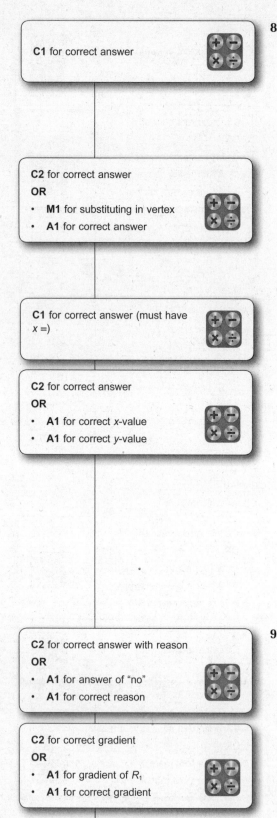

C1 for correct answer

C2 for correct answer
OR
- **M1** for substituting in vertex
- **A1** for correct answer

C1 for correct answer (must have $x =$)

C2 for correct answer
OR
- **A1** for correct x-value
- **A1** for correct y-value

C2 for correct answer with reason
OR
- **A1** for answer of "no"
- **A1** for correct reason

C2 for correct gradient
OR
- **A1** for gradient of R_1
- **A1** for correct gradient

8. (a) $c = -3$

The constant c is the value of the y-intercept. Another way to determine c is to substitute the point $(0, -3)$ in for x and y and solve for c.

$$-3 = a(0)^2 + 4(0) + c$$
$$-3 = c$$

(b) $a = -2$

Now that the constant c has been determined, substitute in the vertex and solve.

$$-1 = a(1)^2 + 4(1) - 3$$
$$-1 = a + 1$$
$$-2 = a$$

(c) $x = 1$

The axis of symmetry passes through the vertex of $(1, -1)$. Thus, the equation is $x = -1$.

(d) $(2, -3)$

Quadratic graphs are symmetric, and you know the y-value of the coordinate is -3. From the axis of symmetry, you know the horizontal distance from point A to the vertex point B is one. Thus, the horizontal distance from point B to point C is also 1. The x-value of point C is 2.

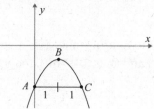

9. (a) **No, because $3(1) - 2(1) \neq 9$**

When x and y are substituted into the linear equation, the two sides of the equation must equal in order for point D to lie on the line.

(b) $-\dfrac{2}{3}$

First, the gradient of line R_1 must be determined.

$$-2y = -3x + 9$$
$$y = \frac{3}{2}x - \frac{9}{2}$$

The gradient of line R_1 is $\dfrac{3}{2}$. Since line R_2 is perpendicular, $\dfrac{3}{2} \times m = -1$. Thus, the gradient of line R_2 is $-\dfrac{2}{3}$.

(c) $-2x - 3y + 5 = 0$ or $2x + 3y - 5 = 0$

The gradient of line R_2 is $-\dfrac{2}{3}$, and it passes through the point (1, 1). Now, substitute the given information into $y = mx + c$ and solve for c.

$$1 = -\frac{2}{3}(1) + c$$

$$\frac{5}{3} = c$$

Thus, $y = -\dfrac{2}{3}x + \dfrac{5}{3}$. The answer must be written in the form $ax + by + d = 0$, where a, b, $d \in \mathbb{Z}$. Set the equation equal to zero and multiply every one to undo the fraction.

$$\left[0 = -\frac{2}{3}x - y + \frac{5}{3}\right] \times 3$$

$$0 = -2x - 3y + 5$$

C2 for correct equation

OR

- A1 for correct $y = mx + c$
- A1 for correct standard form

10. (a) (i) **A to B, C to D, E to F**

   OR $A < x < B$, $C < x < D$, $E < x < F$

   OR $(A, B) \cup (C, D) \cup (E, F)$

C2 for all correct intervals

OR

- A1 for one of the three

The first derivative is greater than zero when the original function is increasing. Looking from left to right on the graph, the intervals are increasing, when the graph is moving uphill.

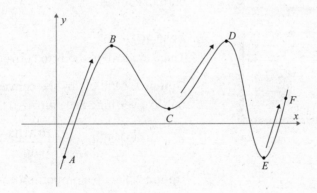

(ii) **B to C, D to E**

   OR $B < x < C, D < x < E$

   OR $(B, C) \cup (D, E)$.

The first derivative is less than zero when the original function is decreasing. Looking from left to right on the graph, the intervals are decreasing, when the graph is moving downhill.

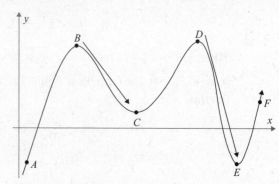

(b) **B, C, D, E**

The first derivative is zero, when the graph has a horizontal tangent. These tangents occur at local maxima or minima.

11. (a) **3 880 euro**

The Mumfords must pay a 3% commission fee: $4\,000(0.03) = 120$. They will then have $4\,000 - 120 = 3\,880$ euro to convert.

(b) **4 268 AUD**

There are two methods to convert:

Option 1: Multiply by the conversion factor. They are only converting 3 880 euro.

$$3\,880 \text{ euro} \times \frac{1.10\,\text{AUD}}{1\,\text{euro}} = 4\,268\,\text{AUD}$$

Option 2: Set up a proportion and solve. They are only converting 3 880 euro.

$$\frac{3\,880 \text{ euro}}{x \text{ AUD}} = \frac{1 \text{ euro}}{1.10\,\text{AUD}}$$

$$x = 4\,268\,\text{AUD}$$

(c) **2 950 euro**

The Mumfords only have 50 AUD remaining, so they spent $4\,268 - 50 = 4\,218$ AUD.

Convert this amount using one of the two methods. Then round to the nearest euro.

Option 1: Multiply by the conversion factor.

$$4\,218\,\text{AUD} \times \frac{1\,\text{euro}}{1.43\,\text{AUD}} = 2\,949.65\,\text{euro} \Rightarrow 2\,950\,\text{euro}$$

Option 2: Set up a proportion and solve.

$$\frac{4218\,\text{AUD}}{x\,\text{euro}} = \frac{1.43\,\text{AUD}}{1\,\text{euro}}$$

$$1.43x = 4\,218$$

$$x = 2\,949.65 \Rightarrow 2\,950\,\text{euro}$$

12. (a) $g'(x) = \dfrac{-2a}{x^3} - 2.$

First, rewrite $g(x)$ using integer exponents, and then apply the first derivative rule to each term.

$$g(x) = ax^{-2} - 2x + 1$$
$$ax^{-2} \Rightarrow -2(ax^{-2-1}) = -2ax^{-3}$$

$-2x$ differentiates to the coefficient -2

The constant 1 differentiates to zero.

(b) $a = 3$

Since the gradient at $x = -1$ is 4, substitute -1 for x and 4 for the first derivative and solve for a.

$$4 = \frac{-2a}{(-1)^3} - 2$$
$$4 = 2a - 2$$
$$6 = 2a$$
$$3 = a$$

(c) **$g(x)$ is increasing because the gradient at $x = -1$ is positive**

When the first derivative is positive, the tangent lines are increasing, and thus the function is increasing.

PRACTICE TEST 3

C2 for correct answer

OR

- **M1** for using correct formula
- **A1** for correct answer

C2 for correct answer

OR

- **M1** for using correct formula
- **A1** for correct answer

C2 for correct answer

OR

- **M1** for using correct formula
- **A1** for correct answer

C1 for correct answer

13. (a) (i) **\$1 579**

Since the interest is paid monthly, $k = 12$. Using the compound interest formula,

$$FV = 1\,200\left(1 + \frac{2.75}{100 \times 12}\right)^{12 \times 10}$$

$$FV = 1\,579$$

(ii) **\$1 582**

Since the interest is paid yearly, $k = 1$. Using the compound interest formula,

$$FV = 1\,200\left(1 + \frac{2.8}{100}\right)^{10}$$

$$FV = 1\,582$$

(b) **19 years**

Option 2 earns interest the quickest. Set up the compound interest formula:

$$2\,000 = 1\,200\left(1 + \frac{2.8}{100}\right)^{n}$$

Let $Y_1 = 1\,200\left(1 + \frac{2.8}{100}\right)^{n}$ and use the table to find when Ashley will have at least \$2 000 in her account.

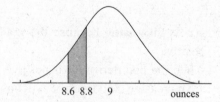

14. (a) (i) **0.157**

Sketch the normal curve and shade the desired area.

Now compute $\boxed{\text{NORMALCDF}}$(8.6, 8.8, 9, 0.25) to get the answer.

(ii) **0.0058**

Sketch the normal curve and shade the desired area.

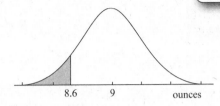

Now compute $\boxed{\text{NORMALCDF}}(-\text{E}99, 8.6, 9, 0.25)$ to get the answer.

(iii) **0.0808**

First, sketch the normal curve and shade the desired area.

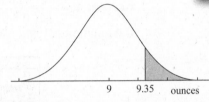

Now compute $\boxed{\text{NORMALCDF}}(9.35, \text{E}99, 9, 0.25)$ to get the answer.

(b) **$w = 9.29$ ounces**

The given probability represents the shaded area, so sketch a diagram.

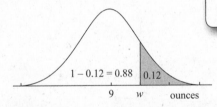

In order to use the $\boxed{\text{INVNORM}}$ option on the calculator, the lower shading must be found.

Now compute $\boxed{\text{INVNORM}}(0.88, 9, 0.25)$ to find the value of w.

15. (a) (i) **40 cm**

(ii) **25.6 cm**

The pendulum moves 40 cm on the first swing, which is the distance traveled on the first swing.

This is a geometric sequence with $u_1 = 40$ and $r = 0.80$. To calculate the distance on the third swing, let $n = 3$.

$$u_3 = 40(0.8)^{3-1}$$
$$u_3 = 25.6$$

C2 for correct answer

OR

- **M1** for using sum formula
- **A1** for correct answer

C2 for correct answer

OR

- **M1** for using sum formula
- **A1** for correct answer

(b) **97.6 cm**

The total distance traveled can be found using the sum of a geometric sequence formula.

$$S_3 = \frac{40(1-0.8^3)}{1-0.8}$$

$$S_3 = 97.6$$

(c) $s = 13$

The total distance traveled is 189, so first set up the sum.

$$189 = \frac{40(1-0.8^n)}{1-0.8}$$

Next, let $Y_1 = \frac{40(1-0.8^x)}{1-0.8}$ and find the value of x when the function is 189.

X	Y₁
11	182.82
12	186.26
13	189
14	191.2
15	192.96
16	194.37
17	195.5

$X = 13$

PAPER 2

1.

Part A

- **M1** for using sequence formula
- **A1** for correct answer

(a) $79 = 3 + (n-1)4$

$20 = n$

You know $u_n = 79$, $u_1 = 3$, and $d = 4$. Substitute those values into the arithmetic sequence formula and solve for n.

$$79 = 3 + (n-1)4$$
$$79 = 4n - 1$$
$$80 = 4n$$
$$20 = n$$

- **M1** for using sum formula
- **A1** for substituting correct values
- **A1** for correct answer

(b) $S_{10} = \frac{10}{2}(2(3)+(10-1)4)$

$S_{10} = 210$

To find the sum of the first 10 terms of the arithmetic sequence, let $n = 10$, $u_1 = 3$, and $d = 4$. Substitute those values into the arithmetic sum formula and simplify.

(c) $\quad 820 = \dfrac{n}{2}(2(3)+(n-1)4)$

$\quad\quad 0 = 4n^2 + 2n - 1\,640$

$\quad\quad n = 20$

$S_n = 820$, $u_1 = 3$, and $d = 4$. Substitute into the arithmetic sum formula and solve for n. The resulting equation is a quadratic. Set it equal to zero and use $\boxed{\text{PLYSMLT}}$.

$$820 = \dfrac{n}{2}(2(3)+(n-1)4)$$

$$820 = \dfrac{n}{2}(6+4n-4)$$

$$1\,640 = n(4n+2)$$

$$1\,640 = 4n^2 + 2n$$

$$0 = 4n^2 + 2n - 1\,640$$

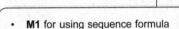

Since n can only be a natural number, ignore the second solution of $-\dfrac{41}{2}$.

Part B

(a) $\quad 216r^2 = 96$

$\quad\quad r^2 = \dfrac{4}{9}$

$\quad\quad r = \dfrac{2}{3}$

$\quad\quad r = 0.667$ is also accepted

You are given $u_3 = 216$ and $u_5 = 96$. The third member and fifth member are two common ratios apart. Therefore, if you multiply 216 by r^2, you will get 96.

(b) $\quad S_{10} = \dfrac{486\left(1-\left(\dfrac{2}{3}\right)^{10}\right)}{1-\dfrac{2}{3}}$

$\quad\quad S_{10} = 1\,433$

To find the sum of the first 10 terms, let $n = 10$, $u_1 = 486$, and $r = \dfrac{2}{3}$. Substitute into the geometric sum formula and simplify. Be sure to round to the nearest integer.

- **M1** for setting sequences equal
- **A1** for correct resulting equation
- **A1** for correct answer

(c) $486\left(\dfrac{2}{3}\right)^{n-1}=\dfrac{1}{16}(4)^{n-1}$

$n=6$

The two sequence formulas are $u_n=486\left(\dfrac{2}{3}\right)^{n-1}$ and $u_n=\dfrac{1}{16}(4)^{n-1}$. They must be equal. Thus, $486\left(\dfrac{2}{3}\right)^{n-1}=\dfrac{1}{16}(4)^{n-1}$.

To solve using the GDC, let $Y_1=486\left(\dfrac{2}{3}\right)^{x-1}$ and $Y_2=\dfrac{1}{16}(4)^{x-1}$. Looking at the table of values, when $x=6$, the function values are equal.

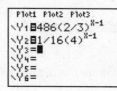 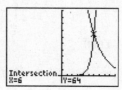

- **A1** for correct hypothesis

2. (a) H₀: Grade average and working status are independent.

The null hypothesis is always that the factors are independent.

- **M1** for using degrees of freedom formula

(b) $df = (2-1)(3-1)$

$df = 2$

Degrees of freedom are found by $df=(\text{rows}-1)(\text{columns}-1)$. There are two and three columns.

- **M1** for multiplying and then dividing by 400
- **A1** for correct answer

(c) $\dfrac{240(139)}{400}=83.4$

The expected value of students who work part time and failed is found by multiplying the total of the "work part time" row by the total of the "fail" column. The product is divided by the total number of students.

- **A1** for *each* correct answer

(d) (i) $\chi^2=5.42$

(ii) $p=0.0666$

Enter the observed data into matrix A and perform the χ^2–test. The calculator output gives the necessary information.

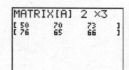 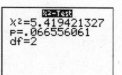

(e) Two possible answers are

Option 1	Option 2
$5.42 < 5.99$	$0.0666 > 0.05$
Accept H_o	Accept H_0
Grade average and working status are independent.	Grade average and working status are independent.

If the χ^2 calculated is smaller than the χ^2 critical, accept the null hypothesis. If the p-value is larger than the significance level, accept the null hypothesis.

(f) (i) $\dfrac{73}{400}$ or 0.1825 or 0.183

There are 73 students who failed AND worked part time out of a total of 400 students.

(ii) $\dfrac{76}{126}$ or $\dfrac{38}{63}$ or 0.603

We know that the selected student has an A/B average and only 126 total students have an A/B average. Therefore, the total number of outcomes is not 400, but 126. Within those 126 students, 76 do not work part time.

(g) $1 - \left(\dfrac{261}{400}\right)\left(\dfrac{260}{399}\right) = 0.575$

To find the probability that at least one student failed, the most direct method is to calculate the complement event of no student failing. You know $1 - P(A') = P(A)$, which means $1 - P(\text{no student failing}) = P(\text{at least 1 fails})$.

The total number of students who did not fail is the combined students with an A/B average and C/D average, $126 + 135 = 261$. There are 400 total students.

Once the first student is selected, there are only 260 students who did not fail out of 399 total students.

3. (a) $96 = a(b^0) + 15$

$81 = a$

The temperature at time $m = 0$ is $T = 96$. When b is raised to the zero power, the result is one.

$$96 = a(b^0) + 15$$
$$\text{Hence, } 96 = a + 15$$
$$81 = a$$

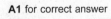

(b) $51 = 81b^{-2} + 15$

$$\frac{4}{9} = b^{-2}$$

$$1.5 = b$$

The temperature at time $m = 2$ is $T = 51$. You already determined $a = 81$, so substitute in the given values.

$$51 = 81b^{-2} + 15$$
$$36 = 81b^{-2}$$
$$\frac{4}{9} = b^{-2}$$

Solve the resulting equation algebraically or graphically.

Algebraically:

$$\frac{4}{9} = b^{-2}$$

$$\frac{4}{9} = \frac{1}{b^2}$$

$$b^2 = \frac{9}{4}$$

$$b = 1.5$$

Graphically:

Let $Y_1 = x^{-2}$ and $Y_2 = \frac{4}{9}$, and find the intersection.

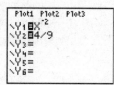 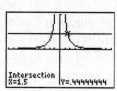

Therefore $b = 1.5$.

(c) $T = 81(1.5)^{-4} + 15$

$T = 31$

Substitute 4 in for m and simplify.

(d) **15°C**

The graph will begin to follow the horizontal asymptote as m gets larger. Thus, the soup will approach a temperature of 15°C.

(e) $18 = 81(1.5)^{-m} + 15$

   $m = 8.1$

To determine m, let $T = 18$ and solve by graphing.

Let $Y_1 = 81(1.5)^{-x} + 15$ and $Y_2 = 18$. Graph and find the intersection.

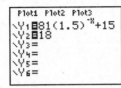 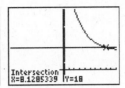

(f) $T = 3.61m + 17.4$

   (also accepted: y = 3.61x + 17.4)

Enter the data into L_1 and L_2. Calculate (LINREG) using the GDC.

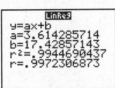

(g) $81(1.5)^{-n} + 15 = 3.61n + 17.4$

   $n = 4$

Since the cooling and the reheating both hit a certain temperature at the same time, set the equations equal and solve by graphing.

Let $Y_1 = 81(1.5)^{-x} + 15$ and $Y_2 = 3.61x + 17.4$; graph and find the intersection.

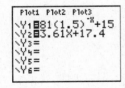 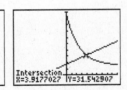

The answer is to be written to the nearest minute; 3.92 rounds to 4 minutes.

- **M1** for using cosine rule
- **A1** for correct answer

4. (a) $AC^2 = 75^2 + 85^2 - 2(75)(85)\cos115°$

$AC^2 = 18238.38$

$AC = 135$

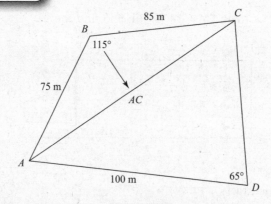

You are given two sides and the included angle, so use the cosine rule to solve. The side opposite the angle is always a, while the angle is A.

$$a^2 = b^2 + c^2 - 2bc\cos A$$

(b) $\dfrac{135}{\sin65°} = \dfrac{100}{\sin C}$

$135\sin C = 100\sin65°$

$\sin C = \dfrac{100\sin65°}{135}$

$C = \sin^{-1}(0.671339)$

$C = 42.2°$

- **M1** for using cosine rule
- **M1** for the use of sine inverse

Looking at the given information, you have angles with their opposite sides. The sine rule is the appropriate method to solve for angle ACD.

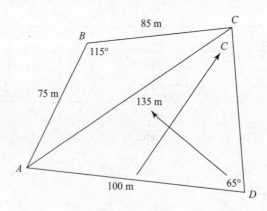

(c) $\hat{CAD} = 180° - 65° - 42.2°$

$\hat{CAD} = 72.8°$

$$\frac{135}{\sin 65°} = \frac{CD}{\sin 72.8°}$$

Thus $\quad CD = \dfrac{135 \sin 72.8°}{\sin 65°}$

$$CD = 142 \text{ m}$$

- **A1** for finding angle *CAD* (could be implied in calculations)
- **M1** for using sine rule
- **A1** for correct answer

To find the length of *CD*, you need the opposite angle. Since all angles add to 180°, subtract the two angles from 180°. Now, set up a sine ratio and solve.

(d) $75 + 85 + 135 + 100 + 142 = 537 \text{ m}$

Add all the sides of the quadrilateral, plus diagonal *AC*.

- **A1** for correct answer

(e) $A = \dfrac{1}{2}(75 \times 85)\sin 115°$ $\qquad A = \dfrac{1}{2}(100 \times 135)\sin 72.8°$

$A = 2\,889$ $\qquad\qquad\qquad A = 6\,448$

Total area $= 2\,889 + 6\,448 = 9\,340 \text{ m}^2$

- **M1** for using area of a triangle formula
- **A1** for correct area of 2 889 (or 2 890)
- **A1** for correct area of 6 448 (or 6 450)
- **A1** for correct total area

To find the total area of the campground, find the area of each triangle. The area of a non-right triangle is found by $A = \dfrac{1}{2}ab\sin C$, where *a* and *b* are included sides to angle *C*.

For triangle *ACD*, any angle with its included sides would be appropriate to use.

(f) $\dfrac{9\,340}{150} = 62 \text{ campers}$

Each camper receives 150 square meters and the campground contains 9 340 square meters. $9\,340 \div 150 = 62.3$. Since decimal amounts of campers do not make sense, the solution is 62 campers.

- **M1** for dividing
- **A1** for correct answer

5. (a)

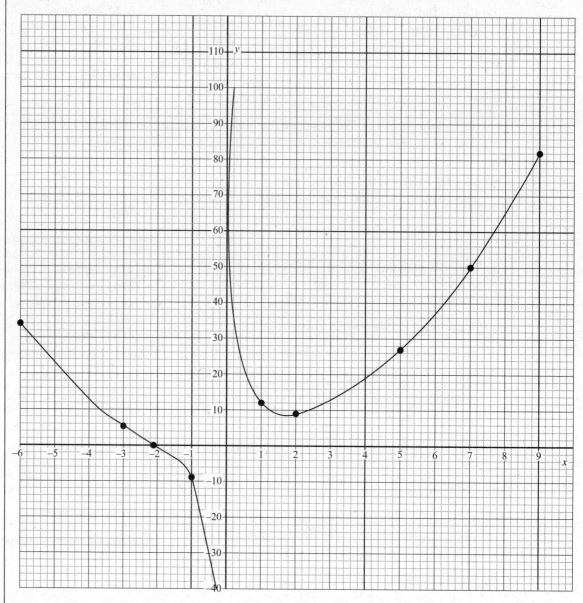

- **A1** for correct scales and labeled axes
- **A1** for *x*-intercept of −2.15
- **A1** for vertical asymptote of *x* = 0
- **A1** for general shape

The axes should be labeled with *x* and *y*, and the provided scale must be used. Since the *x*-intercept was determined in part *b*, the point needs to be marked on the graph. A few points can be used off the table to get the general shape of the graph.

(b) $f(x) = x^2 + 10x^{-1}$

$f'(x) = 2x - \dfrac{10}{x^2}$

First, rewrite the function with a negative exponent. Then, apply the first derivative rule to each term.

$$f(x) = x^2 + 10x^{-1}$$
$$x^2 \Rightarrow 2(x^{2-1}) = 2x$$
$$10x^{-1} \Rightarrow -1(10x^{-1-1}) = -10x^{-2}$$

(c) $2x - \dfrac{10}{x^2} = 0$

$x = 1.71$

Use the GDC to find the zero of $Y_1 = 2x - \dfrac{10}{x^2}$.

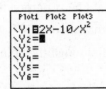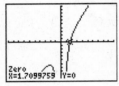

(d) $f(1.71) = (1.71)^2 + \dfrac{10}{1.71}$

(1.71, 8.77)

To find the y-coordinate of the minimum, substitute $x = 1.71$ into the original function.

(e) $\dfrac{3}{2} = 2x - \dfrac{10}{x^2}$

$x = 2$

The gradient of the tangent line is $\dfrac{3}{2}$, so the first derivative must be $\dfrac{3}{2}$. Graph $Y_1 = 2x - \dfrac{10}{x^2}$ and $Y_2 = \dfrac{3}{2}$. Find the intersection.

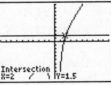

A1 for determining $y = 9$
M1 for finding $c = 6$
A1 for correct equation

(f) $f(2) = (2)^2 + \dfrac{10}{2}$ $9 = \dfrac{3}{2}(2) + c$ $y = \dfrac{3}{2}x + 6$

    $f(2) = 9$ $6 = c$

The tangent line has a gradient of $\dfrac{3}{2}$ and an x-coordinate of 2, as found in part e. To write the equation of the line, the coordinate of y is needed.

Substitute 2 in for x into the original function to determine y. Then, substitute the gradient x and y into $y = mx + c$.

A1 for correct gradient
A1 for correct y-intercept

(g) $y = -\dfrac{2}{3}x + \dfrac{31}{3}$

The normal line is perpendicular to the tangent line. The gradient would be $-\dfrac{2}{3}$, since $\dfrac{3}{2} \times -\dfrac{2}{3} = -1$. The line also passes through the point (2, 9).

$$9 = -\frac{2}{3}(2) + c$$

$$\frac{31}{3} = c$$

Therefore, $y = -\dfrac{2}{3}x + \dfrac{31}{3}$

- **M1** for using compound formula
- **A1** for correct answer

6. (a) $FV = 21\,000\left(1 + \dfrac{4.5}{100 \times 4}\right)^{4 \times 5}$

    $FV = 26\,266$

$PV = 21\,000$, $r = 4.5$, $k = 4$ (quarterly), and $n = 5$.

Substitute the values into the compound interest formula and simplify.

- **M1** for multiplying conversion
- **A1** for correct answer

(b) $26\,266 \text{ USD} \times \dfrac{0.89 \text{ CHF}}{1 \text{ USD}} = 23\,377$

OR $\dfrac{26\,266 \text{ USD}}{x \text{ CHF}} \times \dfrac{1 \text{ USD}}{0.89 \text{ CHF}}$

$x = 23\,377$

George will convert 26 266 USD so either multiply by the conversion factor or set up a proportion and solve.

- **M1** for multiplying by time
- **A1** for correct answer

(c) $2\,800(4) + 700(12 \times 4) = 44\,800 \text{ CHF}$

The son will attend four years where tuition costs 2 800 per year. Room and board is 700 per month for four years.

(d) **21 156 CHF**

George had 23 644 CHF after converting, but he needs 44 800 CHF. The difference is $44\,800 - 23\,644 = 21\,156$ CHF.

(e) $$18\,500\,\text{USD} \times \frac{1\,\text{CHF}}{1.12\,\text{USD}} = 16\,518\,\text{CHF}$$

$$21\,156 = 16\,518\left(1+\frac{r}{100}\right)^{4}$$

$$r = 6.38\%$$

In order to determine the interest rate, convert the 18 500 USD to CHF.

Either multiply by the conversion rate (the values will be divided)	or set up a proportion and solve
$18\,500\,\text{USD} \times \dfrac{1\,\text{CHF}}{1.12\,\text{USD}} = 16\,518\,\text{CHF}$	$\dfrac{18\,500\,\text{USD}}{x\,\text{CHF}} \times \dfrac{1.12\,\text{USD}}{1\,\text{CHF}}$ $1.12x = 18\,500$ $x = 16\,518\,\text{CHF}$

George needs 21 156 CHF, which is FV, $PV = 16\,518$, and $n = 4$

$$21\,156 = 16\,518\left(1+\frac{r}{100}\right)^{4}$$

To solve, let $Y_1 = 21\,156$ and $Y_2 = 16\,518\left(1+\dfrac{x}{100}\right)^{4}$, then find the intersection.

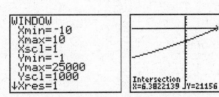

APPENDIX

APPENDIX

TI-Nspire Directions

Many of the calculator operations for the TI-84 and the TI-Nspire are the same. Addition, subtraction, multiplication, division, and operations such as exponents and roots are entered the same for both calculators. However, there are differences when using the TI-Nspire calculator. On the following pages, examples from throughout the book are reworked showing the TI-Nspire (NON-CAS) calculator with operating system 3.6.0.546. The TI-Nspire must be in Press-to-Test mode in order to use on the IB Math Studies exam.

To enter into Press-to-Test mode:

1. Turn the Nspire calculator off.
2. Press and hold the (ESC) and (HOME) keys.
3. A settings screen will appear. Ensure angles are set to degree.

4. Uncheck the restrictions: limit trigonometric functions, disable "$\log_b x$" template and summation functions, and disable polynomial root finder and simultaneous equation solver. These are the only functions allowed by IBO.

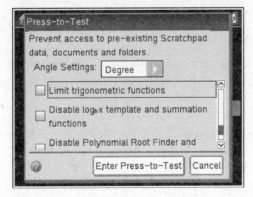

5. Select (ENTER PRESS-TO-TEST). The calculator will restart.

To exit PRESS-TO-TEST mode, connect the calculator to another handheld or computer and go to "My Documents."

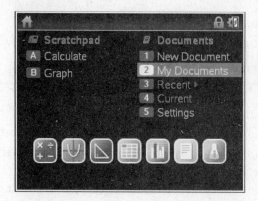 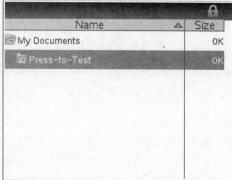

Press the DOC key and select option 9: PRESS-TO-TEST.
Then select option 1: Exit PRESS-TO-TEST.

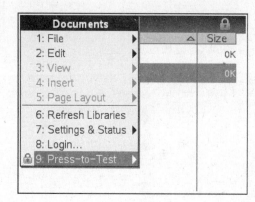 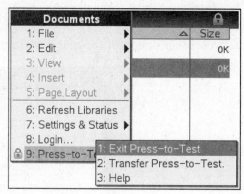

To do simple calculations, the Scratchpad is the easiest option. Most calculations and graphing in this chapter will take place on the scatchpad screen.

Section 1.3 (Standard Form)

To enter a number into the calculator in standard form, use the EE button on the left side of the keyboard.

➡ EXAMPLE 1.31

Let $x = 7.65 \times 10^{-4}$ and $y = 4.97 \times 10^5$. Calculate $\dfrac{10x - y}{x}$, writing the answer in the form $a \times 10^k$, where $1 \le a < 10$ and $k \in \mathbb{Z}$.

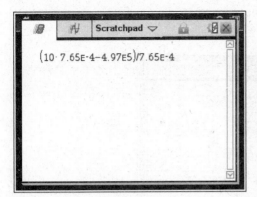
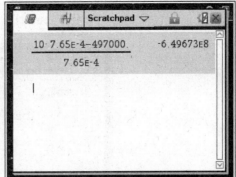

The solution is -6.50×10^8.

Section 1.6 (Solving Linear and Quadratic Equations)

To solve a system of linear equations, access the Algebra menu and select option 7: SOLVE SYSTEM OF EQUATIONS. Select option 2: SOLVE SYSTEM OF LINEAR EQUATIONS.

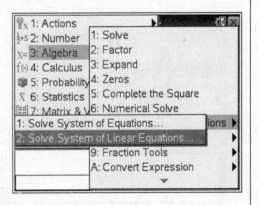

In IB Math Studies, you will only solve 2 equations with 2 variables.

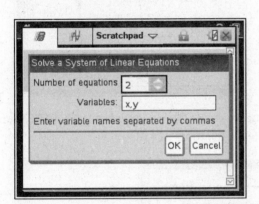

➡ **EXAMPLE 1.60** ——————————————————

Solve

$$2x - 3y = 10.5$$
$$6y - x = -16.5.$$

Input the equations into the solving systems function (as accessed using the above directions). The variable keys are found at the bottom of the keyboard. The equal sign is below the CTRL key.

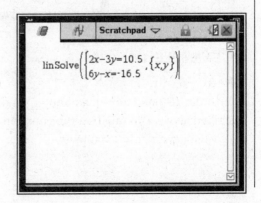

Pressing (ENTER) gives the point of intersection.

The solution is displayed as the point (1.5, –2.5).

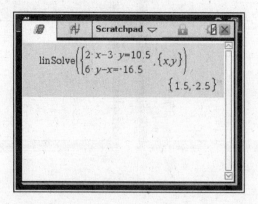

To solve a quadratic equation, you can graph and find the zeros, or you can use the (POLYNOMIAL TOOLS) function found in the Algebra menu. Select option 1: (FIND ROOTS OF POLYNOMIAL).

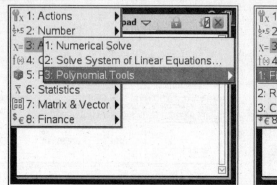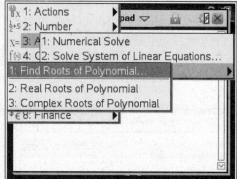

The degree is the value of the highest exponent, so the degree of a quadratic is 2. In IB Math Studies, we only use real numbers, so always leave the roots as real.

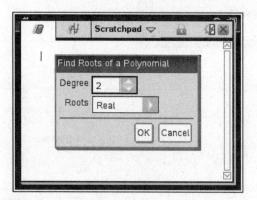

➡ **EXAMPLE 1.61** ———————————————————————————————————

Solve $2x^2 - 5 = 3x$.

In order to enter the quadratic equation into the (FIND ROOTS OF POLYNOMIAL) function as described above, you must set the equation equal to zero. The quadratic should be written in descending order by exponents.

$$2x^2 - 3x - 5 = 0$$

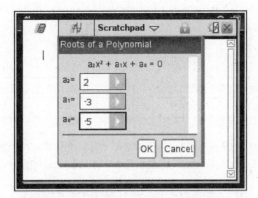

Press ENTER to solve.

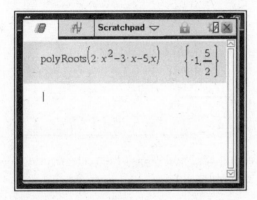

The solutions are displayed as $x = -1$ or $x = \dfrac{5}{2}$

Sections 2.3 and 2.4 (Histograms and Box and Whisker Plots)

To display data, use the LISTS & SPREADSHEETS menu found on the home screen and enter the given data into the first list (A).

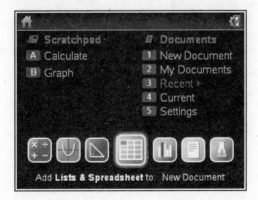

Most calculations can be performed directly in the spreadsheet; however, graphs must be inserted by adding a new page. The directions are given in the example on page 404.

Each list should be named when entering data. In the same cell as the calculator list name (A, B, C, etc), enter a name for the list of data.

The coach of a school's cricket team recorded the number of runs his team earned last season. The data are given below.

150	200	250	195	199	204	210
300	265	288	185	210	232	190
185	240	245	260	288	265	213

Draw a box and whisker plot and histogram to represent the number of runs earned. Before entering the data, label the list as "runs". Then enter the data into the list.

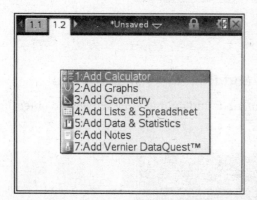

Next, insert a page by pressing (CTRL) and (+PAGE), then select option 5 (ADD DATA & STATISTICS). This gives an initial graph of random points.

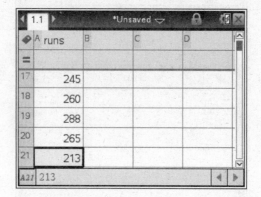 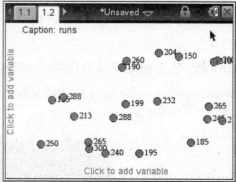

Scroll down to the horizontal axis and select (CLICK TO ADD A VARIABLE). Select (RUNS) as this was the name of our data list. This will change the graph to a dot plot.

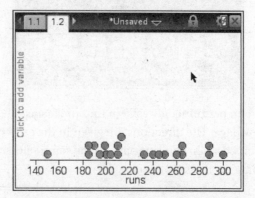

To change the graph to a box and whisker plot, select (MENU), option 1: (PLOT TYPE) and then option 2: (BOX PLOT).

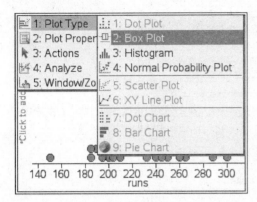
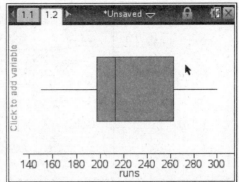

Using the center keypad, move the cursor around to locate the minimum, first quartile, median, third quartile, and maximum.

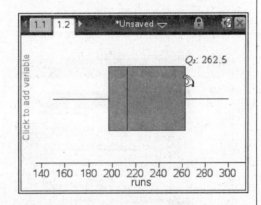

To make a histogram, instead of selecting (BOX PLOT), select option 3: (HISTOGRAM).

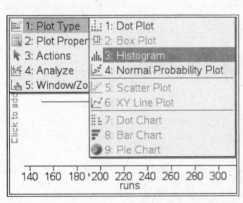
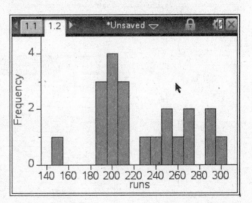

To change the width of the intervals, press (MENU) and select option 2: (PLOT PROPERTIES), then option 2: (HISTOGRAM PROPERTIES).

Select option 2: [BIN SETTINGS], and then option 1: [EQUAL BIN WIDTH]. Now, set the interval width as desired. In our example, a width of 20 is appropriate.

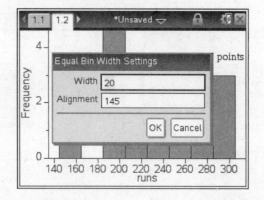

To change the window, press [MENU] and select option 5: [WINDOW/ZOOM]. In this menu, select option 1: [WINDOW SETTINGS]. In our example, the [YMAX] needs to be larger.

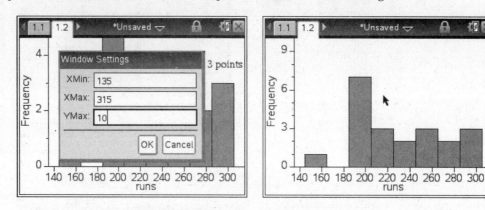

Using the touchpad, scroll over the bars to determine the interval and the height.

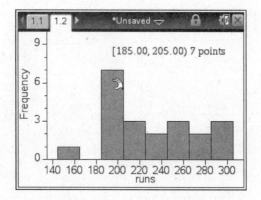

Sections 2.5 and 2.6 (Measures of Central Tendency and Measures of Spread)

Calculating measures of center and spread can be done right in the spreadsheet. After naming the list and entering the data, arrow to the next data column over and calculate [ONE-VARIABLE STATISTICS]. The exact steps are shown on page 407.

➧ **EXAMPLE 2.50**

The marks earned on Paper 1 of the IB Math Studies exam for 24 randomly selected students are recorded below.

85	75	54	77	32	70
44	80	68	53	59	72
81	30	39	47	54	60
77	72	68	71	76	77

Calculate the measures of central tendency and the standard deviation.

Use the [LISTS & SPREADSHEETS] menu found on the home screen and enter the given data into the first list (A) after first naming the list "scores".

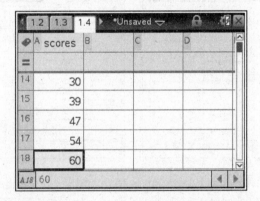

To calculate the one variable statistics, press [MENU] and select option 4: [STATISTICS]. Then select option 1: [STATISTIC CALCULATIONS]. The first option in this menu is the correct one.

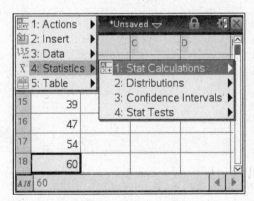
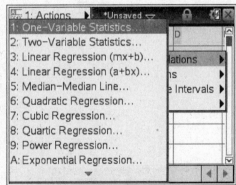

We only have one list, so press [OK].

Under X1 list, select our data named SCORES.
All other options remain in the default setting.
Select OK.

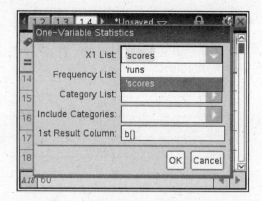

The resulting statistics are displayed in the next two columns of the spreadsheet.

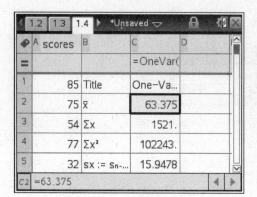
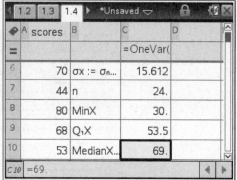

The mean is 63.375, the standard deviation is 15.6, and the 5 number summary is listed at the end of the output: 30, 53.5, 69, 76.5, 85.

To determine the mode, sort the list in ascending order. First, highlight the name of the list, scores. Then press MENU, option 1: ACTIONS, and then option 6: SORT. Select OK.

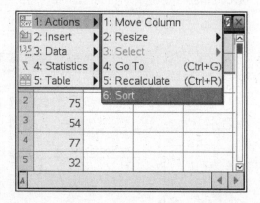

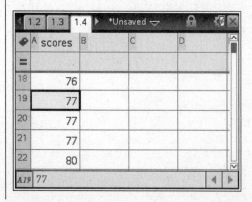

Now, the list is in ascending order, and we can identify the most frequent data value, which is 77.

➡ EXAMPLE 2.51 _____

The weights of 80 patients at a pediatric doctor's office were recorded and are displayed in the table below.

Weight (w)	$10 \leq w < 20$	$20 \leq w < 30$	$30 \leq w < 40$	$40 \leq w < 50$	$50 \leq w < 60$
Frequency	10	18	12	22	18

Using a GDC find an estimate weight for the:

(i) mean
(ii) median
(iii) standard deviation

To find the approximate mean, enter the mid-interval values into List A and the frequencies into List B. Remember to name the lists first. Name List A "weight" and List B "freq".

Arrow over to List C. Select [ONE-VARIABLE STATISTICS] as done in the previous example. We are still only using one list. Select [WEIGHT] as the X1 list and [FREQ] as the frequency list.

The output displays the mean, standard deviation, and 5-number summary.

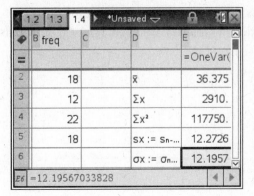 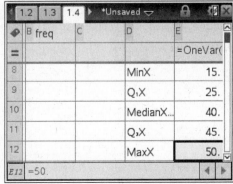

Section 4.1 (Normal Distribution)

To perform calculations using the (NORMAL) distribution, access the Distribution menu by pressing (MENU) and selecting option 5: (PROBABILITY).

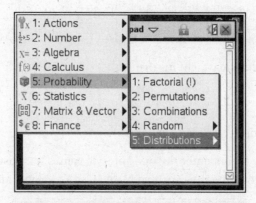

Option 5: Distributions is where the (NORMAL CDF) and (INVERSE NORMAL) will be found.

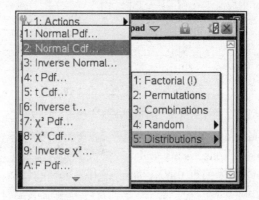

Entering data is the same as the TI-84, except instead of using −E99 for negative infinity or E99 for infinity, the Nspire requires a coefficient for E. The Nspire will use −1E99 and 1E99 for negative or positive infinity.

➡ EXAMPLE 4.13 ───────────────────────────

A local business packages gift boxes of candied almonds. The weights of the boxes are normally distributed with a mean of 16.1 ounces and a standard deviation of 0.25 ounces.

One box is randomly selected from the production line and weighed. Find the probability the box

1. weighs less than 15.7 ounces

Select NORMAL CDF and enter the data. Then press OK.

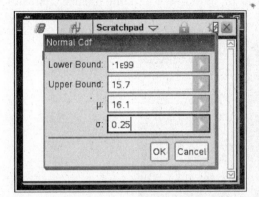 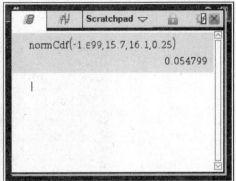

Thus, the probability is 0.0550.

2. weighs more than 16.7 ounces

Enter the data the same as before and select OK.

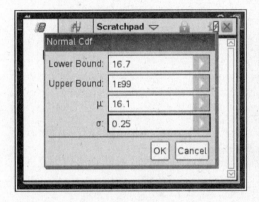 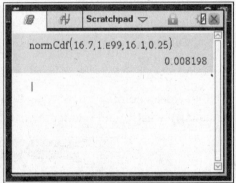

Thus, the probability is 0.00820.

The lightest 5% of the gift boxes weigh w ounces and are discarded.

3. Calculate the value of w.

Select INVERSE NORMAL and enter the data the same as the TI-84.

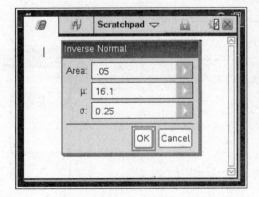 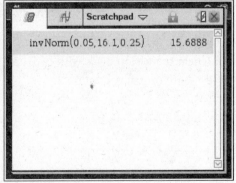

Thus, $w = 15.7$ ounces.

Sections 4.2–4.3 (Bivariate Data and Regression Line)

To work with bivariate data, use the [LISTS & SPREADSHEETS] menu found on the home screen and enter the given data into the list (A) and list (B). Remember to name the lists first with x and y.

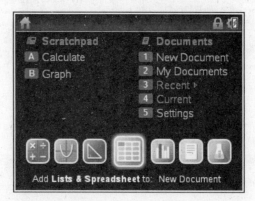

⮕ **EXAMPLE 4.21**

You record the gallons remaining in your fuel tank after driving x miles.

Miles driven	0	40	70	100	130	150	190	240	280	300
Gallons remaining	19	17	14	13	10	9	8	5	3	2

1. Find the mean point.

After naming the lists x and y, enter the data.

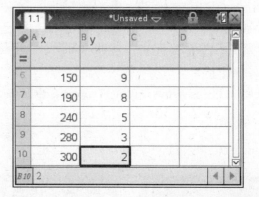

As in chapter 2, access the [STAT CALCULATION] menu. Linear regression deals with two variables; therefore, select option 2.

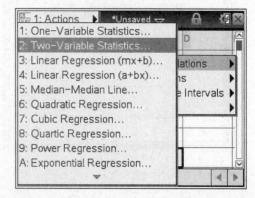

Enter the list names in for x and y and press OK.

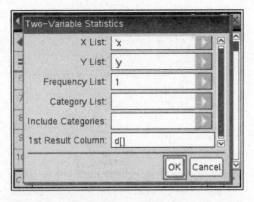

The mean point will be displayed in the statistical output.

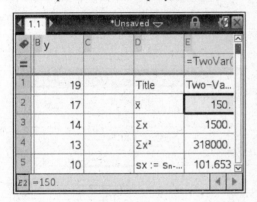
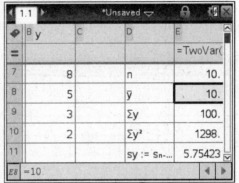

$$\bar{x} = 150 \text{ and } \bar{y} = 10$$

2. Write down the correlation coefficient.

The correlation coefficient is found when the Linear Regression is calculated. This is in the same menu as TWO-VARIABLE STATISTICS.

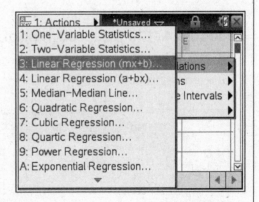

Again, name the x list and the y list, leave the remaining items as given, then press OK.

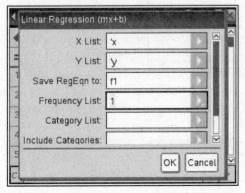

The correlation coefficient is the 4th number in the output.

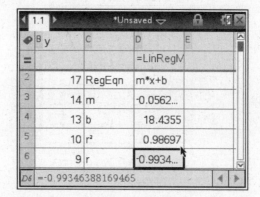

3. Find the equation of the regression line.

The output from part 2 also displays the regression line: $y = -0.0563x + 18.4$.

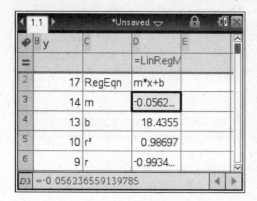

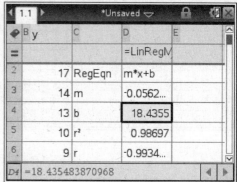

Section 4.4 (Chi-Squared Test of Independence)

To enter a matrix, press MENU while on a sketchpad page. Select option 7: MATRIX & VECTOR.
Then select CREATE and MATRIX

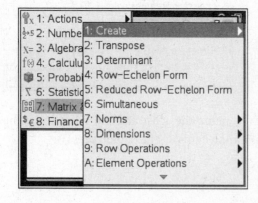

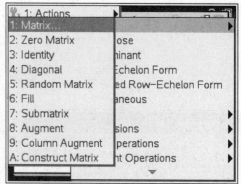

Then, you will select the number of rows and columns to generate the correct sized template.

Ms. Ghali recorded three classes of her students' grades according to where each student sits in class, as shown in the table below.

Seating location		Grade earned			
		A	B	C	D or less
	Front	10	14	7	4
	Middle	6	9	11	9
	Back	7	8	12	8

Ms. Ghali carries out a χ^2 test for independence.

1. Using the GDC, write down the calculated value of χ^2.

First, create a new matrix. There are 3 rows and 4 columns. (Pressing the TAB key easily moves from the rows input to the columns input.)

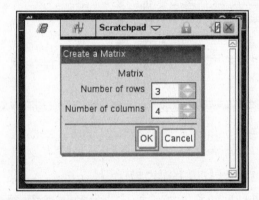

After entering all of the elements, you need to name the matrix. Press CTRL and VAR followed by the letter "a" to name the matrix as a.

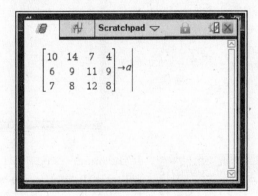

The Chi-Squared test is found in the STATISTICS menu under option 7: STAT TESTS.

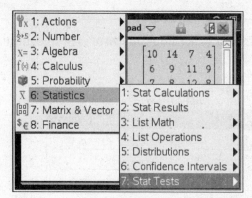
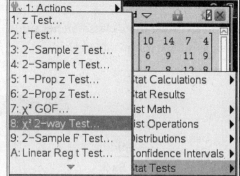

Simply enter *a* as the observed matrix and press ENTER.

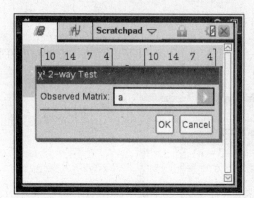
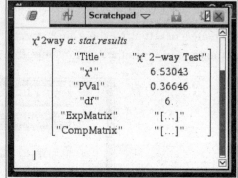

The Chi-Squared calculated value is 6.53 and the *p*-value is 0.366.

Section 5.1 (Trigonometric Ratios)

The calculator should already be set to degree mode when put in PRESS-TO-TEST. To verify, hover the cursor over the gear icon in the upper right-hand corner to display the current setting.

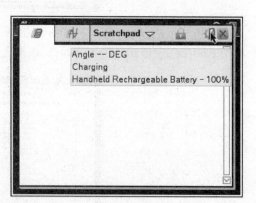

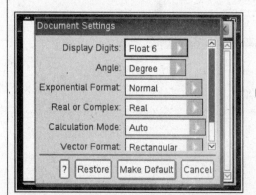

If the mode needs to be changed, press the center of the cursor pad while over the gear icon, then select option 2: DOCUMENT SETTINGS.

Scroll down to ANGLE and change the mode to DEGREE.

➡ EXAMPLE 5.20

Given triangle *RST* below, solve the triangle.

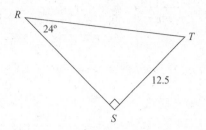

In chapter 5, you found $R\hat{T}S = 66°$ first.

To find side *t*, $\tan 24° = \dfrac{12.5}{t}$.

Thus, $t = \dfrac{12.5}{\tan 24°}$.

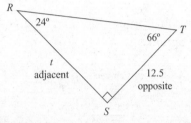

The trig functions are accessed by pressing the TRIG button on the left side of the calculator. Select the correct trig function while entering the calculation.

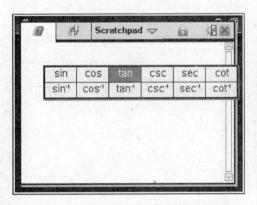
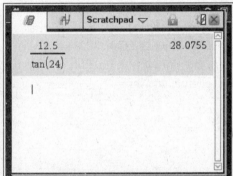

Section 5.3 (Sine and Cosine Rule)

When calculating problems involving the sine rule or cosine rule, you can use a fraction template in order to type in the work.

Press the [⬚|{⬚}] key to open the template menu, then select the first template, the fraction.

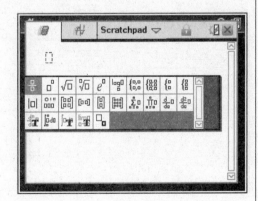

EXAMPLE 5.32

Barbara is making the small garden flag shown.
    Calculate the measurements of all the angles.

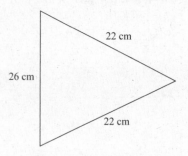

As discussed in chapter 5, finding the largest angle first
is the best option.

$$\cos A = \frac{22^2 + 22^2 - 26^2}{2(22)(22)}$$

$$\cos A = 0.30165$$
$$A = \cos^{-1}(0.30165)$$
$$A = 72.4°$$

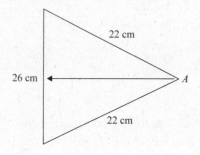

To enter this in the calculator, select the
fraction template and type in the work as
written. The output is in fraction form, but this
is not the final answer.

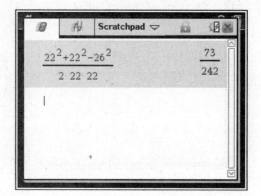

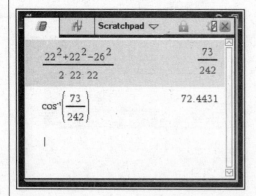

Locate "cos⁻¹" in the trig menu, then press CTRL
and "(−)" for the ANS option.

To continue solving the triangle, use the sine rule.

$$\frac{26}{\sin 72.4°} = \frac{22}{\sin X}$$

$$26 \sin X = 22 \sin 72.4°$$

$$\sin X = \frac{22 \sin 72.4°}{26}$$

$$\sin X = 0.80655$$

$$X = \sin^{-1}(0.80655)$$

$$X = 53.8°$$

Again, the fraction template makes entering the calculations much easier.

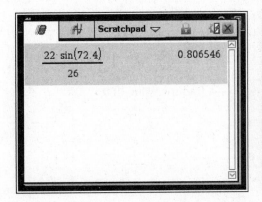 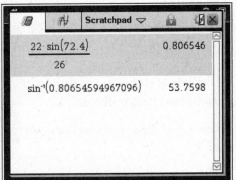

Section 6.3 (Quadratic Models)

To graph a function, use the (GRAPH) option under (SKETCHPAD).

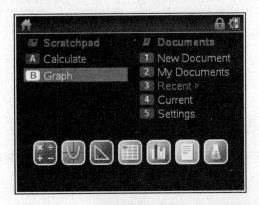

Enter the desired function at the top of the screen.

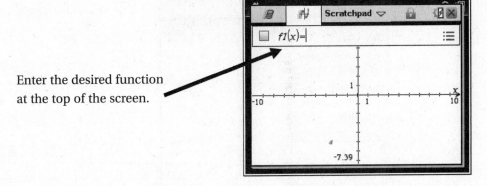

➡ EXAMPLE 6.30 _____

The graph of $y = 2x^2 + bx - 3$ is show below with points A, B, and C on the graph.

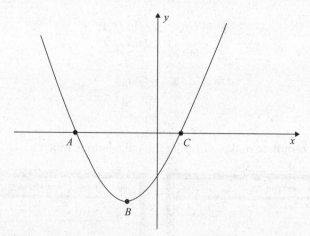

1. If the axis of symmetry for the quadratic is $x = -\dfrac{5}{4}$, find the value of b.
You determined $b = 5$ in chapter 6.

2. Hence, find the coordinates of point:
    (a) A
    (b) B
    (c) C

You need to graph $y = 2x^2 + 5x - 3$ to locate points A, B, and C.

Enter $2x^2 + 5x - 3$ into the $f1(x)$ at the top of the graph screen and press ⎡ENTER⎤.

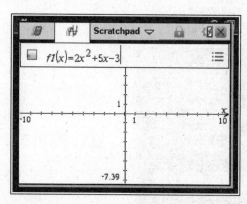
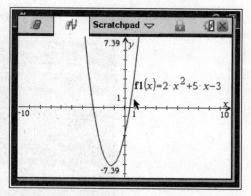

The axes are set on a default setting. To change this, press ⎡MENU⎤ and select option 4: ⎡WINDOW/ZOOM⎤. Then, choose option 1: ⎡WINDOW SETTINGS⎤.

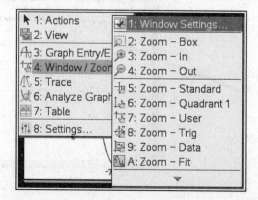

Judging by the original graph, you can narrow the domain while changing the range. You can leave the scale as AUTO or change the scale to suit your needs.

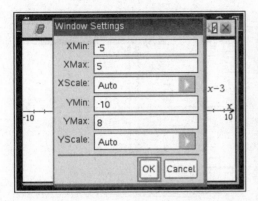 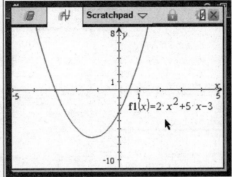

To identify important points on graphs such as zeros, maximum or minimum, or the intersection of two graphs, press MENU and select option 6: ANALYZE GRAPH.

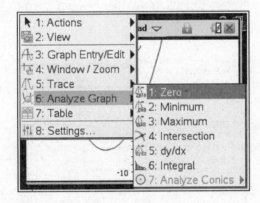

These options are the same as the TI-84.

Point *A* is the zero on the negative *x*-axis, so select option 1: ZERO.

A dotted line with a pointer finger appears and asks for the lower bound. Using the cursor pad in the middle, move the line so it is to the left of the zero and press ENTER. Next, move the line to the right of the zero and press ENTER. The zero is then identified.

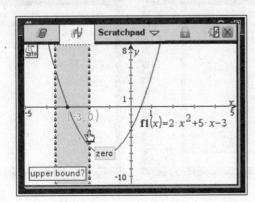

To locate Point B, determine the minimum. Select option 2: MINIMUM and again move the line to the left of the minimum. Press ENTER, and then to the right and press ENTER.

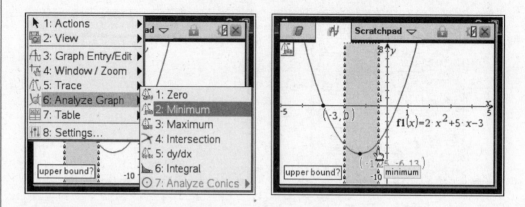

Point C is found by repeating the process to find the zero.

The zeros could also be found using the polynomial root finder. On the calculator sketch-pad, select option 3: POLYNOMIAL TOOLS from the Algebra menu. Then select option 1: FIND ROOTS OF POLYNOMIAL.

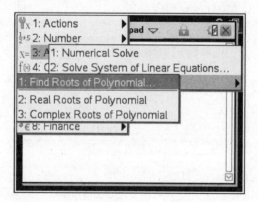

Enter the coefficients and press OK.

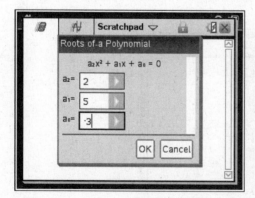
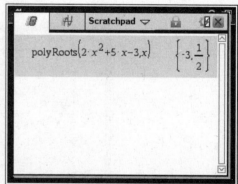

Section 6.4 (Exponential Models)

Graphing an exponential function is the same as a quadratic; however, many exponential functions also involve the use of the table.

➡ **EXAMPLE 6.42** ────────────────────────────

The following graph displays the temperature of a small refrigerator h hours after being turned on. The equation of the graph is $T(h) = 5 + 16(0.65)$, where $T(h)$ is the temperature of the refrigerator in degree Celsius h hours after being turned on.

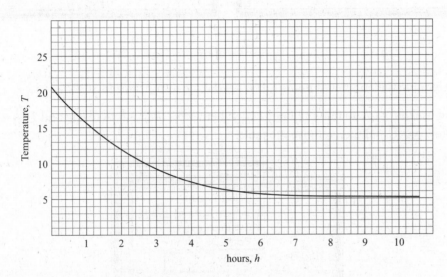

1. Find the temperature of the refrigerator after 2 hours.

Graph the function and set up an appropriate window.

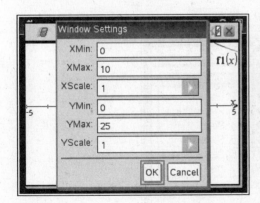 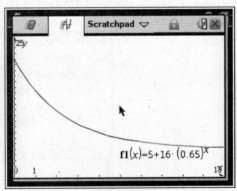

Press ⟨MENU⟩ and select option 7: ⟨TABLE⟩. Now, select option 1: ⟨SPLIT-SCREEN TABLE⟩. This will split the viewing window so half is the graph and half is the table.

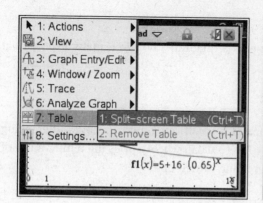
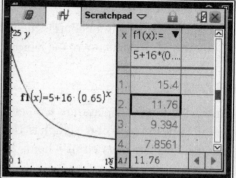

$$T(2) = 11.8$$

2. Determine the number of hours taken to cool the refrigerator to 10°C.

You can use the graph to find the intersection between $T(h) = 5 + 16 (0.65)^t$ and $T(h) = 10$. First press MENU and then remove the table.

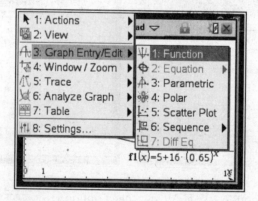

Now, select option 3: GRAPH ENTRY/EDIT and then 1: FUNCTION.

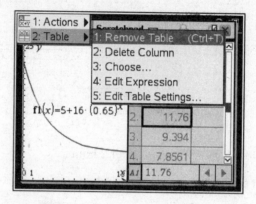

When $f2(x) =$ appears at the top, type "10". Once the second graph appears, select 5: INTERSECTION in the ANALYZE GRAPH menu.

Move the line to the left of the intersection, press ⟨ENTER⟩, and then to the right.

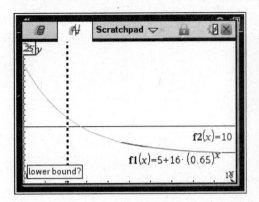
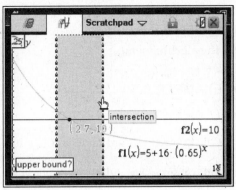

The intersection occurs when $t = 2.7$ minutes.

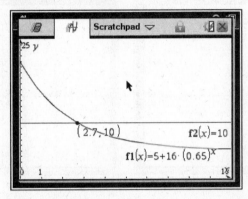

Alternatively, the calculator sketchpad has the option to solve equations. In the Algebra menu, select option 1: ⟨NUMERICAL SOLVE⟩. Enter the equation to be solved followed by a comma and the variable for which it will be solved.

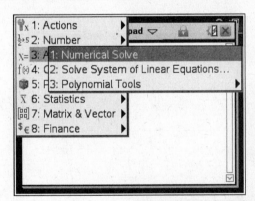
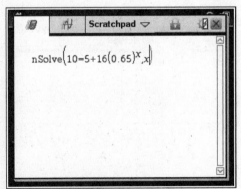

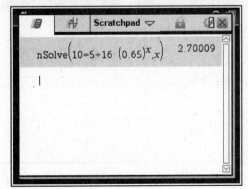

Section 6.4 (Polynomial and Rational Models)

Polynomial and rational functions are graphed in the same way as other functions. The ANALYZE GRAPH menu allows you to find the key features. Another option is the POLYNOMIAL TOOLS from the Algebra menu.

➡ EXAMPLE 6.50 ─────────────────────

Given $f(x) = x^3 - 2x^2 - 4x + 2$.

1. Write down the coordinates of the y-intercept for $f(x)$.

The graph shows the y-intercept of $(0,2)$ and the table confirms the graph.

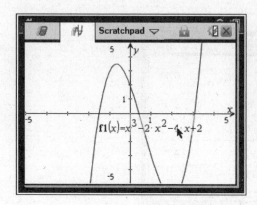
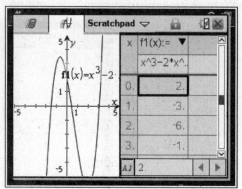

2. Write down the solution to $0 = x^3 - 2x^2 - 4x + 2$ if $x \in \mathbb{R}^-$.

You can find the zero using the graph or solve using the POLYNOMIAL ROOTS OPTION.

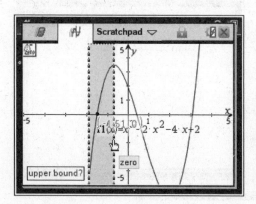
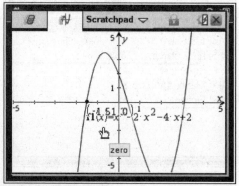

The POLYNOMIAL TOOLS allows us to find all zeros of the cubic function.

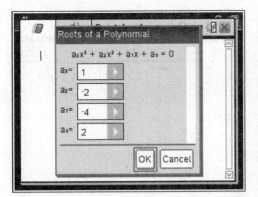
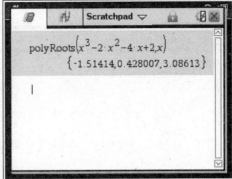

The problem states $x \in \mathbb{R}^-$, so we only include the negative answer.

3. Determine the value of x for which $f(x) = -1$ if $x \in \mathbb{N}$.

You can graph a second function of $f(x) = -1$ and determine the intersection, or find the roots of the polynomial.

In order to find all the roots, write the equation $-1 = x^3 - 2x^2 - 4x + 2$, and then set it equal to zero.

$$0 = x^3 - 2x^2 - 4x + 3$$

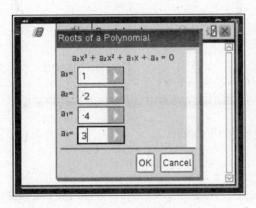
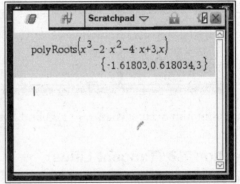

Since $x \in \mathbb{N}$, the only answer is $x = 3$.

➡ EXAMPLE 6.52 ―――――――――――――――――――――――――――――――――――

Consider the function $g(x) = \dfrac{3}{x^2} - 2x$.

1. Write down the x-intercept of $g(x)$ correct to three decimal places.

After graphing the function, find the zero.

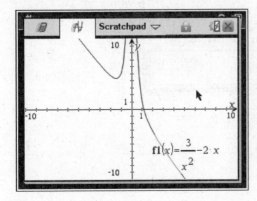
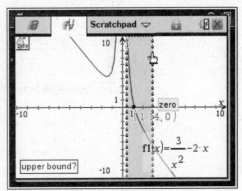

Another option is to use the NSOLVER.

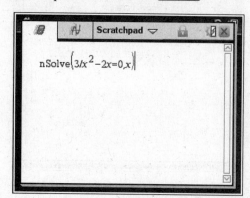
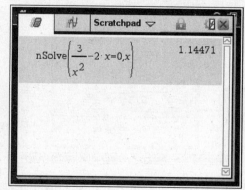

2. Determine the minimum value of $g(x)$, when $x < 0$.

Locate the minimum in the same way as in the Quadratic section.

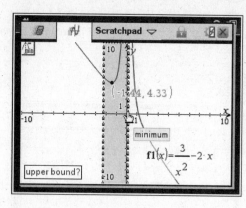
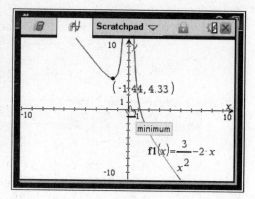

The minimum occurs when $x = -1.33$ and has a value of 4.33.

Section 7.3 (Tangent Lines)

The TI-Nspire can find the gradient of the tangent line at a specific point. On the calculator sketchpad screen, press MENU and select option 4: CALCULUS. The option to find the gradient is option 1: NUMERICAL DERIVATIVE AT A POINT.

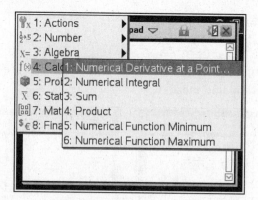

➡ EXAMPLE 7.30 ————————————————————

For each of the following, calculate the gradient of the curve, when $x = -3$.

1. $f(x) = 3x^5 - 2x + 14$

To find the gradient at a given point, access the CALCULUS menu and select option 1: DERIVATIVE AT A POINT. Enter the variable, usually x, and then the value of x given in the problem.

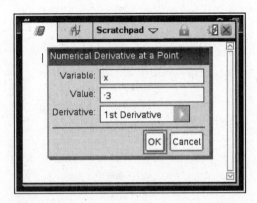

Next, type the original function, and press ENTER.

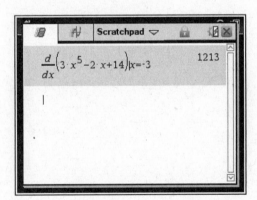

IB Grading Marking Scheme Guide

When the IB Math Studies exam is marked, or graded, different marks are assigned to a student's paper.

On Paper 1, work need not be shown. Students can earn full marks for correct answers. However, if an answer is incorrect, then marks may be assigned to work shown.

On Paper 2, work must be shown, as students will earn marks for correct answers and also their method used.

Types of Marks Earned		
Abbreviation	Meaning	Definition
C	Correct	Correct answer is shown • Paper 1 earns **C** marks for correct answers
A	Accuracy	The answer or values used are accurate • Usually final answers are given **A** marks. • Substituting correct values into an equation can also earn **A** marks.
M	Method	Correct method has been shown • Using the correct formula can earn **M** marks. • Showing an equation equal to a certain value can earn **M** marks. • Questions involving "Show that ..." often only earn **M** marks.
G	Graphics Display Calculator	Full marks earned for correct answers from GDC
R	Reasoning	Explanation or mathematical justification provided • Problems asking "support answer" earn **R** marks. • Chi-Squared tests often earn **R** marks. • Explaining answers earn **R** marks.
FT	Follow Through	A part was incorrect but then used correctly • If a question was missed that was necessary for the next calculation, students can still earn full marks via **FT**. For example, if a student missed part *a* but uses that incorrect answer in part *b* correctly, **FT** marks are awarded for part *b*. • If a student is unsure of how to solve a part needed to continue, make up a reasonable answer and then carry it forward. **FT** marks may be earned.

Formulas Needed

The formulas needed for the IB Math Studies exam are provided throughout this book; however, they are also located here for easy reference.

Name of Formula	Formula	Additional Information
Percent Error	$$\varepsilon = \left| \frac{v_A - v_E}{v_E} \right| \times 100\%$$	v_E is the exact value of v, while v_A is the approximate value.
Arithmetic Sequence	$$u_n = u_1 + (n-1)d$$	u_n is the value at nth term, u_1 is the first term, n is the position in sequence, and d is the common difference.
Arithmetic Sum	$$S_n = \frac{n}{2}[2u_1 + (n-1)d]$$ OR $$S_n = \frac{n}{2}(u_1 + u_n)$$	S_n is the sum of first n term, u_n is the value at nth term, u_1 is the first term, n is the position in sequence, and d is the common difference.
Geometric Sequence	$$u_n = u_1 r^{n-1}$$	u_n is the value at nth term, u_1 is the first term, n is the position in sequence, and r is the rate.
Geometric Sum	$$S_n = \frac{u_1(r^n - 1)}{r - 1}$$ OR $$S_n = \frac{u_1(1 - r^n)}{1 - r}$$	S_n is the sum of first n terms, u_n is the value at nth term, u_1 is the first term, n is the position in sequence, and r is the rate.
Compound Interest	$$FV = PV \times \left(1 + \frac{r}{100k}\right)^{kn}$$	FV is the future value, PV is the present value, n is number of investment years, k is the amount of compounding periods, and r is the annual interest rate.
Interquartile Range	$$IQR = Q_3 - Q_1$$	Q_3 is the third quartile and Q_1 is the first quartile.
Complementary Events	$$P(A') = 1 - P(A)$$	$P\,A$ is the probability of event A.
Combined Events	$$P(A \cup B) = P(A) + P(B) - P(A \cap B)$$	The probability of A **or** B is the probability of A plus the probability of B minus the probability of A **and** B.

Name of Formula	Formula	Additional Information
Mutually Exclusive Events	$P(A \cap B) = 0$	Two events are mutually exclusive if they do not share a common outcome.
Independent Events	$P(A \cap B) = P(A)P(B)$	Independent events are found by multiplying individual probabilities.
Conditional Probability	$P(A|B) = \dfrac{P(A \cap B)}{P(B)}$	Probability of A **given** B is the probability of A and B divided by the probability of B.
Equation of a straight line	$y = mx + c$; $ax + by + d = 0$	m is the gradient and c is the y-intercept.
Gradient	$m = \dfrac{y_2 - y_1}{x_2 - x_1}$	x_1, y_1 and x_2, y_2 are points.
Sine Rule	$\dfrac{a}{\sin A} = \dfrac{b}{\sin B} = \dfrac{c}{\sin C}$	Ratios are comprised of opposite angles with their sides.
Cosine Rule	$a^2 = b^2 + c^2 - 2bc\cos A$ $\cos A = \dfrac{b^2 + c^2 - a^2}{2bc}$	Angle A has the opposite side a.
Area of a triangle	$A = \dfrac{1}{2}ab\sin C$	C is the included angle between the adjacent sides a and b.
Axis of symmetry for a quadratic function	$x = -\dfrac{b}{2a}$	The quadratic function is represented by $y = ax^2 + bx + c$.
Derivative of ax^n	$f(x) = ax^n \Rightarrow f'(x) = nax^{n-1}$	a is the coefficient and n is the exponent.

Solutions to Featured Questions

Featured Question: Topic 1 (page 20)

Part A

(a) **Option 1:** $A = 25\,000 + \dfrac{2\,500 \times 6 \times 3}{100} = 29\,500$

Simple interest is no longer a part of the IB Math Studies curriculum.
The formula for simple interest is $A = Prt$, where A is the final amount, P is the principal, r is the interest rate, and t is time in years.

Option 2: $A = 25\,000\left(1 + \dfrac{5}{100}\right)^{3} = 28\,940.63$

Calculate compound interest by the formula $FV = PV \times \left(1 + \dfrac{r}{100k}\right)^{kn}$.

Using the given information, $PV = 25\,000$, $r = 5\%$, $k = 1$, and $n = 3$.

Option 3: $A = 25\,000\left(1 + \dfrac{4.8}{12(100)}\right)^{3 \times 12} = 28\,863.81$

Using the given information, $PV = 25\,000$, $r = 4.8\%$, $k = 12$, and $n = 3$.

(b) Option 1 is the best investment option.
Option 1 yields the largest amount at the end of 3 years.

Part B

(a) $u_1 = 135 + 7(1)$
$u_1 = 142$
The first term in the sequence occurs when $n = 1$. Substitute 1 for n and simplify.

(b) $u_2 = 135 + 7(2)$ Thus, $d = 149 - 142 = 7$
$u_2 = 149$
To find the common difference, you need two terms. You know $u_1 = 142$, so find u_2.
Then, subtract $u_2 - u_1$ to determine d.

(c) $S_n = \dfrac{n}{2}[2(142) + (n-1)7]$

$S_n = \dfrac{n}{2}(277 + 7n)$

$S_n = 3.5n^2 + 138.5n$

The formula for the sum of an arithmetic sequence is $S_n = \dfrac{n}{2}[2u_1 + (n-1)d]$.

Since $u_1 = 142$ and $d = 7$, substitute in those values and simplify.

(d) $20r^{4-1} = 67.5$

$r^3 = 3.375$

$r = 1.5$

You know $v_1 = 20$ and $v_4 = 67.5$, so substitute the given information into $u_n = u_1 r^{n-1}$ and solve.

$$20r^{4-1} = 67.5$$
$$r^3 = 3.375$$

You can solve algebraically by taking the cube root of both sides: $r = \sqrt[3]{3.375} = 1.5$. Another option is to solve using (PLYSMLT). First, set the equation equal to zero.

$$r^3 - 3.375 = 0$$

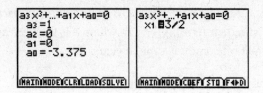

(e) $T_7 = \dfrac{20(1.5^7 - 1)}{1.5 - 1}$

$T_7 = 643$

The sum of a geometric sequence is $S_n = \dfrac{u_1(r^n - 1)}{r - 1}$. You know $v_1 = 20$ and $r = 1.5$, and since you want the sum of the first 7 terms, $n = 7$.

(f) $\dfrac{20(1.5^n - 1)}{1.5 - 1} > 3.5n^2 + 138.5n$

$n = 10$

You need to find when the function T is larger than the function $S (T_n > S_n)$, so substitute the equations in for the functions.

$$\frac{20(1.5^n - 1)}{1.5 - 1} > 3.5n^2 + 138.5n$$

Let $Y_1 = \dfrac{20(1.5^x - 1)}{1.5 - 1}$ and $Y_2 = 3.5x^2 + 138.5x$.

Now, using the table, find when $Y_1 > Y_2$.

X	Y₁	Y₂
6	415.63	957
7	643.44	1141
8	985.16	1332
9	1497.7	1530
10	2266.6	1735
11	3419.9	1947
12	5149.9	2166

X=10

When $x = 10$, the sum of the geometric sequence is larger than the arithmetic. Thus, $n = 10$.

Featured Question: Topic 2 (page 46)

(a) (i) **median = 13 seconds**

(ii) **$Q_3 = 16$ seconds, $Q_1 = 10$ seconds**

(iii) **IQR = 6 seconds**

The median occurs at the halfway point. Since there are 2 000 men in the sample, the median will be at 1 000.

The upper and lower quartile, which are also called the third and first quartile, occur at the 75th percentile and the 25th percentile. Since there are 2 000 men, the third quartile will be at 2 000(0.75) = 1 500 and the first quartile at 2 000(0.25) = 500.

The interquartile range is $Q_3 - Q_1$.

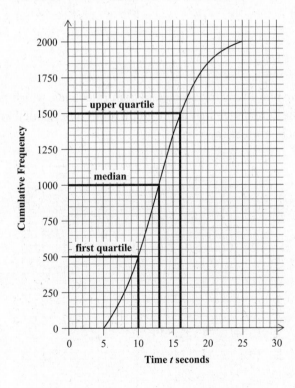

(b) **2 000 − 650 = 1 350 men**

There were 650 men who took 11 seconds or less to perform the task. The question asked for more than 11 seconds, so 650 must be subtracted from 2 000 to determine the men above 11 seconds.

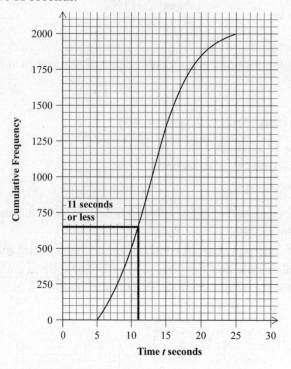

(c) **2 000(0.55) = 1 100**

$p = 13.5$

First, calculate how many men took less than p seconds, which is 2 000(0.55) = 1 100. Now, using the cumulative frequency graph, draw a horizontal line from 1 100 to the curve, and then down to the x-axis.

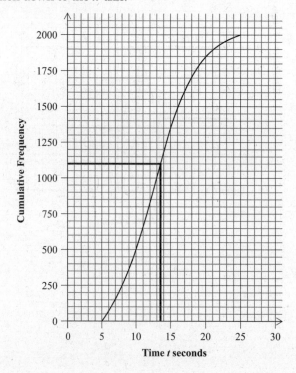

(d) (i) **$a = 500$**

(ii) **$b = 150$**

The values of a and b represent the change from the beginning of the given interval to the end.

For the interval $15 \leq t < 20$, identify the beginning of the interval as 1 350 men and the ending as 1 850 men. Therefore, $a = 1\,850 - 1\,350 = 500$.

Since there are 2 000 men in total, $b = 2\,000 - (500 + 950 + 500) = 150$.

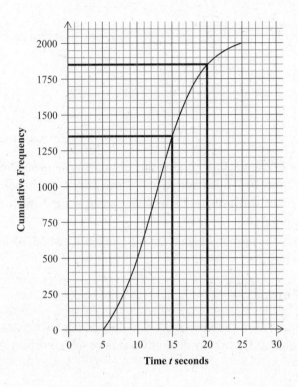

(e) (i) **13.25 seconds (13.3 seconds)**

(ii) **4.41 seconds**

Enter the midinterval values into L_1 and the frequencies into L_2. Then calculate

⌜1-VAR STAT⌝ L_1, L_2 to find the mean and standard deviation.

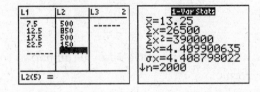

(f) **8.84 < 9.5**

Pedro does not receive the bonus.

One standard deviation below the mean is determined by $13.25 - 4.41 = 8.84$.

To earn the bonus, Pedro must have performed the task in less time than 8.84 seconds. Since he took 9.5 seconds, he does not earn the bonus.

Featured Question: Topic 3 (page 77)

Part A

(a)

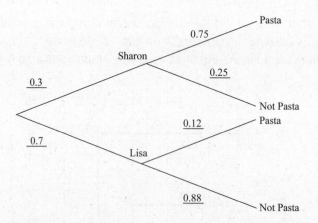

- Sharon cooks dinner three nights out of ten, so the probability she cooks is 0.3. Since the "branches" must add to one, Lisa's probability of cooking is 0.7.
- If Sharon cooks, the probability they have pasta is 0.75, which means 0.25 is left for the event "not pasta".
- If Lisa cooks, the probability they have pasta is 0.12, which means 0.88 is left for the event "not pasta".

(b) $\mathbf{0.7 \times 0.88 = 0.616}$ **or** $\dfrac{\mathbf{77}}{\mathbf{125}}$

To find the probability that Lisa cooks, and they do not have pasta, multiply the probability Lisa cooks, 0.7, by the probability she does not cook pasta, 0.88.

(c) **0.691**

There are two ways for them not to have pasta: Sharon cooks, and they do not have pasta or Lisa cooks, and they do not have pasta.

Find the probability of each event and then add them together.

- P(Sharon cooks, and they do not have pasta) = 0.3(0.25) = 0.075
- P(Lisa cooks, and they do not have pasta) = 0.616 (from part *b*)
- P(they do not have pasta) = 0.075 + 0.616 = 0.691

(d) $\dfrac{\mathbf{0.616}}{\mathbf{0.691}} = \mathbf{0.891}$

You know Sharon and Lisa did not have pasta for dinner, so you must use the conditional probability formula.

$$P(\text{Lisa cooked} \mid \text{did not have pasta}) = \frac{P(\text{Lisa cooked AND did not have pasta})}{P(\text{did not have pasta})}$$

$$P(\text{Lisa cooked} \mid \text{did not have pasta}) = \frac{0.616}{0.691} = 0.891$$

Part B

(a) **3**

$n(R \cap S \cap B)$ means the number of elements that set R, S, and B have in common. This is the number in the very middle of the Venn diagram.

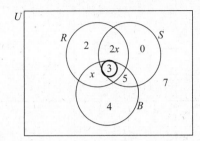

(b) **16**

$n(R')$ means the number of elements **not** in set R. Sum up the numbers outside of the circle containing R.

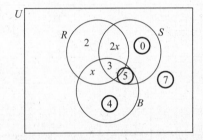

(c) **They like (both) the *Salseros (S)* and they like the *Bluers (B)***

$S \cap B$ contain the students who like Salseros AND Bluers, since $\cap$ = and.

(d) $R \cap B \cap S'$

The statement "the group of pupils who like the *Rockers* **and** the *Bluers* **but** do **not** like the *Salseros*" is represented by the three set names joined with the intersection symbol, $\cap$. Note, that Salseros must be the complement, since the pupils do **not** like the Salseros.

(e) (i) $21 + 3x = 33$

$x = 4$

(ii) **17**

The Venn diagram must add up to 33.

$$33 = 2 + 2x + 3 + x + 5 + 0 + 4 + 7$$
$$33 = 21 + 3x$$
$$12 = 3x$$
$$4 = x$$

The number of pupils who like the Rockers can be calculated when 4 is substituted in for x.

$$2 + 2(4) + 3 + 4 = 17$$

Featured Question: Topic 4 (page 122)

(a)

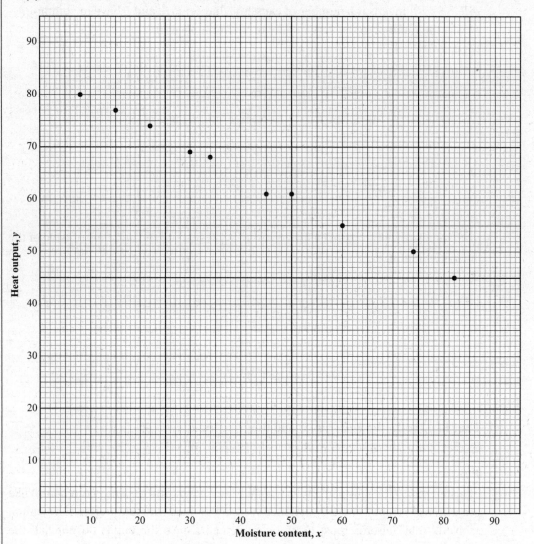

Moisture content, *x*

The scale must be 2 cm blocks to represent 10 both on the *x* and the *y* axes. Each axis must be labeled appropriately. Points from the table should be plotted.

(b) (i) $\bar{x} = 42$

(ii) $\bar{y} = 64$

After entering the data into L_1 and L_2, calculate 2-VAR STATS using the GDC to get the mean *x* and the mean *y*.

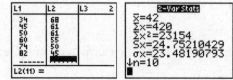

(c)

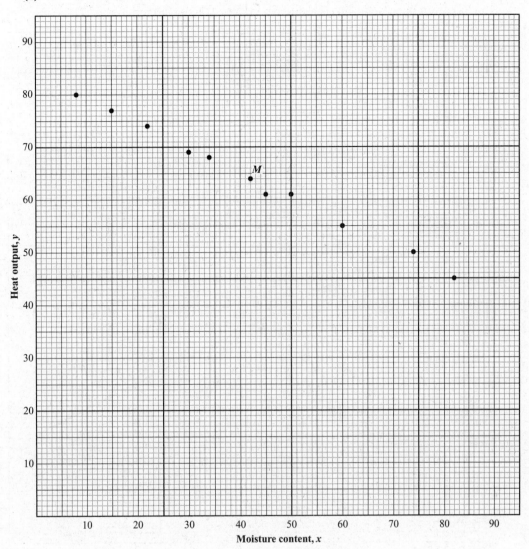

The mean point (42, 64) is plotted and labeled with *M*.

(d) **−0.998**

The correlation coefficient is found when [LINREG] is performed in the GDC.

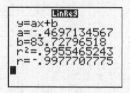

(e)

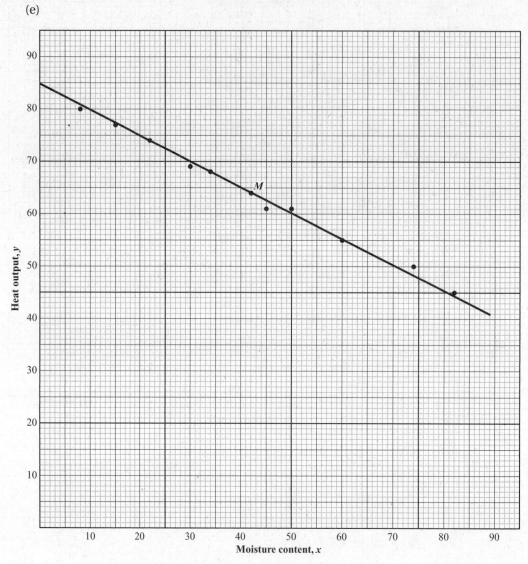

The regression line must pass through the mean point. Since the regression line is given, use the *y*-intercept as the second point on the line.

(f) **72.0 OR 71.95 OR 72**

To estimate the heat output for 25% moisture content, you can either substitute 25 in for x and simplify, or you can use the graph to estimate y.

$$y = -0.470(25) + 83.7$$
$$y = 71.95$$

(g) **Yes, since 25% lies within the data set and r is close to -1**

The correlation suggests a very strong negative, linear relationship, and 25 is within the data range. Therefore, the regression line provides a reasonable estimate for the heat output.

Featured Question: Topic 5 (page 174)

Part A

(a) **$50b + 20c = 260$**

Since Mal bought 50 tins of beans and 20 packets of cereal with a total cost of 260 AUD, the equation would be $50b + 20c = 260$, since the price of the beans together with the price of the cereal equals the total cost.

(b) **$12b + 6c = 66$**

Stephen buys 12 tins of beans and 6 packets of cereal with a total cost of 66 AUD. The equation is written in the same manner as part a.

(c) **$b = 4$**

You can solve simultaneous equations by graphing. First, rewrite each equation into slope intercept form, and then graph. Let $b = x$ and $c = y$.

Mal's Equation	Stephen's Equation
$50b + 20c = 260$	$12b + 6c = 66$
$20c = -50b + 260$	$6c = -12b + 66$
$c = -\dfrac{5}{2}b + 13$	$c = -2b + 11$

The graphs intersect at the point (4, 3). Since we were asked for the value of b, the solution is the x-value.

(d) (i)

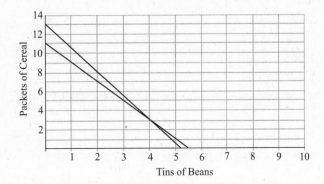

Make sure to label the axes and scale appropriately.

(ii) **(4, 3)**

You found the solution, which is the intersection in part c.

Part B

(a)

The problem states the triangular faces are all inclined at 70° to the base, which means the height of each triangular face creates the 70° angle with the base. Since you must use the letters provided, draw triangle *EGM*.

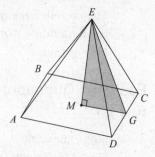

(b) $\tan 70 = \dfrac{h}{5}$

$5 \tan 70 = h$

$13.7 = h$

Using the triangle drawn in part *a*, the height of the pyramid is the height of the triangle.

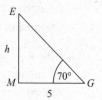

Using the angle of 70°, h is the opposite and 5 is adjacent. Thus, tangent is the correct trigonometric ratio.

(c) (i) $EG^2 = 5^2 + (13.7)^2$

$EG = 14.6$

EG is the hypotenuse of the triangle and can be determined using Pythagoras' theorem.

(ii) $\tan\theta = \dfrac{5}{14.6}$

$\theta = \tan^{-1}\left(\dfrac{5}{14.6}\right)$

$D\hat{E}G = 18.9°$

$D\hat{E}C = 2(18.9°) = 37.8°$

To find $D\hat{E}C$, first calculate $D\hat{E}G$, labeled θ, which is half of the angle needed.

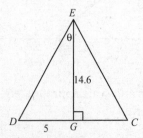

Next, double the value of θ to determine $D\hat{E}C$.

(d) **Surface Area** $= (10 \times 10) + 4\left(\dfrac{1}{2} \times 10 \times 14.6\right)$

Surface Area = 392 cm²

The surface area is the area of the base plus the area of the four triangular faces.

The base is a square with area 10×10.

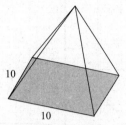

One triangular face has the area $A = \dfrac{1}{2}(10)(14.6)$.

There are four triangular faces, so you must multiply the area by 4.

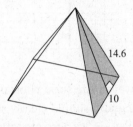

Hence, Surface Area $= (10 \times 10) + 4\left(\dfrac{1}{2} \times 10 \times 14.6\right)$.

(e) $V = \dfrac{1}{3} \times 10 \times 10 \times 13.7$

$V = 457$ **cm³**

The volume of a pyramid is found using the formula $V = \dfrac{1}{3}Ah$. The area of the base is 10×10, and the height was found in part *b*.

The answer of 458 cm³ would also be accepted.

Featured Question: Topic 6 (page 229)

(a)

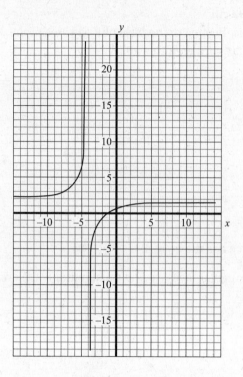

There was no set scale; however, the scale chosen needs to be clearly marked. Both axes should be labeled. The graph should clearly show a vertical asymptote as $x = -4$ and a horizontal asymptote at $y = 2$. These are found using the GDC. The x-intercept and y-intercept should also be graphed.

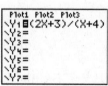

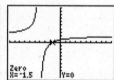

(b) **$x = -4$**

This is found using the table. When $x = -4$, the table shows ⟨ERROR⟩. The graph also splits at $x = -4$. This question is from the older IB Math Studies curriculum. The current curriculum only tests on functions whose vertical asymptote is $x = 0$.

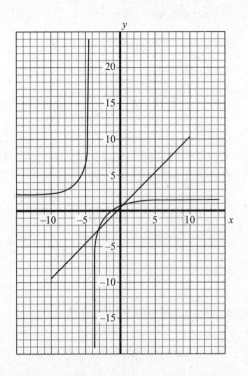

(c) The line $y = x + 0.5$ has a y-intercept of $(0, 0.5)$ and a gradient of 1.

(d) **(−2.85078, −2.35078) OR (0.35078, 0.85078)**

The functions intersect at two points, but you only need to find one.

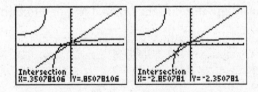

(e) **gradient = 1**

The gradient is the coefficient of x.

(f) **$y = -x - 5$**

Since the lines are perpendicular, the gradient of line L is −1, since $1 \times m = -1$.

To find c, substitute the given point with the perpendicular slope into $y = mx + c$.

$$-3 = -1(-2) + c$$
$$-3 = 2 + c$$
$$-5 = c$$

Thus, the equation of line L is $y = -x - 5$.

Featured Question: Topic 7 (page 268)

(a) $f'(x) = \dfrac{-10}{x^3} + 3$

To find the first derivative, first rewrite the function as $f(x) = 5x^{-2} + 3x + c$. Then, apply the first derivative rule to each term.

$$5x^{-2} \Rightarrow -2(5x^{-2-1}) = -10x^{-3}$$

$3x$ differentiates to the coefficient 3.

c is a constant and differentiates to zero.

(b) $4 = 5 + 3 + c$

$c = -4$

The graph passes through the point $(1,4)$; therefore, $x = 1$ and $y = 4$. You must use the original function, $f(x)$, since you have the value of y. Substitute in x and y, and then solve for c.

(c) (i) **(1.49, 2.72)**

First, set the first derivative equal to zero.

$$0 = \dfrac{-10}{x^3} + 3$$

Now, solve for x by graphing and finding the zero.

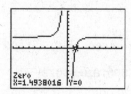

Since $x = 1.49$, substitute this into the original function to find y.

$$f(1.49) = \dfrac{5}{(1.49)^2} + 3(1.49) - 4$$

$$f(1.49) = 2.72$$

(ii) **(0, 1.49) OR $0 < x < 1.49$**

Since the point $(1.49, 2.72)$ is a minimum, the function is first decreasing, and then it switches to increasing. Looking at the graph of $f(x)$, you can see the graph decreases from 0 until 1.49.

(d) (i) $f'(1) = \dfrac{-10}{(1)^3} + 3$

$f'(1) = -10 + 3 = -7$

The first derivative is the gradient function. Substitute 1 in for x and simplify.

(ii) $y = -7x + 11$

Tangent line T passes through the point (1, 4) and has a gradient of -7.

$$4 = -7(1) + c$$
$$4 = -7 + c$$
$$11 = c$$

(e) (−0.5, 14.5)

Graph the original function as Y_1 and the equation of the tangent line as Y_2. Find the second intersection.

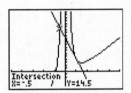

INDEX

NOTES

NOTES

NOTES

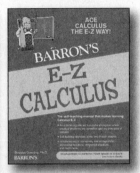
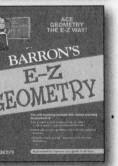
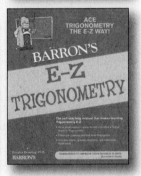